VOLUME ONE HUNDRED AND SIXTEEN

INTERNATIONAL REVIEW OF NEUROBIOLOGY

Brain Transcriptome

INTERNATIONAL REVIEW OF NEUROBIOLOGY

VOLUME 116

SERIES EDITORS

VOLUME ONE HUNDRED AND SIXTEEN

INTERNATIONAL REVIEW OF NEUROBIOLOGY

Brain Transcriptome

Edited by

ROBERT HITZEMANN

Department of Behavioral Neuroscience,
Oregon Health & Science University,
and Research Service, Veterans Affairs Medical Center,
Portland, Oregon, USA

SHANNON MCWEENEY

Oregon Clinical & Translational Research Institute
& Division of Biostatistics,
Public Health & Preventative Medicine,
Oregon Health & Science University,
Portland, Oregon, USA

ELSEVIER

AMSTERDAM • BOSTON • HEIDELBERG • LONDON
NEW YORK • OXFORD • PARIS • SAN DIEGO
SAN FRANCISCO • SINGAPORE • SYDNEY • TOKYO
Academic Press is an imprint of Elsevier

Academic Press is an imprint of Elsevier
32 Jamestown Road, London NW1 7BY, UK
525 B Street, Suite 1800, San Diego, CA 92101-4495, USA
225 Wyman Street, Waltham, MA 02451, USA
The Boulevard, Langford Lane, Kidlington, Oxford OX5 1GB, UK

First edition 2014

ISBN: 978-0-12-801105-8
ISSN: 0074-7742

For information on all Academic Press publications visit our website at store.elsevier.com

CONTENTS

CONTRIBUTORS

Adam Ameur
Department of Immunology, Genetics and Pathology, and Science for Life Laboratory, Uppsala University, Uppsala, Sweden

Daniel Bottomly
Clinical & Translational Research Institute, Oregon Health and Science University, Portland, Oregon, USA

Murray J. Cairns
School of Biomedical Sciences and Pharmacy, Faculty of Health and Medicine, The University of Newcastle, Callaghan, New South Wales, and The Schizophrenia Research Institute, Sydney, Australia

Lucia Cavelier
Department of Immunology, Genetics and Pathology, Uppsala University, and Uppsala University Hospital, Uppsala, Sweden

Alexandre Colville
Department of Behavioral Neuroscience, Oregon Health & Science University, Portland, Oregon, USA

Mark R. Cookson
Cell Biology and Gene Expression Section, Laboratory of Neurogenetics, National Institute on Aging, Bethesda, Maryland, USA

Antony A. Cooper
Garvan Institute of Medical Research, Darlinghurst, and St Vincent's Clinical School and BABS UNSW, Sydney, New South Wales, Australia

Priscila Darakjian
Department of Behavioral Neuroscience, Oregon Health & Science University, Portland, Oregon, USA

Allissa A. Dillman
Cell Biology and Gene Expression Section, Laboratory of Neurogenetics, National Institute on Aging, Bethesda, Maryland, USA

Mary-Anne Enoch
Laboratory of Neurogenetics, National Institute on Alcohol Abuse and Alcoholism, National Institutes of Health, Bethesda, Maryland, USA

Sean P. Farris
Waggoner Center for Alcohol and Addiction Research, The University of Texas at Austin, Austin, TX 78712

Lars Feuk
Department of Immunology, Genetics and Pathology, and Science for Life Laboratory, Uppsala University, Uppsala, Sweden

David Goldman
Laboratory of Neurogenetics, National Institute on Alcohol Abuse and Alcoholism, National Institutes of Health, Bethesda, Maryland, USA

Boris Guennewig
Garvan Institute of Medical Research, Darlinghurst, and St Vincent's Clinical School and BABS UNSW, Sydney, New South Wales, Australia

Robert Hitzemann
Department of Behavioral Neuroscience, Oregon Health & Science University, Portland, Oregon, USA, and Research Service, Veterans Affairs Medical Center, Portland, Oregon, USA

Ovidiu Dan Iancu
Department of Behavioral Neuroscience, Oregon Health & Science University, Portland, Oregon, USA

Sunita Kawane
Clinical & Translational Research Institute, Oregon Health and Science University, Portland, Oregon, USA

R. Dayne Mayfield
Waggoner Center for Alcohol and Addiction Research, The University of Texas at Austin, Austin, TX 78712

Shannon McWeeney
Division of Bioinformatics & Computational Biology, Department of Medical Informatics & Clinical Epidemiology; Oregon Clinical and Translational Research Institute, and Division of Biostatistics, Public Health & Preventative Medicine, Oregon Health & Science University, Portland, Oregon, USA

Michael Mooney
Division of Bioinformatics & Computational Biology, Department of Medical Informatics & Clinical Epidemiology, Oregon Health & Science University, Portland, Oregon, USA

Ashutosh K. Pandey
Department of Genetics, Genomics and Informatics, Center for Integrative and Translational Genomics, University of Tennessee Health Science Center, Memphis, Tennessee, USA

Robert Searles
Integrative Genomics Laboratory, Oregon Health & Science University, Portland, Oregon, USA

Nikki Walter
Department of Behavioral Neuroscience, Oregon Health & Science University, and Research Service, Veterans Affairs Medical Center, Portland, Oregon, USA

Xi Wang
School of Biomedical Sciences and Pharmacy, Faculty of Health and Medicine, The University of Newcastle, Callaghan, New South Wales, Australia

Robert W. Williams
Department of Genetics, Genomics and Informatics, Center for Integrative and Translational Genomics, University of Tennessee Health Science Center, Memphis, Tennessee, USA

Beth Wilmot
Department of Medical Informatics and Clinical Epidemiology, and Clinical & Translational Research Institute, Oregon Health and Science University, Portland, Oregon, USA

Ammar Zaghlool
Department of Immunology, Genetics and Pathology, and Science for Life Laboratory, Uppsala University, Uppsala, Sweden

Christina L. Zheng
Department of Medical Informatics and Clinical Epidemiology, and Knight Cancer Institute, Oregon Health and Science University, Portland, Oregon, USA

Zhifeng Zhou
Laboratory of Neurogenetics, National Institute on Alcohol Abuse and Alcoholism, National Institutes of Health, Bethesda, Maryland, USA

CHAPTER ONE

Introduction to Sequencing the Brain Transcriptome

Robert Hitzemann*,†,1, Priscila Darakjian*, Nikki Walter*,†, Ovidiu Dan Iancu*, Robert Searles‡, Shannon McWeeney§,¶

*Department of Behavioral Neuroscience, Oregon Health & Science University, Portland, Oregon, USA
†Research Service, Veterans Affairs Medical Center, Portland, Oregon, USA
‡Integrative Genomics Laboratory, Oregon Health & Science University, Portland, Oregon, USA
§Oregon Clinical and Translational Research Institute, Oregon Health & Science University, Portland, Oregon, USA
¶Division of Biostatistics, Public Health & Preventative Medicine, Oregon Health & Science University, Portland, Oregon, USA
[1]Corresponding author: e-mail address: hitzeman@ohsu.edu

Contents

Abstract

High-throughput next-generation sequencing is now entering its second decade. However, it was not until 2008 that the first report of sequencing the brain transcriptome appeared (Mortazavi, Williams, Mccue, Schaeffer, & Wold, 2008). These authors compared short-read RNA-Seq data for mouse whole brain with microarray results for the same sample and noted both the advantages and disadvantages of the RNA-Seq approach. While RNA-Seq provided exon level resolution, the majority of the reads were provided by a small proportion of highly expressed genes and the data analysis was exceedingly complex. Over the past 6 years, there have been substantial improvements in both RNA-Seq technology and data analysis. This volume contains 11 chapters that detail various aspects of sequencing the brain transcriptome. Some of the chapters are very methods driven, while others focus on the use of RNA-Seq to study such diverse areas as development, schizophrenia, and drug abuse. This chapter briefly reviews the transition from microarrays to RNA-Seq as the preferred method for analyzing the brain transcriptome. Compared with microarrays, RNA-Seq has a greater dynamic range, detects both coding and noncoding RNAs, is superior for gene network construction,

International Review of Neurobiology, Volume 116
ISSN 0074-7742
http://dx.doi.org/10.1016/B978-0-12-801105-8.00001-1

detects alternative spliced transcripts, and can be used to extract genotype information, e.g., nonsynonymous coding single nucleotide polymorphisms. RNA-Seq embraces the complexity of the brain transcriptome and provides a mechanism to understand the underlying regulatory code; the potential to inform the brain–behavior–disease relationships is substantial.

1. INTRODUCTION

Next-generation sequencing (NGS) refers to a variety of related technologies, often termed massively parallel sequencing. The first NGS platform (Roche 454) was introduced in 2004. Subsequently, other platforms were released by several manufacturers: Illumina (Solexa), Helicos, Pacific Biosciences, and Life Technologies (ABI). Although the instruments differ in the underlying chemistry and technical approach, the platforms are similar in their capability of producing very large numbers of simultaneous reads relative to traditional methods. Thus, it is now possible to sequence whole genomes, exomes, and transcriptomes for a reasonable cost and effort. The technology of transcriptome sequencing, also known as RNA-Seq, has matured to the point that it is reasonable to propose substituting RNA-Seq for microarray-based assessments of global gene expression. Of particular importance to our laboratories are the advantages RNA-Seq has over microarray platforms when analyzing complex rodent crosses, e.g., heterogeneous stocks (HSs). However, the same argument can be made when analyzing any outbred population, including humans. Of particular relevance to the brain transcriptome are the advantages RNA-Seq has over microarrays in analyzing alternative splicing. This chapter provides a starting point for understanding the emergence of RNA-Seq and emphasizes *transcriptome/behavior* relationships.

2. FROM MICROARRAYS TO RNA-Seq

Cirelli and Tononi (1999) were among the first to report genome-wide brain gene expression profiling associated with a behavioral phenotype; both mRNA differential display and cDNA arrays were used to examine the effects of sleep deprivation on rat prefrontal cortex gene expression. Sandberg et al. (2000) used Affymetrix microarrays to detect differences in brain gene expression between two inbred mouse strains (C57BL/6J [B6] and 129SvEv [129; now 129S6/SvEvTac]). Importantly, these authors observed that some differentially expressed (DE) genes were found in chromosomal regions with known behavioral quantitative trait loci (QTLs). For example, *Kcnj9* that encodes for GIRK3, an inwardly rectifying potassium channel, was DE (higher expression in the 129 strain) and is located on distal chromosome 1 in a region where QTLs had been identified for locomotor activity, alcohol and pentobarbital withdrawal, open-field emotionality, and certain aspects of fear-conditioned behavior (see Sandberg et al., 2000). Subsequently, Buck and colleagues (Buck, Milner, Denmark, Grant, & Kozell, 2012; Kozell, Walter, Milner, Wickman, & Buck, 2009) have shown that *Kcnj9* is a quantitative trait gene (QTG) for the withdrawal phenotypes. Over the past decade, this alignment of global brain gene expression data and behavioral QTLs has been reported in numerous publications and discussed in numerous symposia and reviews (e.g., Bergeson et al., 2005; Farris & Miles, 2012; Hoffman et al., 2003; Matthews et al., 2005; Mcbride et al., 2005; Saba et al., 2011; Sikela et al., 2006; Tabakoff et al., 2009). The association gained further support as the focus turned to genes whose expression appeared to be regulated by a factor or factors within the behavioral QTL interval. Web tools have been developed to facilitate integrating behavioral and brain microarray data (e.g., www.genenetwork.org and http://phenogen.ucdenver.edu/PhenoGen/index.jsp; Chapter 8). This integration has been successful in detecting several candidate QTGs for behavioral phenotypes (see, e.g., Hitzemann et al., 2004; Hofstetter et al., 2008; Mulligan et al., 2006; Saba et al., 2011; Tabakoff et al., 2009).

The alignment of DE genes with a behavioral phenotype can be further examined using a variety of secondary analyses, e.g., examining if the DE genes cluster within known gene ontology categories (Pavlidis, Qin, Arango, Mann, & Sibille, 2004) or are part of a known protein–protein interaction network (Bebek & Yang, 2007; Feng, Shaw, Rosen, Lin, & Kibbe, 2012). DE genes can also be grouped on the basis of common transcription factors and other regulatory elements (e.g., Vadigepalli, Chakravarthula, Zak, Schwaber, & Gonye, 2003). In addition to DE genes, microarrays have also facilitated gene coexpression-based analyses, such as the Weighted Gene Coexpression Network Analysis (WGCNA; Horvath et al., 2006; Zhang & Horvath, 2005). The rationale behind these approaches is that coexpressed genes frequently code for interacting proteins, which in turn leads to new insights into protein function(s) and in some cases leads to discovery of protein function (Zhao et al., 2010). Coexpression analysis has been used to analyze differences in functional brain organization between nonhuman primates and humans (Oldham, Horvath, & Geschwind, 2006), regional differences in the functional organization of the human brain (Oldham et al., 2008), and the molecular pathology of autism (Voineagu et al., 2011) and alcoholism (see Chapter 11).

Despite these successes, microarray-based approaches are not without problems. First, differences in brain gene expression among genetically unique individuals or lines selected for behavioral traits are generally small; reported differences of 15–25% are not uncommon. To some extent, these small variations occur because hybridization isotherms for oligonucleotide arrays are frequently not linear due to probe saturation (Pozhitkov, Boube, Brouwer, & Noble, 2010).

A second problem with oligonucleotide arrays is the effect of single nucleotide polymorphisms (SNPs; Duan, Pauley, Spindel, Zhang, & Norgren, 2010; Peirce et al., 2006; Sliwerska et al., 2007; Walter et al., 2009, 2007). Rodent oligonucleotide arrays are based upon the sequence of the B6 mouse or Brown-Norway (BN) rat. Even inbred strains closely related to the B6 or BN strains may differ by several million SNPs (see, e.g., Keane et al., 2011), which in turn can cause significant hybridization artifacts (Walter et al., 2009, 2007). Masking for SNPs can improve this situation but results in deleting probes or even an entire probe set from the analysis. Walter et al. (2009) used NGS to address the SNP problem, building upon the repeated observation that, when comparing gene expression in the B6 and DBA/2J (D2) inbred mouse strains (or crosses and selected lines formed from these strains) and after masking for known SNPs in the D2 strain, there remained an excess of genes showing higher expression in the B6 strain. Similarly, this was also observed in the case of cis-eQTLs showing higher expression associated with the B6 allele (see Mulligan et al., 2006; Peirce et al., 2006; Walter et al., 2007). The two possible explanations for these observations were the following: (a) gene expression was actually higher in the B6 strain or (b) there were many uncharacterized D2 SNPs, which led to decreased binding of D2-derived target on probes containing the SNP locale. Preliminary direct sequencing and quantitative PCR data pointed to missing SNPs. NGS was used to analyze a 3-Mbp region of Chr 1 (171.5–174.5 Mbp) that was enriched in a number of behavioral QTLs and transcripts DE between the B6 and D2 strains. B6 and D2 BAC clones tiled across the region were sequenced using the short-read Illumina IIx and ABI SOLiD 2 platforms. The results obtained (30–100 $\times$ coverage) illustrated that there were 160% more SNPs in the region than previously reported (Walter et al., 2009); these data have been confirmed

(Keane et al., 2011; R. Williams, unpublished observations). The integration of these SNPs to the mask markedly reduced the disparity in DE genes between the B6 and D2 strains.

A third problem with oligonucleotide arrays is the annotation and summarization issues associated with predefined reporters/probes (e.g., Allison, Cui, Page, & Sabripour, 2006; Lu, Lee, Salit, & Cam, 2007). Interestingly, on some arrays, a significant number of the represented transcripts are actually long noncoding RNAs (ncRNAs) (see, e.g., Liao et al., 2011). But tens of thousands of ncRNAs, many of which have important regulatory functions (Mattick, 2011; also see Chapter 7), are not represented on the arrays.

A fourth problem with oligonucleotide arrays is that 3′UTR-orientated microarrays provide relatively little information about alternative splicing, which is particularly high in brain (Johnson et al., 2009; Li, Lee, & Black, 2007; Licatalosi & Darnell, 2006; Mortazavi, Williams, Mccue, Schaeffer, & Wold, 2008). The Affymetrix Mouse 1.0 Exon ST array collects data on alternative splicing, but when used to detect differential alternative splicing, it is particularly sensitive to the "SNP effect" due to the smaller number of probes per probe set (Laderas et al., 2011).

3. NGS PLATFORMS

There are several excellent reviews of the various NGS platforms (e.g., Mardis, 2008, 2011; Martin & Wang, 2011; Metzker, 2010; Ozsolak & Milos, 2011; Rothberg et al., 2011). Understanding in some depth how the platforms work is critical to understanding where errors develop and are propagated from sample preparation to alignment to data analysis. The differences in platforms will not be discussed here. We simply note that for RNA-Seq experiments, the majority have used the Illumina platform (see, e.g., Costa, Angelini, De Feis, & Ciccodicola, 2010). The promise of a high-throughput, high read instrument with minimal library preparation remains a promise. Such an instrument would be particularly welcome for sequencing the brain transcriptome given the diversity of cell types present and the numerous comparisons that could be made.

4. RNA-Seq OVERVIEW

The first and perhaps the most important step of an RNA-Seq experiment is the same as that for a microarray experiment, the isolation of high-quality RNA. Although both RNA-Seq and microarrays can be used on fragmented RNA such as that found in formalin-fixed-paraffin-embedded samples, the biases present in such samples for genome-wide sequencing are difficult to assess. RNA quality is routinely examined on the Agilent BioAnalyzer or a similar instrument; an RNA integrity number (RIN) of ≥ 8 is generally considered high quality. Unfortunately for brain samples, the amount of beginning tissue may be very small, and obtaining a reliable RIN or even accurately measuring the amount of RNA may be difficult. Even within very discrete brain regions, there are multiple cell types, and some experiments need to focus on a specific subset of cells or even a single cell. Eberwine and colleagues at the University of Pennsylvania have pioneered techniques for the linear amplification of small amounts of RNA; an online audio describing the procedures when beginning with only fentograms of material is available (Morris, Singh, & Eberwine, 2011). Many RNA-Seq experiments begin with postmortem material that has been stored, often under variable conditions, including differences in the postmortem interval (PMI). Depending on the length of the PMI, the RNA in a sample may be moderately to significantly degraded as assessed by the RIN and other *Q/C* measures. For samples with integrity numbers <6, one should consider ribosome depletion as opposed to a polyA+preparation. Ribosome-depleted samples also have the advantage of including coding and ncRNAs which are not polyadenylated; tiling array data suggest that more than 40% of transcripts are not polyadenylated (Cheng et al., 2005). Cui et al. (2010) have compared RNA-Seq of RiboMinus (rmRNA) and poly(A)-selected (mRNA) samples; the starting total RNA was extracted from BALB/c mouse whole brain. The authors found (on a percent basis) that there were marked read distribution differences between samples. The percentage of known exon reads was twice as high in the mRNA sample (60%), while the percentage of both intronic and intergenic reads was twice as high in the rmRNA sample (25% and 44%, respectively). Both samples detected reads in essentially the same population of RefSeq-defined genes, i.e., there was not a substantial read bias. So the use of rRNA-depleted or poly(A)-selected RNA depends on the questions being asked and the estimated read density per sample. Data collected in our laboratory and elsewhere (Bottomly et al., 2011; Marioni, Mason, Mane, Stephens, & Gilad, 2008; Mortazavi et al., 2008) have found that 20–40 million reads are generally adequate for most estimates of gene expression. If the goal is to quantitatively measure expression at the exon level, then the read density must be increased significantly, perhaps by an order of magnitude (see Labaj et al., 2011; Lee, Mayfield, & Harris, 2014). Such exon level measurements are obviously best suited for poly(A)-selected samples, especially when one is dealing with multiple biological replicates and assuming resources are reasonably limited; i.e., it is very likely that it will be necessary to multiplex samples. But if one is only interested in gene expression and can maintain total exonic read density at 20–40 million, then rmRNA could be used, and significant information on ncRNAs and mRNAs without a poly(A) tail can be obtained. Cui et al. (2010) also used a procedure that facilitates both the quantification of transcripts derived from opposite strands and determining the directionality of transcription (Costa et al., 2010; Martin & Wang, 2011; see also Chapter 2). Using the strand-specific data, Cui et al. (2010) made several salient observations: (a) 99.9% of the junction reads are in the sense orientation; (b) nearly all expressed genes have natural antisense transcripts (the proportion may be as high as 70% of expressed genes [Katayama et al., 2005]); (c) poorly expressed genes tend to have more pronounced antisense transcription; and (d) the antisense transcripts are enriched in the promoter and terminal transcript regions. This enrichment is likely the result of divergent transcription initiation of RNA polymerase II (Core, Waterfall, & Lis, 2008; Preker et al., 2008).

Samples from very discrete brain regions are often prepared by laser capture microdissection (LCM). Given the steps involved in preparing the LCM samples, including staining and dehydration, care needs to be taken to maintain RNA quality. Chen et al. (2011) appear to be the first to couple LCM and RNA-Seq to examine brain gene expression. They examined rat GABAergic neurons projecting from the nucleus accumbens to the ventral pallidum. Cells were labeled using the retrograde tracer, Fluorogold. Approximately 1500 cells were labeled and isolated by LCM in each of four animals; this in turn produced ~4 ng of RNA per animal, and the average RIN was 8.1. Samples were independently amplified for microarray and RNA-Seq; for genes detected on both platforms, the correlation for gene expression was ~0.7. Not surprisingly, the correlation was better for the highly expressed genes. We have used LCM to examine gene expression in discrete regions of the mouse brain (prelimbic cortex, nucleus accumbens shell, and central nucleus of the amygdala; Colville, AM & Hitzemann, RJ unpublished observations). Sufficient high-quality RNA was obtained from each sample (>100 ng) that amplification was not necessary. Although the samples were only used for RNA-Seq, the data obtained for the nucleus accumbens shell appear at the gene level to be very similar to data previously obtained for the ventral striatum when using microarrays (e.g., Iancu et al., 2010).

In addition to examining gene expression in discrete brain regions and discrete cell types, for some applications, it is desirable to assess the synaptic transcriptome (see, e.g., Eipper-Mains, Eipper, & Mains, 2012). A key mechanism of synaptic plasticity is the local synthesis of proteins from synaptic mRNA. Techniques for isolating synaptosomes from adult brains and growth cones from developing brains are well established using gradient centrifugation (e.g., Hitzemann & Loh, 1978). Synaptoneurosomes are prepared by filtration of tissue homogenate through a series of filters to obtain a fraction that is enriched in pinched-off dendritic spines (Lugli & Smalheiser, 2013). Regardless of preparation, once isolated, these fractions can be subjected to sequencing as outlined earlier (e.g., Eipper-Mains et al., 2011). A key to the use of these fractions will be assessments of subcellular contamination.

The next step in an RNA-Seq experiment involves the synthesis of high-quality double-stranded (ds) cDNA. The most widely used procedure fragments the RNA before reverse transcription, followed by second-strand synthesis. This approach has the advantage of minimizing the effects of secondary RNA structure on first-strand synthesis. If the adapters needed for the sequencing are added after the ds cDNA is formed, information on strandedness is lost. There are several procedures, including ligating adapters to the fragmented RNA, that will maintain strand information (Ingolia, Ghaemmaghami, Newman, & Weissman, 2009; Li et al., 2008; Parkhomchuk et al., 2009). The alternative to using fragmented RNA is to synthesize the cDNA from intact RNA and then fragment. This approach has a clear advantage for platforms that are capable of long to very long reads. For the Illumina, SOLiD, and 454 platforms, the final step prior to the actual sequencing is the clonal amplification of the fragmented cDNA. Both 454 and SOLiD use emulsion PCR on a bead surface, while Illumina uses enzymatic amplification on a glass surface (flow cell). The sequencing and detection methods differ among the three platforms (see Mardis, 2011 and Metzker, 2010 for details). The 454 sequencer use a polymerase-mediated incorporation of unlabeled nucleotides; detection is via light emitted by secondary reactions with the released PPi. Illumina also uses a polymerase-mediated sequencing but uses end-blocked fluorescent nucleotides in a protocol similar to traditional Big Dye sequencing; detection comes from following the incorporation of the nucleotide attached fluorescent tags. SOLiD sequencing uses the ligase-mediated addition of 2-base encoded fluorescent oligonucleotides; detection is from fluorescent emission of the incorporated oligonucleotides. The SOLiD system differs from Illumina and 454 in that each base is determined twice. The quality of the base calls for all three platforms is very good. Quality is measured in terms of a Phred

Score (Q), which was originally developed to assess base calls for the human genome project (Ewing, Hillier, Wendl, & Green, 1998). A Q score of 20 indicates a 99% accuracy rate, and a score of 30 indicates a 99.9% accuracy rate. Q30 values are routinely obtained for NGS platforms. Typically, the Q value decreases with increasing read length.

5. RNA-Seq AND DATA ANALYSIS

Before commenting on the analysis of RNA-Seq data, it is useful to recount the analysis controversies that arose with the introduction of microarrays. In 1999, *Nature Genetics* devoted an entire issue (volume 21—January) to microarrays. Cautionary concerns were raised around issues of data analysis (Lander, 1999). Microarray experiments, at the time, were generally expensive, limiting sample sizes. Small sample sizes and thousands of independent observations per sample were seen as a prescription for statistical disaster. Initial attempts to deal with this problem frequently involved using a nonstatistical threshold for a meaningful difference, e.g., a twofold difference in expression. This approach frequently worked well in some applications, e.g., when comparing cancerous and noncancerous tissue; however, this approach was destined not to work well in brain, where differences in expression among experimental groups were much smaller. Initially, journal reviewers, editors, and study sections panned microarray experiments as being "fishing expeditions," with no clear hypothesis. The idea of discovery science as a valuable strategy was a minority opinion.

Despite the obstacles, microarray experiments eventually flourished; technology and analysis methods improved. One might have predicted that the microarray experience would have laid the groundwork for the acceptance of NGS. However, the introduction of the 454 sequencer (Margulies et al., 2005) was met with a similar resistance; the argument was made that the data sets were so large that only one of the established genome centers would have the necessary bioinformatics expertise. But as NGS technology improved so did the analytic approaches, such that by 2007/2008, RNA-Seq data appeared from several different laboratories (Marioni et al., 2008; Mortazavi et al., 2008; Sugarbaker et al., 2008; Torres, Metta, Ottenwalder, & Schlotterer, 2008; Weber, Weber, Carr, Wilkerson, & Ohlrogge, 2007). Workflows emerged that addressed the measurement of not only DE genes but also differential alternative splicing and the detection of novel transcripts (Marioni et al., 2008). Bullard, Purdom, Hansen, and Dudoit (2010) examined a number of statistical issues associated with using

RNA-Seq to detect DE genes. Similar to Marioni et al. (2008), they found that most sources of technical variation had only small effects on detecting DE transcripts. The most significant effect on DE transcripts was data normalization. Bullard et al. (2010) concluded that their "main novel finding is the extent to which normalization affects differential expression results: sensitivity varies more between normalization procedures than between test statistics. . .we propose scaling gene counts by a quantile of the gene count distribution (the upper-quartile)."

This volume contains several chapters that address in some detail the analysis of RNA-Seq data (see Chapters 2, 3, and 11); these chapters especially emphasize the evolution of RNA-Seq analysis over the past 3–4 years. In addition to improvements in analysis strategy, sample power has in general improved with decreasing costs and the ability to multiplex samples with adequate read depth (at least at a level sufficient for gene summarization statistics). If one is interested in quantifying alternative splicing, then substantially greater read depth is required (see, e.g., Lee et al., 2014).

RNA-Seq data have some unique properties that affect the strategies for data analysis (Garber, Grabherr, Guttman, & Trapnell, 2011). First, unlike microarray data where the output is fluorescence intensity (more or less a continuous measure), the output from an RNA-Seq experiment is digital in the form of read counts. For the microarray experiment, familiar statistics such as a t-test or ANOVA are appropriate (assuming variances are equal); for RNA-Seq data, these statistics are not directly applicable. Robinson, Mccarthy, and Smyth (2009) proposed the use of the empirical analysis of digital gene expression in R (edgeR), a variant of a procedure used to analyze SAGE data. edgeR models count data using an overdispersed Poisson model and use an empirical Bayes' procedure to moderate the degree of overdispersion across genes; the overdispersion reflects the biological variation among samples (Robinson et al., 2009). An implementation of edgeR to mouse brain RNA-Seq data is found in Bottomly et al. (2011).

Second, RNA-Seq data are biased in several important ways. First, the majority of the counts are produced by a small number (<10% of the total) of very highly expressed genes. Thus, many genes of interest may have only moderate to low counts. Also, for genes with equal levels of expression, the long genes will be overrepresented, distorting the relative expression among genes. Similarly, within a given gene, long exons are overrepresented. Normalization and weighting algorithms can be used to address these issues, but they in turn may introduce new biases (Bullard et al., 2010).

Third, RNA-Seq provides a substantial amount of data with very low read counts, which will be quite variable (see, e.g., Cui et al., 2010), and thus, regardless of the analytic strategy, makes detecting DE genes difficult.

Fourth, RNA-Seq data includes multireads, i.e., reads that map equally well to multiple genomic locations. The multireads arise predominantly from conserved domains in paralogous genes and from repeats (Costa et al., 2010). Mortazavi et al. (2008) found that, in the mouse brain, 76% of the 25-bp transcriptome sequence segments uniquely mapped; 6% mapped 2–10 times in the genome; and the remainder mapped more than 10 times. Depending on the gene model used and assuming a high-read density, ignoring the multireads may only have a minimal effect on detecting DE genes. But one can easily contrive a situation involving alternative splicing and multireads where this would not be the case.

Fifth, RNA-Seq collects data across splice junctions that (a) are ignored by many alignment tools and (b) may be unknown. While there are <25,000 known protein-coding genes in the mammalian genome, the number of gene-related transcripts may well be 10–20 times higher (Pan, Shai, Lee, Frey, & Blencowe, 2008; Johnson et al., 2009). Given the heterogeneous nature of brain tissue, the complexity problem is significantly amplified. Tools are available that detect splice junctions and will estimate the minimum number of gene isoforms that account for the observed data (Guttman et al., 2010; Katz, Wang, Airoldi, & Burge, 2010; Trapnell, Pachter, & Salzberg, 2009; Trapnell et al., 2012, 2010). Roberts, Pimentel, Trapnell, and Pachter (2011) illustrate a procedure that makes use of annotated model organism genomes, such as those available for the laboratory mouse and rat. For both correctly aligning multireads and splice junctions, paired-end sequencing is a useful approach. The downside is the added expense of sequencing the cDNA fragment from both ends.

Sixth, RNA-Seq data can be used to detect allele-specific expression and both synonymous and nonsynonymous SNPs within gene-coding sequences. This application may be particularly useful in complex crosses such as the HS-CC (Iancu et al., 2010) where RNA-Seq can provide detailed genotype information. In the RNA-Seq context, the advantages of using model organisms with a well-annotated genome cannot be underestimated (Martin & Wang, 2011). Reference genome alignment is computationally simpler and faster as the problem is reduced from assembling millions of reads to assembling a much smaller number of reads to known loci. For both the mouse and rat, the reference genomic sequence was obtained using tiled BAC clones, and thus, there are essentially no gaps.

But if one believes that there are a substantial number of missed exons, then some combination of reference-based and *de novo* alignment may be the most effective approach (Martin & Wang, 2011). The Mouse Genomes Project (Keane et al., 2011) released genomic sequence data for 17 inbred strains; the data are aligned to the B6 reference strain. It is important to note that these data are not equivalent to the reference genome. The data were acquired using a short-read NGS platform (Illumina), which naturally means that in regions of high repeats/low genetic complexity, it is not possible to correctly align the sequence data. For the standard laboratory strains, this effect is most notable on the proximal aspect of chromosome 7 (Keane et al., 2011). RNA-Seq data are also available for six tissues from a B6D2 F_1 hybrid and for whole brain transcriptome data from 15 strains. These data sets can be freely downloaded and provide an excellent training set for RNA-Seq analysis.

6. SEQUENCING THE BRAIN TRANSCRIPTOME

PubMed lists 2702 RNA-Seq publications (6/1/14) with the first appearing in June 2008 (Nagalakshmi et al., 2008); the number has steadily increased from 11 in 2008, to 34 in 2009, to 127 in 2010, to 339 in 2011, to 639 in 2012, and to 1123 in 2013. Of these publications, 162 are also coded as "RNA-Seq and Brain" (~6% of total). However, this number most certainly represents a low estimate of the number of publications where RNA-Seq is used to assess the brain transcriptome or brain surrogates such as induced pluripotent stem cells. Nonetheless, sequencing the brain transcriptome is still an emerging area. The first publication using RNA-Seq to compare brain gene expression between two inbred mouse strains appeared in 2011 (Bottomly et al., 2011). The first application of RNA-Seq to brain WGCNA appeared in 2012 (Iancu et al., 2012). Iancu and colleagues extend this network approach to cosplicing in Chapter 4 building upon the earlier work of Dai, Li, Liu, & Zhou (2012) and Aschoff et al. (2013). Mudge et al. (2008) is an early example of using RNA-Seq in a neuropsychiatric context (schizophrenia) but as noted by Wang and Cairns in Chapter 6 most of the work in this area has appeared within the last 2 years. Chapter 7 details just how quickly our understanding of the functional roles of the ncRNAs has changed due to the introduction of RNA-Seq; further, Guennewig and Copper make compelling arguments for the roles of the ncRNAs in both normal brain function and disease states. Alternative splicing is higher in the brain as compared to other tissues

(Johnson et al., 2009); RNA-Seq facilitates a genome-wide assessment of alternative splicing which is key to understanding both brain development (Dillman and Cookson—Chapter 9) and normal brain function (Zaghlool et al.—Chapter 5). Lewohl et al. (2000) were among the first to use microarrays to study the human brain transcriptome, comparing alcoholics and matched controls. Zhou et al. (Chapter 10) and Farris and Mayfield (Chapter 11) illustrate how readily investigators in the fields of alcoholism and drug abuse research have adopted RNA-Seq to examine human samples. Although still in the preliminary data stage, RNA-Seq is being extensively used to examine the brain transcriptome in nonhuman primates chronically exposed to alcohol (Grant KA, Hitzemann RJ, Darakjian P, & Iancu OD, unpublished observations).

For many investigators, the interest in RNA-Seq and the brain transcriptome is not matched by available funding. Williams and Pandey (Chapter 8) describe a number of freely available mouse resources that allows one to interrogate the relationship(s) between phenotypes and RNA-Seq data. A key element to these resources has been the use of mouse reference populations such as the BXD recombinant inbred series and the Collaborative Cross (Churchill et al., 2004).

RNA-Seq has many applications outside of those mentioned in this volume. One area where it proven to have particular value has been in the examination of the brain transcriptome in nonmodel organisms. Frequently, these organisms have a significant behavioral and/or evolutionary value. A *de novo* assembly of the data can be used in the absence of high-quality genomic sequence data by aligning the reads to conserved protein sequence and/or the annotated genomes of closely related organisms. Four examples are described. Fraser, Weadick, Janowitz, Rodd, and Hughes (2011) assembled brain transcriptome data from the guppy (*Poecilia reticulata*) and were able to detect both sex-specific expression and the effect of predator (*Rivulus hartii*) exposure. Malik et al. (2011) examined the brain transcriptome of blind subterranean mole rat (*Spalax galili*); some modest differences in brain gene expression were found after prolonged exposure to low oxygen concentrations (a normally occurring condition in the underground tunnels). Tzika, Helaers, Schramm, and Milinkovitch (2011) used RNA-Seq in an evolutionary context to compare brain transcriptomes of four divergent reptilian and one reference avian species: the Nile crocodile, the corn snake, the bearded dragon, the red-eared turtle, and the chicken. Somewhat surprisingly, the data suggest that the turtle was evolutionarily closer to the crocodile than was expected. All three of these examples used the Roche

454 platform for sequencing; the longer reads compared with other instruments facilitated the *de novo* transcriptome assembly. Balakrishnan et al. (2014) used RNA-Seq to examine the relationships among the brain transcriptome, avian vocal communication, and social behavior. Brain transcriptomes were sequenced for three emberizid model systems, song sparrow *Melospiza melodia*, white-throated sparrow *Zonotrichia albicollis*, and Gambel's white-crowned sparrow *Zonotrichia leucophrys* gambelii. Each of the assemblies covered fully or in part, over 89% of the previously annotated protein-coding genes in the zebra finch *Taeniopygia guttata*, with 16,846, 15,805, and 16,646 unique BLAST hits in song, white-throated and white-crowned sparrows, respectively. As in previous studies, these authors found tissue of origin (auditory forebrain versus hypothalamus and whole brain) as an important determinant of the expression profile.

7. CONCLUSIONS

Historically, the main arguments against using RNA-Seq (as opposed to using microarrays) have been cost and difficulties with data analysis. Over the past 6 years, technical improvements have and will continue to reduce costs; if the primary goal is gene-wide summarization, transcriptome samples can now be multiplexed and sequenced at adequate depth for less than $200/ sample (not including the cost of library preparation). RNA-Seq data analysis remains substantially more complex than a comparable microarray analysis. The data sets are much larger and are generally not suitable for analysis on a personal computer. While the analysis of RNA-Seq data could still be described as not for the "faint of heart," a rapidly improving data analysis trajectory is clear as indicated by the numerous reports described in this volume. RNA-Seq has several distinct advantages over microarray-based approaches to transcriptome analysis. RNA-Seq data have a significantly greater dynamic range (there are no probe saturation effects); the gene expression data are not biased to the 3′UTR (although there is a bias to the most highly expressed and longest genes) and data are collected on both alternative splicing and inter- and intragenic ncRNAs. Overall, RNA-Seq embraces the complexity of the transcriptome and provides a mechanism to understand the underlying regulatory code.

ACKNOWLEDGMENTS

This study was supported in part by grants MH 51372, AA 11034, AA 13484, and a grant from the Veterans Affairs Research Service. The authors want to thank Dr. Kristin Demarest and Kris Thomason for editorial assistance in preparing the review.

REFERENCES

Allison, D. B., Cui, X., Page, G. P., & Sabripour, M. (2006). Microarray data analysis: From disarray to consolidation and consensus. *Nature Reviews. Genetics*, *7*, 55–65.

Aschoff, M., Hotz-Wagenblatt, A., Glatting, K. H., Fischer, M., Eils, R., & König, R. (2013). SplicingCompass: differential splicing detection using RNA-seq data. *Bioinformatics*, *29*, 1141–1148.

Balakrishnan, C. N., Mukai, M., Gonser, R. A., Wingfield, J. C., London, S. E., Tuttle, E. M., et al. (2014). Brain transcriptome sequencing and assembly of three songbird model systems for the study of social behavior. *PeerJ*, *2*, e396.

Bebek, G., & Yang, J. (2007). Pathfinder: Mining signal transduction pathway segments from protein-protein interaction networks. *BMC Bioinformatics*, *8*, 335.

Bergeson, S. E., Berman, A. E., Dodd, P. R., Edenberg, H. J., Hitzemann, R. J., Lewohl, J. M., et al. (2005). Expression profiling in alcoholism research. *Alcoholism: Clinical and Experimental Research*, *29*, 1066–1073.

Bottomly, D., Walter, N. A., Hunter, J. E., Darakjian, P., Kawane, S., Buck, K. J., et al. (2011). Evaluating gene expression in C57BL/6J and DBA/2J mouse striatum using RNA-Seq and microarrays. *PLoS One*, *6*, e17820.

Buck, K. J., Milner, L. C., Denmark, D. L., Grant, S. G., & Kozell, L. B. (2012). Discovering genes involved in alcohol dependence and other alcohol responses: Role of animal models. *Alcohol Research: Current Reviews*, *34*(3), 367–374.

Bullard, J. H., Purdom, E., Hansen, K. D., & Dudoit, S. (2010). Evaluation of statistical methods for normalization and differential expression in mRNA-Seq experiments. *BMC Bioinformatics*, *11*, 94.

Chen, H., Liu, Z., Gong, S., Wu, X., Taylor, W. L., Williams, R. W., et al. (2011). Genome-wide gene expression profiling of nucleus accumbens neurons projecting to ventral pallidum using both microarray and transcriptome sequencing. *Frontiers in Neuroscience*, *5*, 98.

Cheng, J., Kapranov, P., Drenkow, J., Dike, S., Brubaker, S., Patel, S., et al. (2005). Transcriptional maps of 10 human chromosomes at 5-nucleotide resolution. *Science*, *308*, 1149–1154.

Churchill, G. A., Airey, D. C., Allayee, H., Angel, J. M., Attie, A. D., & Beatty, J. (2004). The Collaborative Cross, a community resource for the genetic analysis of complex traits. *Nature Genetics*, *36*, 1133–1137.

Cirelli, C., & Tononi, G. (1999). Differences in brain gene expression between sleep and waking as revealed by mRNA differential display and cDNA microarray technology. *Journal of Sleep Research*, *8*(Suppl. 1), 44–52.

Core, L. J., Waterfall, J. J., & Lis, J. T. (2008). Nascent RNA sequencing reveals widespread pausing and divergent initiation at human promoters. *Science*, *322*, 1845–1848.

Costa, V., Angelini, C., De Feis, I., & Ciccodicola, A. (2010). Uncovering the complexity of transcriptomes with RNA-Seq. *Journal of Biomedicine & Biotechnology*, *2010*, 853916.

Cui, P., Lin, Q., Ding, F., Xin, C., Gong, W., Zhang, L., et al. (2010). A comparison between ribo-minus RNA-sequencing and polya-selected RNA-sequencing. *Genomics*, *96*, 259–265.

Dai, C., Li, W., Liu, J., & Zhou, X. J. (2012). Integrating many co-splicing networks to reconstruct splicing regulatory models. *BMC Systems Biology*, *16*, S1–S17.

Duan, F., Pauley, M. A., Spindel, E. R., Zhang, L., & Norgren, R. B., Jr. (2010). Large scale analysis of positional effects of single-base mismatches on microarray gene expression data. *BioData Mining*, *3*, 2.

Eipper-Mains, J. E., Eipper, B. A., & Mains, R. E. (2012). Global approaches to the role of miRNAs in drug-induced changes in gene expression. *Frontiers in Genetics*, *3*, 109.

Eipper-Mains, J. E., Kiraly, D. D., Palakodeti, D., Mains, R. E., Eipper, B. A., & Graveley, B. R. (2011). MicroRNA-Seq reveals cocaine-regulated expression of striatal microRNAs. *RNA*, *17*(8), 1529–1543.

Ewing, B., Hillier, L., Wendl, M. C., & Green, P. (1998). Base-calling of automated sequencer traces using phred. I. Accuracy assessment. *Genome Research*, *8*, 175–185.

Farris, S. P., & Miles, M. F. (2012). Ethanol modulation of gene networks: Implications for alcoholism. *Neurobiology of Disease*, *45*, 115–121.

Feng, G., Shaw, P., Rosen, S. T., Lin, S. M., & Kibbe, W. A. (2012). Using the bioconductor gene answers package to interpret gene lists. *Methods in Molecular Biology*, *802*, 101–112.

Fraser, B. A., Weadick, C. J., Janowitz, I., Rodd, F. H., & Hughes, K. A. (2011). Sequencing and characterization of the guppy (Poecilia reticulata) transcriptome. *BMC Genomics*, *12*, 202.

Garber, M., Grabherr, M. G., Guttman, M., & Trapnell, C. (2011). Computational methods for transcriptome annotation and quantification using RNA-Seq. *Nature Methods*, *8*, 469–477.

Guttman, M., Garber, M., Levin, J. Z., Donaghey, J., Robinson, J., Adiconis, X., et al. (2010). Ab initio reconstruction of cell type-specific transcriptomes in mouse reveals the conserved multi-exonic structure of lincRNAs. *Nature Biotechnology*, *28*, 503–510.

Hitzemann, R. J., & Loh, H. H. (1978). High-affinity GABA and glutamate transport in developing nerve ending particles. *Brain Research*, *159*(1), 29–40.

Hitzemann, R., Reed, C., Malmanger, B., Lawler, M., Hitzemann, B., Cunningham, B., et al. (2004). On the integration of alcohol-related quantitative trait loci and gene expression analyses. *Alcoholism: Clinical and Experimental Research*, *28*, 1437–1448.

Hoffman, P. L., Miles, M., Edenberg, H. J., Sommer, W., Tabakoff, B., Wehner, J. M., et al. (2003). Gene expression in brain: A window on ethanol dependence, neuroadaptation, and preference. *Alcoholism: Clinical and Experimental Research*, *27*, 155–168.

Hofstetter, J. R., Hitzemann, R. J., Belknap, J. K., Walter, N. A., Mcweeney, S. K., & Mayeda, A. R. (2008). Characterization of the quantitative trait locus for haloperidol-induced catalepsy on distal mouse chromosome 1. *Genes, Brain, and Behavior*, *7*, 214–223.

Horvath, S., Zhang, B., Carlson, M., Lu, K. V., Zhu, S., Felciano, R. M., et al. (2006). Analysis of oncogenic signaling networks in glioblastoma identifies ASPM as a molecular target. *Proceedings of the National Academy of Sciences of the United States of America*, *103*, 17402–17407.

Iancu, O. D., Darakjian, P., Walter, N. A., Malmanger, B., Oberbeck, D., Belknap, J., et al. (2010). Genetic diversity and striatal gene networks: Focus on the heterogeneous stock-collaborative cross (hs-cc) mouse. *BMC Genomics*, *11*, 585.

Iancu, O. D., Kawane, S., Bottomly, D., Searles, R., Hitzemann, R., & McWeeney, S. (2012). Utilizing RNA-Seq data for de novo coexpression network inference. *Bioinformatics*, *28*(12), 1592–1597.

Ingolia, N. T., Ghaemmaghami, S., Newman, J. R., & Weissman, J. S. (2009). Genome-wide analysis in vivo of translation with nucleotide resolution using ribosome profiling. *Science*, *324*, 218–223.

Johnson, M. B., Kawasawa, Y. I., Mason, C. E., Krsnik, Z., Coppola, G., Bogdanovic, D., et al. (2009). Functional and evolutionary insights into human brain development through global transcriptome analysis. *Neuron*, *62*, 494–509.

Katayama, S., Tomaru, Y., Kasukawa, T., Waki, K., Nakanishi, M., Nakamura, M., et al. (2005). Antisense transcription in the mammalian transcriptome. *Science*, *309*, 1564–1566.

Katz, Y., Wang, E. T., Airoldi, E. M., & Burge, C. B. (2010). Analysis and design of RNA sequencing experiments for identifying isoform regulation. *Nature Methods*, *7*, 1009–1015.

Keane, T. M., Goodstadt, L., Danecek, P., White, M. A., Wong, K., Yalcin, B., et al. (2011). Mouse genomic variation and its effect on phenotypes and gene regulation. *Nature*, *477*, 289–294.

Kozell, L. B., Walter, N. A., Milner, L. C., Wickman, K., & Buck, K. J. (2009). Mapping a barbiturate withdrawal locus to a 0.44 Mb interval and analysis of a novel null mutant identify a role for Kcnj9 (GIRK3) in withdrawal from pentobarbital, zolpidem, and ethanol. *Journal of Neuroscience*, *29*(37), 11662–11673.

Labaj, P. P., Leparc, G. G., Linggi, B. E., Markillie, L. M., Wiley, H. S., & Kreil, D. P. (2011). Characterization and improvement of RNA-Seq precision in quantitative transcript expression profiling. *Bioinformatics*, *27*, i383–i391.

Laderas, T. G., Walter, N. A., Mooney, M., Vartanian, K., Darakjian, P., Buck, K., et al. (2011). Computational detection of alternative exon usage. *Frontiers in Neuroscience*, *5*, 69.

Lander, E. S. (1999). Array of hope. *Nature Genetics*, *21*(Suppl. 1), 3–4.

Lee, C., Mayfield, R. D., & Harris, R. A. (2014). Altered gamma-aminobutyric acid type B receptor subunit 1 splicing in alcoholics. *Biological Psychiatry*, *75*, 765–773.

Lewohl, J. M., Wang, J., Miles, M. F., Zhang, L., Dodd, P. R., & Harris, R. A. (2000). Gene expression in human alcoholism: microarray analysis of frontal cortex. *Alcohol, Clinical and Experimental Research*, *24*, 1873–1882.

Li, Q., Lee, J. A., & Black, D. L. (2007). Neuronal regulation of alternative pre-mRNA splicing. *Nature Reviews. Neuroscience*, *8*, 819–831.

Li, H., Lovci, M. T., Kwon, Y. S., Rosenfeld, M. G., Fu, X. D., & Yeo, G. W. (2008). Determination of tag density required for digital transcriptome analysis: Application to an androgen-sensitive prostate cancer model. *Proceedings of the National Academy of Sciences of the United States of America*, *105*, 20179–20184.

Liao, Q., Liu, C., Yuan, X., Kang, S., Miao, R., Xiao, H., et al. (2011). Large-scale prediction of long non-coding RNA functions in a coding-non-coding gene co-expression network. *Nucleic Acids Research*, *39*(9), 3864–3878.

Licatalosi, D. D., & Darnell, R. B. (2006). Splicing regulation in neurologic disease. *Neuron*, *52*, 93–101.

Lu, J., Lee, J. C., Salit, M. L., & Cam, M. C. (2007). Transcript-based redefinition of grouped oligonucleotide probe sets using AceView: High-resolution annotation for microarrays. *BMC Bioinformatics*, *8*, 108.

Lugli, G., & Smalheiser, N. R. (2013). Preparing synaptoneurosomes from adult mouse forebrain. *Methods in Molecular Biology*, *936*, 173–179.

Malik, A., Korol, A., Hubner, S., Hernandez, A. G., Thimmapuram, J., Ali, S., et al. (2011). Transcriptome sequencing of the blind subterranean mole rat, Spalax galili: Utility and potential for the discovery of novel evolutionary patterns. *PLoS One*, *6*, e21227.

Mardis, E. R. (2008). The impact of next-generation sequencing technology on genetics. *Trends in Genetics*, *24*, 133–141.

Mardis, E. R. (2011). A decade's perspective on DNA sequencing technology. *Nature*, *470*, 198–203.

Margulies, M., Egholm, M., Altman, W. E., Attiya, S., Bader, J. S., Bemben, L. A., et al. (2005). Genome sequencing in microfabricated high-density picolitre reactors. *Nature*, *437*, 376–380.

Marioni, J. C., Mason, C. E., Mane, S. M., Stephens, M., & Gilad, Y. (2008). RNA-Seq: An assessment of technical reproducibility and comparison with gene expression arrays. *Genome Research*, *18*, 1509–1517.

Martin, J. A., & Wang, Z. (2011). Next-generation transcriptome assembly. *Nature Reviews. Genetics*, *12*, 671–682.

Matthews, D. B., Bhave, S. V., Belknap, J. K., Brittingham, C., Chesler, E. J., Hitzemann, R. J., et al. (2005). Complex genetics of interactions of alcohol and CNS function and behavior. *Alcoholism: Clinical and Experimental Research*, *29*, 1706–1719.

Mattick, J. S. (2011). The central role of RNA in human development and cognition. *FEBS Letters*, *585*, 1600–1616.

Mcbride, W. J., Kerns, R. T., Rodd, Z. A., Strother, W. N., Edenberg, H. J., Hashimoto, J. G., et al. (2005). Alcohol effects on central nervous system gene expression in genetic animal models. *Alcoholism: Clinical and Experimental Research*, *29*, 167–175.

Metzker, M. L. (2010). Sequencing technologies—The next generation. *Nature Reviews. Genetics*, *11*, 31–46.

Morris, J., Singh, J. M., & Eberwine, J. H. (2011). Transcriptome analysis of single cells. *Journal of Visualized Experiments*, *50*, pii: 2634.

Mortazavi, A., Williams, B. A., Mccue, K., Schaeffer, L., & Wold, B. (2008). Mapping and quantifying mammalian transcriptomes by RNA-Seq. *Nature Methods*, *5*, 621–628.

Mudge, J., Miller, N. A., Khrebtukova, I., Lindquist, I. E., May, G. D., Huntley, J. J., et al. (2008). Genomic convergence analysis of schizophrenia: mRNA sequencing reveals altered synaptic vesicular transport in post-mortem cerebellum. *PLoS One*, *3*, e3625.

Mulligan, M. K., Ponomarev, I., Hitzemann, R. J., Belknap, J. K., Tabakoff, B., Harris, R. A., et al. (2006). Toward understanding the genetics of alcohol drinking through transcriptome meta-analysis. *Proceedings of the National Academy of Sciences of the United States of America*, *103*, 6368–6373.

Nagalakshmi, U., Wang, Z., Waern, K., Shou, C., Raha, D., Gerstein, M., et al. (2008). The transcriptional landscape of the yeast genome defined by RNA sequencing. *Science*, *320*, 1344–1349.

Oldham, M. C., Horvath, S., & Geschwind, D. H. (2006). Conservation and evolution of gene coexpression networks in human and chimpanzee brains. *Proceedings of the National Academy of Sciences of the United States of America*, *103*, 17973–17978.

Oldham, M. C., Konopka, G., Iwamoto, K., Langfelder, P., Kato, T., Horvath, S., et al. (2008). Functional organization of the transcriptome in human brain. *Nature Neuroscience*, *11*, 1271–1282.

Ozsolak, F., & Milos, P. M. (2011). RNA sequencing: Advances, challenges and opportunities. *Nature Reviews. Genetics*, *12*, 87–98.

Pan, Q., Shai, O., Lee, L. J., Frey, B. J., & Blencowe, B. J. (2008). Deep surveying of alternative splicing complexity in the human transcriptome by high-throughput sequencing. *Nature Genetics*, *40*, 1413–1415.

Parkhomchuk, D., Borodina, T., Amstislavskiy, V., Banaru, M., Hallen, L., Krobitsch, S., et al. (2009). Transcriptome analysis by strand-specific sequencing of complementary DNA. *Nucleic Acids Research*, *37*, e123.

Pavlidis, P., Qin, J., Arango, V., Mann, J. J., & Sibille, E. (2004). Using the gene ontology for microarray data mining: A comparison of methods and application to age effects in human prefrontal cortex. *Neurochemical Research*, *29*, 1213–1222.

Peirce, J. L., Li, H., Wang, J., Manly, K. F., Hitzemann, R. J., Belknap, J. K., et al. (2006). How replicable are mRNA expression QTL? *Mammalian Genome*, *17*, 643–656.

Pozhitkov, A. E., Boube, I., Brouwer, M. H., & Noble, P. A. (2010). Beyond Affymetrix arrays: Expanding the set of known hybridization isotherms and observing pre-wash signal intensities. *Nucleic Acids Research*, *38*, e28.

Preker, P., Nielsen, J., Kammler, S., Lykke-Andersen, S., Christensen, M. S., Mapendano, C. K., et al. (2008). RNA exosome depletion reveals transcription upstream of active human promoters. *Science*, *322*, 1851–1854.

Roberts, A., Pimentel, H., Trapnell, C., & Pachter, L. (2011). Identification of novel transcripts in annotated genomes using RNA-Seq. *Bioinformatics*, *27*, 2325–2329.

Robinson, M. D., Mccarthy, D. J., & Smyth, G. K. (2009). Edger: A bioconductor package for differential expression analysis of digital gene expression data. *Bioinformatics*, *26*, 139–140.

Rothberg, J. M., Hinz, W., Rearick, T. M., Schultz, J., Mileski, W., Davey, M., et al. (2011). An integrated semiconductor device enabling non-optical genome sequencing. *Nature*, *475*, 348–352.

Saba, L. M., Bennett, B., Hoffman, P. L., Barcomb, K., Ishii, T., Kechris, K., et al. (2011). A systems genetic analysis of alcohol drinking by mice, rats and men: Influence of brain GABAergic transmission. *Neuropharmacology, 60*, 1269–1280.

Sandberg, R., Yasuda, R., Pankratz, D. G., Carter, T. A., Del Rio, J. A., Wodicka, L., et al. (2000). Regional and strain-specific gene expression mapping in the adult mouse brain. *Proceedings of the National Academy of Sciences of the United States of America, 97*, 11038–11043.

Sikela, J. M., Maclaren, E. J., Kim, Y., Karimpour-Fard, A., Cai, W. W., Pollack, J., et al. (2006). DNA microarray and proteomic strategies for understanding alcohol action. *Alcoholism: Clinical and Experimental Research, 30*, 700–708.

Sliwerska, E., Meng, F., Speed, T. P., Jones, E. G., Bunney, W. E., Akil, H., et al. (2007). SNPs on chips: The hidden genetic code in expression arrays. *Biological Psychiatry, 61*, 13–16.

Sugarbaker, D. J., Richards, W. G., Gordon, G. J., Dong, L., De Rienzo, A., Maulik, G., et al. (2008). Transcriptome sequencing of malignant pleural mesothelioma tumors. *Proceedings of the National Academy of Sciences of the United States of America, 105*, 3521–3526.

Tabakoff, B., Saba, L., Printz, M., Flodman, P., Hodgkinson, C., Goldman, D., et al. (2009). Genetical genomic determinants of alcohol consumption in rats and humans. *BMC Biology, 7*, 70.

Torres, T. T., Metta, M., Ottenwalder, B., & Schlotterer, C. (2008). Gene expression profiling by massively parallel sequencing. *Genome Research, 18*, 172–177.

Trapnell, C., Pachter, L., & Salzberg, S. L. (2009). TopHat: Discovering splice junctions with RNA-Seq. *Bioinformatics, 25*, 1105–1111.

Trapnell, C., Roberts, A., Goff, L., Pertea, G., Kim, D., Kelley, D. R., et al. (2012). Differential gene and transcript expression analysis of RNA-Seq experiments with TopHat and Cufflinks. *Nature Protocols, 7*, 562–578.

Trapnell, C., Williams, B. A., Pertea, G., Mortazavi, A., Kwan, G., Van Baren, M. J., et al. (2010). Transcript assembly and quantification by RNA-Seq reveals unannotated transcripts and isoform switching during cell differentiation. *Nature Biotechnology, 28*, 511–515.

Tzika, A. C., Helaers, R., Schramm, G., & Milinkovitch, M. C. (2011). Reptilian-transcriptome v1.0, a glimpse in the brain transcriptome of five divergent Sauropsida lineages and the phylogenetic position of turtles. *EvoDevo, 2*, 19.

Vadigepalli, R., Chakravarthula, P., Zak, D. E., Schwaber, J. S., & Gonye, G. E. (2003). Paint: A promoter analysis and interaction network generation tool for gene regulatory network identification. *OMICS: A Journal of Integrative Biology, 7*, 235–252.

Voineagu, I., Wang, X., Johnston, P., Lowe, J. K., Tian, Y., Horvath, S., Mill, J., Cantor, R. M., Blencowe, B. J., & Geschwind, D. H. (2011). Transcriptomic analysis of autistic brain reveals convergent molecular pathology. *Nature, 474*, 380–384.

Walter, N. A., Bottomly, D., Laderas, T., Mooney, M. A., Darakjian, P., Searles, R. P., et al. (2009). High throughput sequencing in mice: A platform comparison identifies a preponderance of cryptic SNPs. *BMC Genomics, 10*, 379.

Walter, N. A., Mcweeney, S. K., Peters, S. T., Belknap, J. K., Hitzemann, R., & Buck, K. J. (2007). SNPs matter: Impact on detection of differential expression. *Nature Methods, 4*, 679–680.

Weber, A. P., Weber, K. L., Carr, K., Wilkerson, C., & Ohlrogge, J. B. (2007). Sampling the Arabidopsis transcriptome with massively parallel pyrosequencing. *Plant Physiology, 144*, 32–42.

Zhang, B., & Horvath, S. (2005). A general framework for weighted gene co-expression network analysis. *Statistical Applications in Genetics and Molecular Biology, 4*, Article17.

Zhao, W., Langfelder, P., Fuller, T., Dong, J., Li, A., & Hovarth, S. (2010). Weighted gene coexpression network analysis: State of the art. *Journal of Biopharmaceutical Statistics, 20*, 281–300.

CHAPTER TWO

Analysis Considerations for Utilizing RNA-Seq to Characterize the Brain Transcriptome

Christina L. Zheng[*,†,1], **Sunita Kawane**[‡], **Daniel Bottomly**[‡], **Beth Wilmot**[*,‡]

[*]Department of Medical Informatics and Clinical Epidemiology, Oregon Health and Science University, Portland, Oregon, USA
[†]Knight Cancer Institute, Oregon Health, Oregon Health and Science University, Portland, Oregon, USA
[‡]Clinical & Translational Research Institute, Oregon Health and Science University, Portland, Oregon, USA
[1]Corresponding author: e-mail address: zheng@ohsu.edu

Contents

Abstract

RNA-Seq allows one to examine only gene expression as well as expression of noncoding RNAs, alternative splicing, and allele-specific expression. With this increased sensitivity and dynamic range, there are computational and statistical considerations that need to be contemplated, which are highly dependent on the biological question being asked. We highlight these to provide an overview of their importance and the impact they can have on downstream interpretation of the brain transcriptome.

International Review of Neurobiology, Volume 116
ISSN 0074-7742
http://dx.doi.org/10.1016/B978-0-12-801105-8.00002-3

1. INTRODUCTION

The utilization of RNA-Seq studies is rapidly increasing (over 2500 publications in PubMed as of 4/2014 with first publication in 2008). We note that approximately 7% of those publications are focused on RNA-Seq applications using brain or neuronal tissue. With the decreasing costs and rapid changes in technology, it is reasonable to expect more studies routinely using RNA-Seq, particularly in neuroscience. Given this, this chapter focuses on some of the key considerations in the analysis of an RNA-Seq experiment that should be reviewed at the design of the experiment and in the context of the primary research hypothesis. We refer the reader background papers on general experimental design considerations (e.g., Fang & Cui, 2011; McIntyre et al., 2011; Robles et al., 2012; and citations therein) and instead focus on three areas: defining and quantifying transcript/gene expression, detecting differential expression (DE) and frameworks for interpretation (Fig. 2.1).

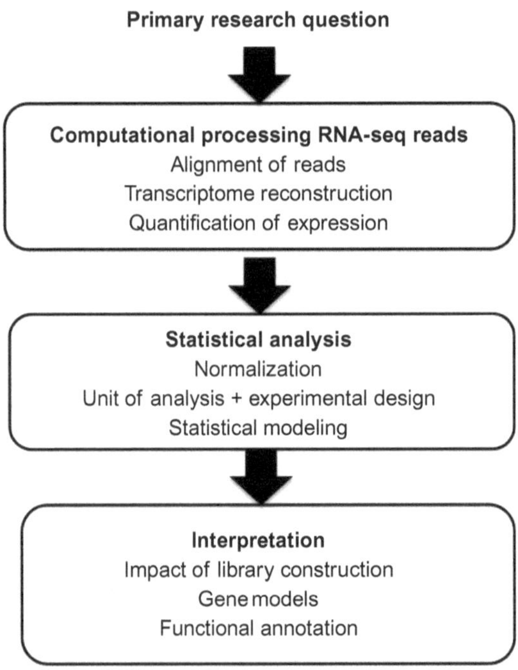

Figure 2.1 Overview of the key components for consideration in the analysis of RNA-Seq data. The choices made are determined by the primary research question of interest.

2. DEFINING AND QUANTIFYING TRANSCRIPT/GENE EXPRESSION

The ultimate goal of transcriptome profiling is to define and quantify a precise map of the expressed transcripts and genes in a given sample. RNA-Seq profiling of a transcriptome results in short sequenced segments (~100 bps), known as short reads, of expressed transcripts. The computational challenge is to analytically "reassemble" these short reads to define the transcript of origin and subsequently quantify its expression. Depending on the biological questions and the genomic resources available for a given organism, there are several scenarios for processing reads to define and quantify expression (Fig. 2.2). For example, if your organism of interest has an annotated reference genome and your focus is on quantifying previously annotated genes, transcripts, or exons, the available genomic/transcriptomic information can be leveraged to greatly increase the sensitivity of your RNA-Seq experiment. A typical protocol in this case would involve aligning the RNA-Seq reads to the reference genome followed by quantifying (counting) the aligned reads within the annotated transcriptome to obtain expression levels. The aligned reads may also be used to define and discover as yet annotated transcripts in well-curated (and less well curated) transcriptomes (Roberts, Pimentel, Trapnell, & Pachter, 2011). Alternatively, one may have an RNA-Seq experiment involving an organism that is lacking a reference genome or transcriptome sequence. In this scenario, the RNA-Seq reads would first be used to define and then subsequently quantify the transcriptome. Although these scenarios highlight the diversity

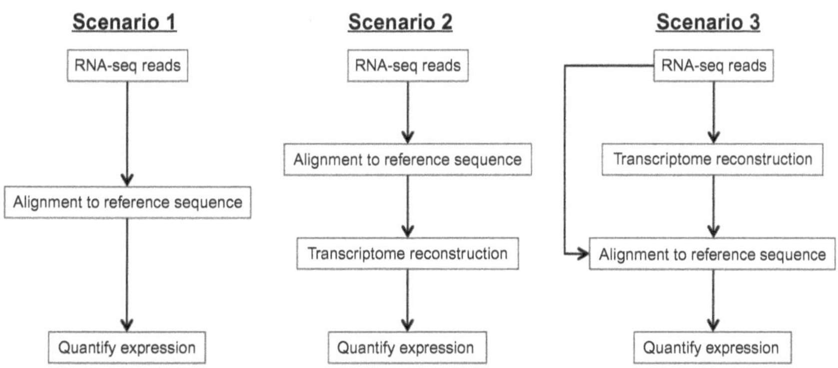

Figure 2.2 Scenarios for processing RNA-Seq reads which are dependent upon the biological question(s) and the genomic resources available for a given organism.

Table 2.1 Overview of common RNA-processing tools

Tool	Type	Source
BedTools	Q.E.	https://code.google.com/p/bedtools/
Bowtie	Alignment (N)	http://bowtie-bio.sourceforge.net/index.shtml
BWA	Alignment (N)	http://bio-bwa.sourceforge.net/
Cufflinks	T.R. (RG)/ Q.E.	http://cufflinks.cbcb.umd.edu/
GSNAP	Alignment (S)	http://research-pub.gene.com/gmap/
HTSeq	Q.E.	http://www-huber.embl.de/users/anders/HTSeq/doc/overview.html
Oases	T.R. (GI)	https://www.ebi.ac.uk/~zerbino/oases/
RSEM	Q.E.	http://deweylab.biostat.wisc.edu/rsem/
Scripture	T.R. (RG)/ Q.E.	http://www.broadinstitute.org/software/scripture/
SOAPdenovo-Trans	T.R. (RI)	http://soap.genomics.org.cn/SOAPdenovo-Trans.html
STAR	Alignment (S)	https://code.google.com/p/rna-star/
Subread	Alignment (N)	http://bioinf.wehi.edu.au/subread/
TopHat	Alignment (S)	http://tophat.cbcb.umd.edu/
Trinity	T.R. (GI)	http://trinityrnaseq.sourceforge.net/

Alignment, short read aligner; N, non-splice-aware; S, splice-aware; T.R., transcript reconstruction; RG, reference guided; RI, reference independent; Q.E., quantify expression.

and power of RNA-Seq profiling, it also helps to highlight the three unifying steps involved in computationally processing RNA-Seq reads to define and quantify transcriptome expression: (1) alignment of RNA-Seq reads to a reference sequence, (2) transcriptome reconstruction, and (3) quantification of expression. Table 2.1 summarizes common RNA-processing tools used in each of these three steps.

2.1. Step 1: Alignment of RNA-Seq reads to a reference sequence

In an effort to identify the transcript/gene origin of the RNA-Seq reads, the short reads may be mapped to either a reference transcriptome or a reference genome sequence. Alignment of sequences based on sequence similarity is a classic problem in bioinformatics (Pearson, 2013). The mapping of RNA-Seq reads to a reference sequence is analogous to the mapping of expressed sequenced tags (ESTs) to a reference genome (Kent, 2002; Wu & Watanabe, 2005); however, its short length (~36–125 nucleotides), higher sequence error rates, and its sheer volume, currently as many as hundreds of millions, introduce a host of additional challenges. Thus, the development of RNA-Seq alignment tools is an active area of research (Engstrom et al., 2013; Fonseca, Rung, Brazma, & Marioni, 2012; Lindner & Friedel, 2012) resulting in a great number of alignment tools currently available for RNA-Seq. This then leads many researchers asking the question, which is the best aligner to use. The answer ultimately depends on the biological question and the genomic resources available for your organism of interest. Different alignment tools have been developed to answer specific biological questions such as the identification of novel splice sites (Dobin et al., 2013; Forster, Finkel, Gould, & Hertzog, 2013; Huang et al., 2011; Kim et al., 2013) or gene fusions (Carrara et al., 2013; Kim et al., 2013; Liu, Ma, Chang, & Zhou, 2013). Alignment tools have also been designed to specifically cater to the needs of specific sequencing platforms (Cloonan et al., 2009; De Bona, Ossowski, Schneeberger, & Ratsch, 2008; Trapnell, Pachter, & Salzberg, 2009). Below we introduce a few key features of different alignment approaches and discuss their potential impacts. Please refer to recent reviews for a more comprehensive comparison and evaluation of different alignment tools (Engstrom et al., 2013; Fonseca et al., 2012).

2.1.1 Splice-aware aligners

One of the most distinctive features of RNA-Seq alignment tools is its ability to accommodate large sequence gaps corresponding to spliced exon–exon junctions (i.e., introns). Splice-aware aligners such as STAR (Dobin et al., 2013), GSNAP (Wu & Nacu, 2010), and TopHat (Trapnell et al., 2009) should be targeted to alignments with genomic sequences, particularly with biological questions pertaining to the identification of novel or annotated spliced junctions while nonsplice-aware aligners such as Bowtie (Langmead, Trapnell, Pop, & Salzberg, 2009) and BWA (Li & Durbin, 2009) should be targeted toward alignments to transcriptome sequences. To increase their sensitivity in identifying spliced junctions, many splice-aware aligners use information from previously annotated exon–exon junctions (Dobin et al., 2013; Trapnell et al., 2009; Wu & Nacu, 2010).

Interestingly, a number of spliced aligners are adaptations/extensions of nonsplice-aware aligners. For example, TopHat is an adaptation of Bowtie.

TopHat first uses Bowtie to align reads to the genome in a nonsplice-aware manner and then all reads which were not mapped (unmapped) are broken into smaller segments and remapped to the genome in search of candidate splice sites. This method is termed the exon-first method as exons are identified first. Another major category of splice-aware aligners are the "seed-and-extend" methods. Seed and extend methods such as GSNAP, STAR, and Subread (Liao, Smyth, & Shi, 2013) segment the reads into shorter segments which are then placed on the genome to localize the alignment. Candidate genomic regions are then extended and merged with initial seeds. With this approach, all reads are aligned at the same time, thus will not be biased toward continuous alignments such as exon-first approaches are. However, due to their more efficient approach of only remapping unmapped reads, exon-first approaches are computationally faster.

2.1.2 Sequence variations between the short read and reference sequence

To better accommodate potential sequencing errors or sequence variations among individuals (SNPs, single nucleotide polymorphisms) or different animal strains, in addition to exact matches, alignment tools must also allow approximate matches of the reads to the reference sequence. To do so, alignment tools allow for varying degrees of sequence mismatches or insertions/deletions (indels) between the short read and the reference sequence. Many tools also allow sequences of the read to be trimmed (clipped), particularly within the ends of the reads, where sequencing quality is known to suffer (Dohm, Lottaz, Borodina, & Himmelbauer, 2008). However, it is important to note that different alignment tools handle and tolerant each of these cases very differently. For example, although GSNAP and Subread allow for sequence mismatches, the maximum number of mismatches allowed is dictated by the overall score of the alignment, while other programs such as STAR and TopHat allow users to define the maximum number allowed. In an attempt to improve computational efficiency, tools such as BWA impose constraints on the maximum number of mismatches and indels allowed, while other tools such as Bowtie once again impose alignment score thresholds in place of imposing constraints on the number of each allowed. As expected, alignment tools which automatically trim reads result in overall higher numbers of aligned reads however at the price of decreased coverage of the aligned reads (Engstrom et al., 2013). Understanding the individual characteristics of how different alignment tools computationally handle sequence variations should be an important consideration in the final choice of the aligner.

2.1.3 Uniquely mapping or multimapping reads

When a read is aligned to a reference sequence, the read may align uniquely to one position or may align to multiple positions (multimapped) in the reference sequence. Multimapped reads may be a result of the repetitive nature of the original transcriptome or genome sequence, multiple isoforms of a gene when aligning to a transcriptome, or potential sequencing errors. Once again different aligners process and report multimapped reads differently. For example, GSNAP and STAR allow for the option of reporting the uniquely mapped reads separately from the multimapped reads, while other aligners such as Bowtie and TopHat can be tuned to report the alignment(s) with the best score up to a user-defined limit of alignments. The use of either uniquely or multimapped reads has potential consequences in downstream analyses (discussed below) and should be a factor when considering the aligner of choice.

2.2. Step 2: Transcriptome reconstruction

The ultimate goal of RNA-Seq profiling is to reconstruct a precise map of all transcripts and isoforms expressed in a given sample. However, the complex nature of the transcriptome (i.e., multiple isoforms for one gene, gene expression differences spanning orders of magnitude across genes, and the mix of processed and unprocessed transcripts) and the short length of RNA-Seq reads make reconstructing the transcriptome a very difficult computational task. There are two main approaches to transcriptome construction, genome guided and genome independent. Genome-guided approaches reconstruct transcripts by assembling spatially adjacent or overlapping reads previously mapped to a genome reference sequence, while genome-independent approaches reconstruct transcripts by assembling reads together with direct sequence (*de novo*) similarity sans alignment to a reference genome. Each is described more below.

2.2.1 Genome guided

Two of the most widely used genome-guided approaches are Scripture (Guttman et al., 2010) and Cufflinks (Trapnell et al., 2010). Both methods utilize spliced reads to directly guide their reconstructions, but they differ in their approach to the problem. Scripture approaches the reconstruction problem as a segmentation problem and attempts to identify significant transcript paths from a graph topology of all possible connections of bases in the transcriptome. Cufflinks approaches the problem with more of an exon focus and connects aligned reads into a graph based on the locations of their spliced alignments. Conceptually, the two approaches build similar graphs; however, Scriptures strength is in identifying all isoforms compatible with the read data (sensitivity), while Cufflinks focuses on reporting the minimal number of compatible isoforms (specificity). Both tools have been found to assemble similar transcripts at high expression levels, however differ substantially for lower expressed transcripts (Garber, Grabherr, Guttman, & Trapnell, 2011).

2.2.2 Genome independent

Genome-independent approaches, however, aim to reconstruct the transcriptome directly from the short reads themselves, thus bypassing the alignment of the reads to a reference genome particularly when a reference genome is not available. A commonly used strategy of genome-independent approaches such as Oases (Schulz, Zerbino, Vingron, & Birney, 2012), Trinity (Grabherr et al., 2011), and SOAPdenovo-Trans (Xie et al., 2014) is the use of de Bruijn graphs to model overlapping subsequences (k-mers) of the reads, thus reducing the complexity of millions of reads to a fixed number of all possible k-mers originating from the reads. The overlaps of k-1 bases between the k-mers constitute the graph of all possible transcript paths. The short reads are then used to identify transcript paths which are either supported or not supported by the reads. Genome-independent transcript reconstruction and *de novo* assembly of whole genomes from short reads (Zhang et al., 2011) share many computational strategies; however, genome-independent transcriptome reconstruction has the added challenges introduced by differential read distribution across genes (differential gene expression), nonuniform read coverage within individual transcripts (sequencing biases), and the added complexity of multiple isoforms for one gene making this a very active area of research (Li et al., 2013; Martin & Wang, 2011).

The method of reconstruction chosen should once again be dictated by the biological question and also the availability of a reference genome. Genome-independent methods are an obvious choice for organisms lacking a reference genome, whereas the increased sensitivity of genome-guided approaches is recommended for organisms with reference genomes. Genome-guided approaches may also help to guide the discovery of novel or as yet annotated transcripts (Roberts et al., 2011). However in cases where genomic rearrangements are suspected such as in cancer or among different strains of a model organism (i.e., laboratory mouse strains), a hybrid approach incorporating both genome-guided and genome-independent approaches might be a powerful avenue in capturing both known and novel variations.

2.3. Step 3: Quantification of expression levels

The alignment of the RNA-Seq reads to a reference sequence provides a digital count of the expressed transcripts in a given sample. Once again depending on the biological question and the genomic annotation available, digital counts can be obtained at the level of exons, transcripts, and/or genes. One common method of obtaining digital counts at the gene level is to sum up reads mapped to all annotated exons within a gene. Digital counts at the exon level are commonly obtained in a similar fashion for both constitutively and alternatively expressed exons (Anders, Reyes, & Huber, 2012; Griffith et al., 2010). Tools such as HTSeq (Anders, Pyl, & Huber, 2014) and BEDTools (Quinlan & Hall, 2010) allow users to indicate how the read should be "counted" in cases where the read maps to a genomic location with multiple annotations (i.e., overlapping genes) or on a different strand (i.e., stranded RNA-Seq libraries).

Due to the added complexity of shared exons across different isoforms of the same gene, some reads cannot be unequivocally assigned to individual isoforms; thus, obtaining digital counts for individual isoforms of a gene becomes a much more complicated task. Some approaches attempt to simplify the problem by obtaining counts for only unique portions of an isoform (Griffith et al., 2010); however, this approach is limited for genes that do not contain unique exon(s) or exon portions for a particular isoform. Alternative methods such as Cufflinks (Trapnell et al., 2010) and RSEM (Li & Dewey, 2011) approach the read assignment uncertainty in a probabilistic manner by constructing a maximum likelihood estimation of isoform abundance that best explains the short reads. Thus, the abundance estimates are greatly impacted by the coverage of, the number of, and any incorrectly annotated/assembled isoforms.

For the digital counts to be meaningful and comparable across different samples and different genes, two main sources of variability need to be accounted for. The first sources of variability result from differences in the total number of mapped reads per sample (i.e., library size), while the second variability results from differences in transcript/gene lengths. Larger library sizes and longer transcripts/genes are expected to have greater numbers of reads than their smaller and shorter, respectively, counter parts. One popular approach to account for these variables is to normalize the digital counts by reads or fragments (for paired-end reads) per kilobase of transcript per million mapped reads resulting in the metric reads per kilobase per million (RPKM) reads (Mortazavi, Williams, McCue, Schaeffer, & Wold, 2008) or FPKM (Trapnell et al., 2010), respectively. These metrics account for both differences in library sizes across samples and differences in transcript/gene lengths within samples. Additional approaches for accounting for these and additional variables will be described in Section 3.

Irrespective of the chosen quantification method, the choice of the type of mapped reads (i.e., uniquely or multimapped) used greatly impact transcript quantification and its interpretation. For example, the use of only uniquely mapping reads has become common practice when quantifying at the exon and gene level, while the exclusion of multimapped reads has been warned to skew and misguide results particularly when quantifying at the transcript level (Li & Dewey, 2011; Trapnell et al., 2010).

3. DETECTING DIFFERENTIAL EXPRESSION

The primary goal of many RNA-Seq studies is gene expression profiling between samples. This can be between two or more groups or an analysis of subgroups or outliers relative to a main group or groups, depending on the biological question/primary hypothesis of interest. For smaller studies, these analyses can be done on available desktop workstations. Approaches such as Myrna are being introduced allowing large datasets to be analyzed utilizing cloud computing or on local Hadoop clusters (Langmead, Hansen, & Leek, 2010).

3.1. The need for normalization

Prior to sequencing, mRNA is fragmented to obtain read coverage throughout the length of the transcript. Therefore, longer transcripts will have more reads than shorter transcripts with similar expression. Total read count is proportional to the expression level of a gene times the length of the gene. This results in more power to detect DE for longer genes than for shorter genes and creates bias in comparing genes between libraries with differences in sequencing depth. Normalization is a means to correct for any systematic errors in the counts based on various technical factors and is critical for both within sample gene comparisons and across sample comparisons. Factors such as library preparation, personnel, sequencing depth, read length, exon length, GC content, and other technical differences result in differing read counts per gene that are not related to differences in the particular comparison of interest. Normalization enables accurate comparisons of expressions levels between and within samples. Many approaches have been developed to investigate appropriate methods for analyzing RNA-Seq data and improve the accuracy of the final DE estimates. We discuss several of these approaches below.

The first normalization method proposed, RPKM, divides the summarized gene counts by the length of the gene (Mortazavi et al., 2008). Although an intuitive solution, the RPKM approach introduces a bias in the lower abundant genes because only a small number of genes make up a large proportion of the total reads in a library. The proportion of expression for each gene in a library is related to the expression level of all the genes, and small changes in highly expressed genes affect the proportion of lower abundance genes to a greater extent (Finotello et al., 2014; Ramskold, Wang, Burge, & Sandberg, 2009). RPKM may also overcorrect for exon length bias (Finotello et al., 2014). Similarly, accounting for the RNA fragments observed in a library as is used for paired-end RNA-Seq experiments in the method, fragments per kilobase per million (FPKM) reads from the Tuxedo suite of RNA-processing tools (Trapnell et al., 2010) suffers from the same biases.

Another way to correct for differences in sequencing depth between libraries is simply to normalizing by the total counts where gene counts are divided by the total number of mapped reads (or library size) and multiplied by the mean total counts across all samples of the dataset. This method also suffers from the same bias as RPKM due to the greater proportion of small numbers of highly expressed genes in a library and has relatively lower sensitivity for detecting DE (Bullard, Purdom, Hansen, & Dudoit, 2010).

Quantile normalization, a method first developed in the context of microarray data, has been extended to RNA-Seq data (Bullard et al., 2010). A reference distribution is created from all samples in the data set by sorting the read counts from each sample and computing the median counts across all of the sorted samples. The distribution of each sample is matched to this reference and rounded to produce integer values. This method ensures that all samples have the same distribution of counts and are implemented in the Bioconductor package limma (Smyth, 2004). In some instances, this procedure can increase the intragroup variability (Dillies et al., 2013).

Methods that compute scaling factors have an advantage that the raw count data are not transformed and scaling factors can be estimated from the data. One approach is to match libraries on the upper 25% quartile (Bullard et al., 2010). Another way to estimate scaling factors, implemented in the Bioconductor package DESeq, uses the median of scaled counts under the assumption that most genes are not differentially expressed (Anders & Huber, 2010). Another method that assumes most genes are not differentially expressed first excludes highly expressed genes and highly variable genes and calculates a trimmed mean of the log expression values (TMM) (Robinson & Oshlack, 2010). TMM is implemented in the poplar R package edgeR (Robinson & Smyth, 2007).

The PoissonSeq package (Li, Witten, Johnstone, & Tibshirani, 2012) determines a least variable gene set through goodness-of-fit estimation, which is then used as the reference. The library scaling factors for each sample are computed using the ratio of the sum of the counts of the reference gene set for that sample and the sum of the reference gene set counts for all samples.

Several multiple step procedures have been suggested (Kadota, Nishiyama, & Shimizu, 2012; Risso, Schwartz, Sherlock, & Dudoit, 2011; Trapnell et al., 2013). GC content biases seem to be driven by library preparation (Risso et al., 2011) and can be reduced through a procedure where a within lane normalization is followed by a between-lane distributional normalization. However, depending on the counting method used, this approach may not perform better than scaling factors (Finotello et al., 2014). A different approach is where genes potentially different are removed in the first step and then scaling factors for the final normalization are computed using the TMM normalization method (Kadota et al., 2012). This method has been expanded to include multiple combinations of other normalization methods and DE analysis methods (Sun, Nishiyama, Shimizu, & Kadota, 2013).

To account for transcript isoform differences, Cuffdiff (Trapnell et al., 2013) uses a two-step scaling procedure first by within a condition and then between conditions. An additional transcript-level normalization that estimates isoform abundance is implemented.

It is unclear that more complex normalization methods perform better than scaling factors (Bullard et al., 2010; Dillies et al., 2013). However, there is no one solution for all challenges with RNA-Seq (Dillies et al., 2013; Finotello et al., 2014) and the best normalization procedure will depend on both the data set-specific issues and the methods used for counting features (Finotello et al., 2014). Careful exploratory data analysis such as plots of the data distributions is critical to determining the best normalization procedure for a specific data set. In addition, the effect of different normalization procedures on the downstream distributions of the DE *p*-values should be evaluated. The method chosen should reflect the least biases and more uniform distributions of the final DE *p*-values.

3.2. Inferring putative DE

The goal of a DE analysis is to highlight genes that are significantly different in abundance across experimental conditions. This is a problem of assigning a probability to whether, in reality, each gene has different number of reads mapped to it. Results from a DE analysis are therefore probabilities that the genes are differently expressed at some level between the conditions of interest. Therefore, obtaining reliable and accurate estimates of the variability inherent in each condition of interest is a key factor in determining the probability that the gene is differently expressed in the groups. Due to the inherent variability among biological samples, detection of a transcript that is differentially expressed is challenging because high overdispersion (Standard deviation (STD) is greater than the mean of the distribution) across the samples and the conditions means that only studies with many samples and high coverage can be used to detect DE reliably (Anders & Huber, 2010). Therefore, the usual assumption is that the number of conditions is small compared to the number of biological samples.

A number of algorithms have been developed to test for DE between two or more groups and are approached using either parametric or nonparametric methods. Parametric algorithms often model the data using either the Poisson or negative binomial (NB) distributions. Algorithms using nonparametric approaches model the noise distribution based on the actual data and therefore do not depend on the assumptions associated with a known probability distribution model (Li & Tibshirani, 2013; Tarazona, García-Alcalde, Dopazo, Ferrer, & Conesa, 2011). We discuss several of the major parametric approaches below.

RNA-Seq data are discrete count data and as such multiple methods have been developed using the discrete probability distributions Poisson and NB. Early RNA-Seq studies using a single biological sample and technical replicates showed that the distribution of read counts fit well to a Poisson distribution (Bullard, 2010; Marioni, Mason, Mane, Stephens, & Gilad, 2008). However, the assumption that the variance is equal to the mean that defines a

Poisson distribution is violated when using biological replicates, given the extensive variability across biological replicates in RNA-Seq (overdispersion). This underestimates the sampling error and results in greater false-positive rates (Anders & Huber, 2010). RNA-Seq has a high dynamic range (Anders & Huber, 2010), making the gene-specific dispersion estimation a critical and challenging problem.

Many modifications to the simple Poisson DE test have been developed to address this challenge. Assuming that only a subset of genes demonstrate overdispersion, a two-stage Poisson model has been proposed (Auer & Doerge, 2011). In the first stage, each gene is first tested for overdispersion relative to a Poisson model. A Poisson quasi-likelihood approach was developed to test DE for the overdispersed genes. The other genes are tested using Poisson model. The false discovery rate (FDR; Benjamini & Hochberg, 1995) is controlled separately on the two lists of genes. This method was shown to correct for severe overdispersion only and so leads to greater false-positive rates for less variable genes (Lund, Nettleton, McCarthy, & Smyth, 2012). QuasiSeq (Lund et al., 2012) implements two improvements to the quasi-likelihood approach by incorporating more flexibility into the variance estimation using an *F*-test and sharing information across genes to estimate gene-specific error variances similar to the approach developed for microarray analysis (Smyth, 2004).

Another method develops a generalized Poisson (GP) model (Srivastava & Chen, 2010) that adds an additional parameter to model the position-level read counts. This model takes into consideration the potential positional bias in a DE analysis by accounting for reads mapped to each position of an exon. When there is no sequencing bias, the model reduces to the Poisson model. Likelihood ratio tests are used to identify differentially expressed genes by position-level read counts.

In PossionSeq, the normalized library size and the correlation of the gene expression with the condition are modeled by a log-linear relationship. If a gene is not significantly correlated with the condition based on score statistics, there is no DE. A novel permutation method for obtaining the *p*-value distribution appears to result in a more accurate FDR (Li et al., 2012).

The NB distribution specifically characterizes the feature variation and, as such, is a natural extension to the Poisson distribution. The observed read counts are modeled by the Poisson distribution, while the unobserved true expression levels in each biological sample follow a gamma distribution. The NB distribution allows greater flexibility in modeling the mean–variance relationship through the addition of a dispersion parameter modeled by the gamma random variable. The difference among these NB methods is how the dispersion, or biological variation, is modeled, estimated, and used in inference. Accurate estimates of this critical parameter determine whether a DE signal can be found above the natural noise of biological variation inherent in the samples. Improving the model fit to the data also increases the ability to identify DE genes across all levels of abundance.

edgeR (Robinson & Smyth, 2007) and DESeq (Anders & Huber, 2010) are the most widely used methods for DE analysis and compute p-values for the tests based on exact test or approximation of exact test derived from the probabilities. Both methods utilize information across the genes to generate dispersion estimates. edgeR was based on methods for small sample sizes developed for SAGE data (Robinson & Smyth, 2007) where a common dispersion parameter across all genes was suggested. This can be measured accurately as all the data are used in the computation but depends on the assumption is that all genes have similar biological variance. In practice, genes have different variabilities and a method for estimating gene-specific dispersion by borrowing information across genes similar to that used in microarray data (Smyth, 2004) is also implemented. In addition, methods similar to the QuasiSeq algorithms have been implemented.

DESeq assumes a locally linear relationship between the variance and mean expression levels. The dispersion estimates are generated by pooling data from genes with similar expression levels. Since these methods use the dispersion parameter as a fixed, known constant, they can result in greater false-positive rates in some instances from not considering possible uncertainty in the parameter estimates. In particular, genes with higher biological variability are more likely to be reported as DE (Lund et al., 2012; Wu, Phan, & Wang, 2013; Wu, Wang, & Wu, 2013). The methods implemented in QuasiSeq have also been shown improve the accuracy of the NB models (Lund et al., 2012).

A recent approach takes a Bayesian approach by modeling the dispersion parameter using a log-normal prior and an NB likelihood (Wu, Phan, et al., 2013; Wu, Wang, et al., 2013) and is deployed in the Bioconductor package DSS (http://www.bioconductor.org/packages/release/bioc/html/DSS.html).

Another Bayesian method, baySeq, also models an NB distribution estimating the prior probability parameters by sampling from the data under the assumption that similar samples should fit the same distribution (Hardcastle & Kelly, 2010). The result of a baySeq analysis is the posterior likelihood of a DE model, given the data for each gene. There is some inherent variability in these estimates from repeated analyses due to the resampling approach. Fold changes or test statistics are not given making this method difficult to compare to others and the direction of expression cumbersome to determine.

Parametric methods are powerful when the distributional assumptions hold but will fail as the data deviate from the specified distribution. In addition, for parametric methods, the number of significant DE genes is affected by the sequencing depth where increasing library read depth resulted in increasing false-positive rates due to increased power to detect smaller count differences between groups (Tarazona et al., 2011) and a sensitivity to outliers (Li & Tibshirani, 2013). Methods using nonparametric methods have no such explicit assumptions about the data distribution. A nonparametric approach that models the noise distribution in the data, NOISeq, has been shown to be unaffected by library size (Tarazona et al., 2011). The noise distribution is determined by comparing all within-group log ratios and absolute count differences in a pairwise fashion. DE genes are determined by the odds of the gene being DE above the noise distribution. Since it does not estimate model parameters, NOISeq performs well without replicates. In this case, it estimates the noise distribution by simulating technical replicates from the data.

In general, gene-specific variability is higher for genes with higher read counts and this phenomenon has recently been exploited to improve the estimate of the mean–variance relationship (Law, Chen, Shi, & Smyth, 2014). The method used in variance modeling at the observational level (voom) uses a LOWESS regression of the log counts per million ($\log_2$cpm) to estimate precision weights for each observation nonparametrically from the data. The transformed read counts are used for linear modeling utilizing the widely used limma pipeline (Smyth, 2004, 2005), and the many techniques originally developed for modeling DE in gene expression microarrays can be applied to RNA-Seq data. The voom/limma pipeline has been shown to improve the accuracy of the type I error rate compared to other methods particularly when the sequencing depths for each sample are different (Law et al., 2014).

In conclusion, there is no one-size-fits-all analysis procedure for testing for DE in a particular dataset (Guo, Li, Ye, & Shyr, 2013; Rapaport et al., 2013). Caution should be exercised and extensive evaluation of different methods should be conducted before choosing a final analysis method for the analysis of the data. It is noted that the challenge of accurately estimating the dispersion across all genes highlights the importance of planning for biological replicates in the study design.

3.3. Outliers, subgroups, and individual expression

For designed experiments, one determines *a priori* the variables of importance and can fit a model testing the significance of all or part of the model corresponding to the hypotheses of interest. These procedures have been described above in the context of RNA-Seq. A simple example would be the comparison of the expression levels between cases and controls. Significance of this model for a given gene would imply that the expression pattern for the cases is higher or lower than the controls relative to the observed variation. However, it need not be true that all the cases are expressed in a similar fashion. It was shown in Tomlins et al. (2005) that the use of methodology that ranked genes on the basis of whether some genes had "outlier" microarray gene expression profiles could be used to find genes involved in fusion events in the context of prostate cancer. Their method, cancer outlier profile analysis (COPA), was the first of many such approaches which looked for biologically meaningful outliers or subgroups in gene expression datasets. For instance, the outlier sum (Tibshirani & Hastie, 2007) and the mCOPA (Wang, Sun, Ji, Xing, & Liang, 2012; Wang, Taciroglu, et al., 2012) methods are refinements on the COPA approach and involve detection of outliers after a robust transformation. If both cases and controls are present, a number of methods have been proposed to find genes differentially expressed in a subset of patients including those similar to traditional DE approaches (Gadgil, 2008; Ghosh, 2010; Gleiss, Sanchez-Cabo, Perco, Tong, & Heinze, 2011; Hu, 2008; Ji et al., 2010; Karrila, Lee, & Tucker-Kellogg, 2011; Lian, 2008; Liu & Wu, 2007; Pinese et al., 2009; Wang & Rekaya, 2010; Wang, Sun, et al., 2012; Wang, Taciroglu, et al., 2012; Wang, Wu, Ji, Wang, & Liang, 2011; Wu, 2007).

Other approaches focus on the observation that the outliers detected by COPA were really indicative of the expression distribution in the cases being bi- or multimodal and were amendable to mixture models (Ghosh & Chinnaiyan, 2009; Wang, Wen, Symmans, Pusztai, & Coombes, 2009). As detection of outliers and subgroups is not only of interest in genomics, other approaches have been suggested based on approaches used in other fields such as variations in the outlying degree (Bottomly, Ryabinin, et al., 2013; Bottomly, Wilmot, et al., 2013), as well as the gene tissue index (Mpindi et al., 2011). Other related approaches include the antiprofile method (Bravo, Pihur, McCall, Irizarry, & Leek, 2012). However, all of these methods were devised for microarray analysis and although it has been suggested that some microarray array methodology may be applicable to

RNA-Seq data after some correction (Law et al., 2014), methodological developments and assessment are necessary with several having been carried out so far. The first is an extension of the bimodality index (Wang et al., 2009) called SIBER (Tong, Chen, Su, & Coombes, 2013). The SIBER method allows the count data of RNA-Seq to be modeled using mixtures of the NB distribution, GP, or a normal mixture after using a box–cox transformation. Interestingly, they found that the best performer was modeling the counts after the use of a log transformation (which was what the box–cox transformation suggested empirically for many genes). Importantly, incorporation of scaling normalization (Bullard et al., 2010; Dillies et al., 2013) procedures was allowed providing a mechanism for accounting for technical variation. However, one of the issues with mixture model-based approaches is the reliance on large sample sizes (>50 samples) for accurate parameter estimation and the computationally intensive procedures involved in expectation–maximization or Markov chain Monte Carlo methods.

The second approach to assessment of outliers/subgroups in RNA-Seq (termed OASIS in their paper) assesses and builds on previous work on the theory of spacings developing several methods call MAST and MIST and compares these methods to other statistical approaches as well as SIBER (Pawlikowska et al., 2014). No method is found to perform best overall, each with its strengths in certain situations. All of these approaches appear more computationally efficient than SIBER and perform similarly in their assessment, if not better. However, they utilize their own custom normalization procedure though do not provide evidence that it removes the effect of technical noise as has been done with other normalization procedures in the context of DE (Bullard et al., 2010; Dillies et al., 2013). One of the most salient arguments in the paper is that the choice of method should be dictated by the underlying biological hypothesis. For instance, a single outlier may be important in some contexts where it is expected that a given sample is drastically different than the others in the cohort. In other contexts, mutually exclusive bimodal subgroups may be expected such as in Tomlins et al. (2005). Also, it is important to correct for relevant covariates which could lead to false positives. For instance, the presence of gender or ethnicity differences in the cohort could lead to the formation of multiple expression subgroups in the cases where none would be expected. An ideal method would be able to adjust for technical artifacts and confounders to increase power and limit false positives.

While the application of these methods has been primarily in cancer, there is exciting opportunity for psychiatric disorders and pharmacogenomic applications.

Being able to detect single sample outliers as well as multiple sample subgroups has implications for the integration of genomics data in personalized medicine. Approaches to discern differences among patients can potentially be useful to assign patients to drugs based on their subgroup or outlier status or to tease apart less common etiology in the research context. Given the difficulty of this problem and the fact that no method performs optimally in all situations, it is likely to remain a research question of interest for the foreseeable future.

3.4. Isoform-specific DE

Much more complex than gene-level expression, isoforma-specific DE requires the use of sophisticated statistical models in order to estimate, rather than count, expression levels of the transcript isoforms (Leng et al., 2013; Li & Dewey, 2011; Trapnell et al., 2013, 2010). DEXseq uses the differential exon usage based on read counts per exon and applying the DESeq normalization (Anders et al., 2012). The above GP method models position-level counts by exon in addition to gene-level counts. NOISeq will determine DE for exons or transcripts as well as genes. While gene-level DE is fairly well established, solving the challenges surrounding transcript-level DE is still in development. Given the recent studies showing complicated forms of gene expression across the genome (Djebali et al., 2012), this is likely to become a more critical aspect of RNA-Seq analysis.

4. FRAMEWORKS FOR INTERPRETATION

Throughout this chapter, we have focused on the importance of being guided by our biological questions throughout the experiment. In doing so, we also need to be cognizant of the impact these different choices have on the ultimate interpretation of our biological results. We highlight a few examples below.

4.1. RNA-Seq library construction

Many steps during the preparation of RNA-Seq libraries could affect or even bias the interpretation of RNA-Seq experiments (Van Dijk, Jaszczyszyn, & Thermes, 2014). For example, the choice of strand-specific protocol (Levin et al., 2010) and RNA fragmentation method (Wery, Descrimes, Thermes, Gautheret, & Morillon, 2013) was found to affect the coverage, assembly, and quantification of analyzed transcripts. Multiple studies have also reported on the increased transcript diversity and more uniform transcript coverage obtained through polyA selection as compared to other protocols (Beane et al., 2011; Sun, Asmann, et al., 2013; Tariq, Kim, Jejelowo, & Pourmand, 2011). Moreover, a recent study highlighted the overarching effects initiated by the design choices made during library construction on the downstream analyses of exon/transcript/gene quantification, SNV detection, and DE (Sun, Asmann, et al., 2013). Both for primary analysis/study design and for secondary analysis of public data, library construction must be evaluated for its impact on downstream interpretation.

4.2. Gene-model databases

There are numerous ongoing efforts annotating gene models including Ensembl (Gencode) (Harrow et al., 2012), NCBI RefSeq (Pruitt, Tatusova, Klimke, & Maglott, 2009), and AceView (Thierry-Mieg & Thierry-Mieg, 2006). Each effort utilizes different annotation strategies; thus, high variability exists among these different gene-model sources. Gencode annotation is guided by manual curations of transcriptional evidence of cDNAs, ESTs, and protein sequences; AceView uses heuristics to closely reproduce manual curation in an automated fashion, while Refseq utilizes a combination of both manual and automatic curation. A recent study found RefSeq to have the fewest number of annotated transcripts and genes when compared to Ensembl and AceView (Chen et al., 2013). The choice in gene model was found to have a significant effect on the analysis and interpretation of RNA-Seq results (Wu, Phan, et al., 2013) with less complex (smaller number of transcripts/genes) gene models having more reproducible and robust gene estimates as compared to more complex gene models which were better suited for exploratory/novel transcriptional or regulatory mechanisms.

Interestingly, a more comprehensive transcriptomic and genetic analysis of RNA-Seq was obtained by combining multiple gene annotation databases (Chen et al., 2013). However, even with the combination of multiple gene annotation databases, the majority (over 95%) of the genome is not included in known gene models. A more comprehensive genome-wide annotation framework is needed for annotating the genome in its entirety. Within this framework, basic individual genomic features could be created for genes, exons, introns, and intergenic regions. To preserve the double-stranded nature of DNA, each genomic feature can be further annotated on both strands (i.e., exon, opposite an exon, intron, opposite an intron). By preserving the strandedness of each genomic feature, one would be able

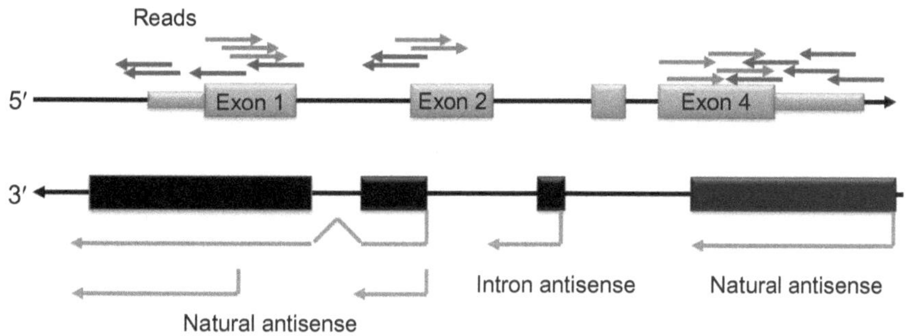

Figure 2.3 Schematic highlighting ambiguity for reads that overlap two genes. Stranded libraries can allow assignment of reads to the appropriate gene along with appropriate annotation frameworks.

to disambiguate and subsequently accurately quantify overlapping genes and antisense noncoding/regulatory RNAs, two genomic features that are commonly ignored or inaccurately counted (Fig. 2.3).

For gene-level annotation, all of the exon features from the ENSEMBL GTF file can be examined and gene-level ranges (start of first exon to end of last exon, including introns) can be calculated for each gene. Gene–gene overlap type can then be annotated (see Table 2.2) and used by any number of read-counting methods. Gene-model annotation is an area where the impact on downstream discovery and interpretation can be quite high, warranting serious evaluation during study design.

4.3. Functional annotation databases

A wealth of functional genomic annotation is currently available to help interpret RNA-Seq results. Annotation efforts such as the Kyoto Encyclopedia of Genes and Genomes (Kanehisa, Goto, Furumichi, Tanabe, & Hirakawa, 2010), Reactome (Croft et al., 2014), Database of Interacting Proteins (Salwinski et al., 2004), Gene Ontology (GO) (Gene Ontology Consortium, 2010), and Encyclopedia of DNA Elements (ENCODE) (Qu & Fang, 2013) have provided invaluable knowledge bases at the level of biological pathways, protein–protein interactions, individual genes, and all the way down to individual nucleotides (i.e., SNPs). Functional annotation analysis using GO terms and biological pathways has become routine with respect to the interpretation of RNA-Seq results (Hung, 2013). Functional annotations provided by ENOCDE are also heavily used to interpret and further understand RNA-Seq experiments with regard to splicing, gene

Table 2.2 Annotation framework to capture gene-level overlap to guide interpretation

Gene-level overlap categories	Description
ggNonOlap	Genes that do not overlap any other gene
ggOlapDiffFuncMultStr	Overlapping genes that have different functions and are in the same and opposite directions, i.e., a lincRNA overlaps one protein-coding gene in the opposite direction and also overlaps miRNA in the same direction
ggOlapDiffFuncOppStr	Overlapping genes that have different functions and are in the opposite direction
ggOlapDiffFuncSameStr	Overlapping genes that have different functions and are in thesame direction
ggOlapMultFunc	Same gene overlaps two other genes, one with same function and the other with different functions (same gene has two overlaps: 1. Protein_coding to protein coding. 2. Protein_coding to lincRNA)
ggOlapSameFuncMultStr	Overlapping genes that have the same function and are in the same and opposite directions, i.e., a protein-coding gene overlaps one protein-coding gene in the same direction and the other protein-coding gene in the opposite direction
ggOlapSameFuncOppStr	Overlapping genes have same function and are in the opposite direction
ggOlapSameFuncSameStr	Overlapping genes have same function and are in the same direction

regulations, and epigenetic factors (Hart, Komori, LaMere, Podshivalova, & Salomon, 2013; Mitra, Das, & Chakrabarti, 2013; Ye et al., 2014).

However, before using any knowledge base, we need to be aware of potential limitations and/or biases which may affect the final interpretation of our RNA-Seq results. For example, pathways with the same name were found to be inconsistent across independent knowledge bases (Bauer-Mehren, Furlong, & Sanz, 2009; Mitrea et al., 2013); thus, the use of different pathway knowledge bases may lead to different interpretations of the same result. Furthermore, although many knowledge bases are at the resolution of individual genes, meaning that annotations of isoform/exon results obtained from RNA-Seq are going to be limited or misleading, within these same knowledge bases many genes either have very limited or no annotation available (Khatri, Sirota, & Butte, 2012; Mitrea et al., 2013). Moreover, many knowledge bases are curated by experiments performed in different cell types at different times point under different conditions; however, these details are not always adequately captured, thus resulting in inaccurate and/or redundant annotations (Khatri et al., 2012). Interestingly, as many knowledge bases have been criticized for not being comprehensive enough, ENCODE annotations has been criticized for maybe being too comprehensive. From the production of 1640 datasets, using 24 experimental platforms on 147 different cell types, the ENCODE consortium reports that 80.4% of the human genome displays some biochemical functionality in at least one cell type (Qu & Fang, 2013). However, many question if ENCODE's definition of a functional element is too lenient, thus leading to a very high false-positive rate of functional annotations (Doolittle, 2013; Eddy, 2012). This highlights how assessment of functional annotation prior to its incorporation is key to ensure confidence in the downstream interpretation. We note again that this is influenced by the primary question of interest and trade-off between discovery and hypothesis testing.

5. SUMMARY

Throughout this chapter, we have focused on three major components in the analysis of RNA-Seq data (Fig. 2.1). The choices made regarding computational processing, statistical analysis, and frameworks for interpretation should be driven by the primary research question. It is of course critical that these decisions are made early during the initial experimental design to ensure best chances for the success of the study. With the rapid advances in technology and algorithms, the complexity will only increase, adding additional considerations in each of the three components. It is therefore critical for the full promise of routine brain transcriptome profiling to be met that continual reevaluation and assessment of these components are made.

REFERENCES

Anders, S., & Huber, W. (2010). Differential expression analysis for sequence count data. *Genome Biology*, *11*, R106. http://dx.doi.org/10.1186/gb-2010-11-10-r106.

Anders, S., Pyl, P. T., & Huber, W. (2014). HTSeq—A Python framework to work with high-throughput sequencing data. *bioRxiv*. http://dx.doi.org/10.1101/002824.

Anders, S., Reyes, A., & Huber, W. (2012). Detecting differential usage of exons from RNA-seq data. *Genome Research*, *22*, 2008–2017. http://dx.doi.org/10.1101/gr.133744.111.

Auer, P. L., & Doerge, R. W. (2011). A two-stage Poisson model for testing RNA-Seq data. *Statistical Applications in Genetics and Molecular Biology*, *10*, 1–26. Retrieved from: http://www.degruyter.com/view/j/sagmb.2011.10.issue-1/sagmb.2011.10.1.1627/sagmb.2011.10.1.1627.xml.

Bauer-Mehren, A., Furlong, L. I., & Sanz, F. (2009). Pathway databases and tools for their exploitation: Benefits, current limitations and challenges. *Molecular Systems Biology*, *5*, 290. http://dx.doi.org/10.1038/msb.2009.47.

Beane, J., Vick, J., Schembri, F., Anderlind, C., Gower, A., Campbell, J., et al. (2011). Characterizing the impact of smoking and lung cancer on the airway transcriptome using RNA-Seq. *Cancer Prevention Research (Philadelphia, PA)*, *4*, 803–817. http://dx.doi.org/10.1158/1940-6207.CAPR-11-0212.

Benjamini, Y., & Hochberg, Y. (1995). Controlling the false discovery rate: A practical and powerful approach to multiple testing. *Journal of the Royal Statistical Society: Series B*, *57*, 289–300.

Bottomly, D., Ryabinin, P. A., Tyner, J. W., Chang, B. H., Loriaux, M. M., Druker, B. J., et al. (2013). Comparison of methods to identify aberrant expression patterns in individual patients: Augmenting our toolkit for precision medicine. *Genome Medicine*, *5*, 1–10.

Bottomly, D., Wilmot, B., Tyner, J. W., Eide, C. A., Loriaux, M. M., Druker, B. J., et al. (2013). HitWalker: Variant prioritization for personalized functional cancer genomics. *Bioinformatics*, *29*, 509–510.

Bravo, H. C., Pihur, V., McCall, M., Irizarry, R., & Leek, J. (2012). Gene expression anti-profiles as a basis for accurate universal cancer signatures. *BMC Bioinformatics*, *13*, 272.

Bullard, J. H., Purdom, E., Hansen, K. D., & Dudoit, S. (2010). Evaluation of statistical methods for normalization and differential expression in mRNA-Seq experiments. *BMC Bioinformatics*, *11*, 94. http://dx.doi.org/10.1186/1471-2105-11-94.

Carrara, M., Beccuti, M., Lazzarato, F., Cavallo, F., Cordero, F., Donatelli, S., et al. (2013). State-of-the-art fusion-finder algorithms sensitivity and specificity. *BioMed Research International*, *2013*, 340620. http://dx.doi.org/10.1155/2013/340620.

Chen, G., Wang, C., Shi, L., Qu, X., Chen, J., Yang, J., et al. (2013). Incorporating the human gene annotations in different databases significantly improved transcriptomic and genetic analyses. *RNA (New York, N.Y.)*, *19*, 479–489. http://dx.doi.org/10.1261/rna.037473.112.

Cloonan, N., Xu, Q., Faulkner, G. J., Taylor, D. F., Tang, D. T., Kolle, G., et al. (2009). RNA-MATE: A recursive mapping strategy for high-throughput RNA-sequencing data. *Bioinformatics (Oxford, England)*, *25*, 2615–2616. http://dx.doi.org/10.1093/bioinformatics/btp459.

Croft, D., Mundo, A. F., Haw, R., Milacic, M., Weiser, J., Wu, G., et al. (2014). The reactome pathway knowledgebase. *Nucleic Acids Research*, *42*, D472–D477. http://dx.doi.org/10.1093/nar/gkt1102.

De Bona, F., Ossowski, S., Schneeberger, K., & Ratsch, G. (2008). Optimal spliced alignments of short sequence reads. *Bioinformatics (Oxford, England)*, *24*, i174–i180. http://dx.doi.org/10.1093/bioinformatics/btn300.

Dillies, M.-A. A., Rau, A., Aubert, J., Hennequet-Antier, C., Jeanmougin, M., Servant, N., et al. (2013). A comprehensive evaluation of normalization methods for Illumina high-throughput RNA sequencing data analysis. *Briefings in Bioinformatics*, *14*, 671–683. http://dx.doi.org/10.1093/bib/bbs046.

Djebali, S., Davis, C. A., Merkel, A., Dobin, A., Lassmann, T., Mortazavi, A., et al. (2012). Landscape of transcription in human cells. *Nature*, *489*, 101–108. http://dx.doi.org/10.1038/nature11233.

Dobin, A., Davis, C. A., Schlesinger, F., Drenkow, J., Zaleski, C., Jha, S., et al. (2013). STAR: Ultrafast universal RNA-seq aligner. *Bioinformatics, 29*(1), 15–21.

Dohm, J. C., Lottaz, C., Borodina, T., & Himmelbauer, H. (2008). Substantial biases in ultra-short read data sets from high-throughput DNA sequencing. *Nucleic Acids Research, 36*, e105. http://dx.doi.org/10.1093/nar/gkn425.

Doolittle, W. F. (2013). Is junk DNA bunk? A critique of ENCODE. *Proceedings of the National Academy of Sciences of the United States of America, 110*, 5294–5300. http://dx.doi.org/10.1073/pnas.1221376110.

Eddy, S. R. (2012). The C-value paradox, junk DNA and ENCODE. *Current Biology, 22*, R898–R899. http://dx.doi.org/10.1016/j.cub.2012.10.002.

Engstrom, P. G., Steijger, T., Sipos, B., Grant, G. R., Kahles, A., Consortium, R., et al. (2013). Systematic evaluation of spliced alignment programs for RNA-seq data. *Nature Methods, 10*, 1185–1191. http://dx.doi.org/10.1038/nmeth.2722.

Fang, Z., & Cui, X. (2011). Design and validation issues in RNA-seq experiments. *Briefings in Bioinformatics, 12*, 280–287. http://dx.doi.org/10.1093/bib/bbr004.

Finotello, F., Lavezzo, E., Bianco, L., Barzon, L., Mazzon, P., Fontana, P., et al. (2014). Reducing bias in RNA sequencing data: A novel approach to compute counts. *BMC Bioinformatics. 15*. http://dx.doi.org/10.1186/1471-2105-15-S1-S7.

Fonseca, N. A., Rung, J., Brazma, A., & Marioni, J. C. (2012). Tools for mapping high-throughput sequencing data. *Bioinformatics (Oxford, England), 28*, 3169–3177. http://dx.doi.org/10.1093/bioinformatics/bts605.

Forster, S. C., Finkel, A. M., Gould, J. A., & Hertzog, P. J. (2013). RNA-eXpress annotates novel transcript features in RNA-seq data. *Bioinformatics (Oxford, England), 29*, 810–812. http://dx.doi.org/10.1093/bioinformatics/btt034.

Gadgil, M. (2008). A population proportion approach for ranking differentially expressed genes. *BMC Bioinformatics, 9*, 380.

Garber, M., Grabherr, M. G., Guttman, M., & Trapnell, C. (2011). Computational methods for transcriptome annotation and quantification using RNA-seq. *Nature Methods, 8*, 469–477. http://dx.doi.org/10.1038/nmeth.1613.

Gene Ontology Consortium. (2010). The Gene Ontology in 2010: Extensions and refinements. *Nucleic Acids Research, 38*, D331–D335. http://dx.doi.org/10.1093/nar/gkp1018.

Ghosh, D. (2010). Discrete nonparametric algorithms for outlier detection with genomic data. *Journal of Biopharmaceutical Statistics, 20*, 193–208.

Ghosh, D., & Chinnaiyan, A. M. (2009). Genomic outlier profile analysis: Mixture models, null hypotheses, and nonparametric estimation. *Biostatistics, 10*, 60–69.

Gleiss, A., Sanchez-Cabo, F., Perco, P., Tong, D., & Heinze, G. (2011). Adaptive trimmed t-statistics for identifying predominantly high expression in a microarray experiment. *Statistics in Medicine, 30*, 52–61.

Grabherr, M. G., Haas, B. J., Yassour, M., Levin, J. Z., Thompson, D. A., Amit, I., et al. (2011). Full-length transcriptome assembly from RNA-Seq data without a reference genome. *Nature Biotechnology, 29*, 644–652. http://dx.doi.org/10.1038/nbt.1883.

Griffith, M., Griffith, O. L., Mwenifumbo, J., Goya, R., Morrissy, A. S., Morin, R. D., et al. (2010). Alternative expression analysis by RNA sequencing. *Nature Methods, 7*, 843–847. http://dx.doi.org/10.1038/nmeth.1503.

Guo, Y., Li, C.-I., Ye, F., & Shyr, Y. (2013). Evaluation of read count based RNAseq analysis methods. *BMC Genomics, 14*(Suppl. 8), S2. http://dx.doi.org/10.1186/1471-2164-14-S8-S2.

Guttman, M., Garber, M., Levin, J. Z., Donaghey, J., Robinson, J., Adiconis, X., et al. (2010). Ab initio reconstruction of cell type-specific transcriptomes in mouse reveals the conserved multi-exonic structure of lincRNAs. *Nature Biotechnology, 28*, 503–510. http://dx.doi.org/10.1038/nbt.1633.

Hardcastle, T. J., & Kelly, K. A. (2010). baySeq: Empirical Bayesian methods for identifying differential expression in sequence count data. *BMC Bioinformatics*, *11*, 422. http://dx.doi.org/10.1186/1471-2105-11-422.

Harrow, J., Frankish, A., Gonzalez, J. M., Tapanari, E., Diekhans, M., Kokocinski, F., et al. (2012). GENCODE: The reference human genome annotation for The ENCODE Project. *Genome Research*, *22*, 1760–1774. http://dx.doi.org/10.1101/gr.135350.111.

Hart, T., Komori, H. K., LaMere, S., Podshivalova, K., & Salomon, D. R. (2013). Finding the active genes in deep RNA-seq gene expression studies. *BMC Genomics*, *14*, 778. http://dx.doi.org/10.1186/1471-2164-14-778.

Hu, J. (2008). Cancer outlier detection based on likelihood ratio test. *Bioinformatics*, *24*, 2193–2199.

Huang, S., Zhang, J., Li, R., Zhang, W., He, Z., Lam, T. W., et al. (2011). SOAPsplice: Genome-wide ab initio detection of splice junctions from RNA-Seq data. *Frontiers in Genetics*, *2*, 46. http://dx.doi.org/10.3389/fgene.2011.00046.

Hung, J. H. (2013). Gene set/pathway enrichment analysis. *Methods in Molecular Biology (Clifton, N.J.)*, *939*, 201–213. http://dx.doi.org/10.1007/978-1-62703-107-3_13.

Ji, Z., Wang, Y., Wu, C., Wu, X., Xing, C., & Liang, Y. (2010). Mean, median and tri-mean based statistical detection methods for differential gene expression in microarray data. *Image and Signal Processing (CISP), 3rd International Congress on. Vol. 7. IEEE.*

Kadota, K., Nishiyama, T., & Shimizu, K. (2012). A normalization strategy for comparing tag count data. *Algorithms for Molecular Biology*, *7*. Retrieved from: http://www.almob.org/content/7/1/5/abstract.

Kanehisa, M., Goto, S., Furumichi, M., Tanabe, M., & Hirakawa, M. (2010). KEGG for representation and analysis of molecular networks involving diseases and drugs. *Nucleic Acids Research*, *38*, D355–D360. http://dx.doi.org/10.1093/nar/gkp896.

Karrila, S., Lee, J. H. E., & Tucker-Kellogg, G. (2011). A comparison of methods for data-driven cancer outlier discovery, and an application scheme to semisupervised predictive biomarker discovery. *Cancer Informatics*, *10*, 109–120.

Kent, W. J. (2002). BLAT—The BLAST-like alignment tool. *Genome Research*, *12*, 656–664. http://dx.doi.org/10.1101/gr.229202. Article published online before March 2002.

Khatri, P., Sirota, M., & Butte, A. J. (2012). Ten years of pathway analysis: Current approaches and outstanding challenges. *PLoS Computational Biology*, *8*, e1002375. http://dx.doi.org/10.1371/journal.pcbi.1002375.

Kim, D., Pertea, G., Trapnell, C., Pimentel, H., Kelley, R., & Salzberg, S. L. (2013). TopHat2: Accurate alignment of transcriptomes in the presence of insertions, deletions and gene fusions. *Genome Biology*, *14*, R36. http://dx.doi.org/10.1186/gb-2013-14-4-r36.

Langmead, B., Hansen, K. D., & Leek, J. T. (2010). Cloud-scale RNA-sequencing differential expression analysis with Myrna. *Genome Biology*, *11*, r83. http://dx.doi.org/10.1186/gb-2010-11-8-r83.

Langmead, B., Trapnell, C., Pop, M., & Salzberg, S. L. (2009). Ultrafast and memory-efficient alignment of short DNA sequences to the human genome. *Genome Biology*, *10*(3), r25. http://dx.doi.org/10.1186/gb-2009-10-3-r25.

Law, C. W., Chen, Y., Shi, W., & Smyth, G. K. (2014). Voom! Precision weights unlock linear model analysis tools for RNA-seq read counts. *Genome Biology*, *15*, R29. http://dx.doi.org/10.1186/gb-2014-15-2-r29.

Leng, N., Dawson, J. A., Thomson, J. A., Ruotti, V., Rissman, A. I., Smits, B. M. G., et al. (2013). EBSeq: An empirical Bayes hierarchical model for inference in RNA-seq experiments. *Bioinformatics*, *29*, 1035–1043. http://dx.doi.org/10.1093/bioinformatics/btt087.

Levin, J. Z., Yassour, M., Adiconis, X., Nusbaum, C., Thompson, D. A., Friedman, N., et al. (2010). Comprehensive comparative analysis of strand-specific RNA sequencing methods. *Nature Methods*, *7*, 709–715. http://dx.doi.org/10.1038/nmeth.1491.

Li, B., & Dewey, C. N. (2011). RSEM: Accurate transcript quantification from RNA-Seq data with or without a reference genome. *BMC Bioinformatics*, *12*, 323. http://dx.doi.org/10.1186/1471-2105-12-323.

Li, H., & Durbin, R. (2009). Fast and accurate short read alignment with Burrows-Wheeler transform. *Bioinformatics (Oxford, England)*, *25*, 1754–1760. http://dx.doi.org/10.1093/bioinformatics/btp324.

Li, H. Z., Gao, X., Li, X. Y., Chen, Q. J., Dong, J., & Zhao, W. C. (2013). Evaluation of assembly strategies using RNA-seq data associated with grain development of wheat (Triticum aestivum L.). *PLoS One*, *8*, e83530. http://dx.doi.org/10.1371/journal.pone.0083530.

Li, J., & Tibshirani, R. (2013). Finding consistent patterns: A nonparametric approach for identifying differential expression in RNA-Seq data. *Statistical Methods in Medical Research*, *22*, 519–536. http://dx.doi.org/10.1177/0962280211428386.

Li, J., Witten, D. M., Johnstone, I. M., & Tibshirani, R. (2012). Normalization, testing, and false discovery rate estimation for RNA-sequencing data. *Biostatistics*, *13*(3), 523–538. http://dx.doi.org/10.1093/biostatistics/kxr031.

Lian, H. (2008). MOST: Detecting cancer differential gene expression. *Biostatistics*, *9*, 411–418.

Liao, Y., Smyth, G. K., & Shi, W. (2013). The Subread aligner: Fast, accurate and scalable read mapping by seed-and-vote. *Nucleic Acids Research*, *41*, e108. http://dx.doi.org/10.1093/nar/gkt214.

Lindner, R., & Friedel, C. C. (2012). A comprehensive evaluation of alignment algorithms in the context of RNA-seq. *PLoS One*, *7*, e52403. http://dx.doi.org/10.1371/journal.pone.0052403.

Liu, C., Ma, J., Chang, C. J., & Zhou, X. (2013). FusionQ: A novel approach for gene fusion detection and quantification from paired-end RNA-Seq. *BMC Bioinformatics*, *14*, 193. http://dx.doi.org/10.1186/1471-2105-14-193.

Liu, F., & Wu, B. (2007). Multi-group cancer outlier differential gene expression detection. *Computational Biology and Chemistry*, *31*(2), 65–71.

Lund, S. P., Nettleton, D., McCarthy, D. J., & Smyth, G. K. (2012). Detecting differential expression in RNA-sequence data using quasi-likelihood with shrunken dispersion estimates. *Statistical Applications in Genetics and Molecular Biology*, *11*(5). http://dx.doi.org/10.1515/1544-6115.1826.

Martin, J. A., & Wang, Z. (2011). Next-generation transcriptome assembly. *Nature Reviews. Genetics*, *12*, 671–682. http://dx.doi.org/10.1038/nrg3068.

Marioni, J. C., Mason, C. E., Mane, S. M., Stephens, M., & Gilad, Y. (2008). RNA-seq: An assessment of technical reproducibility and comparison with gene expression arrays. *Genome Research*, *18*, 1509–1517.

McIntyre, L. M., Lopiano, K. K., Morse, A. M., Amin, V., Oberg, A. L., Young, L. J., et al. (2011). RNA-seq: Technical variability and sampling. *BMC Genomics*, *12*, 293. http://dx.doi.org/10.1186/1471-2164-12-293.

Mitra, S., Das, S., & Chakrabarti, J. (2013). Systems biology of cancer biomarker detection. *Cancer Biomarkers: Section A of Disease Markers*, *13*, 201–213. http://dx.doi.org/10.3233/CBM-130363.

Mitrea, C., Taghavi, Z., Bokanizad, B., Hanoudi, S., Tagett, R., Donato, M., et al. (2013). Methods and approaches in the topology-based analysis of biological pathways. *Frontiers in Physiology*, *4*, 278. http://dx.doi.org/10.3389/fphys.2013.00278.

Mortazavi, A., Williams, B. A., McCue, K., Schaeffer, L., & Wold, B. (2008). Mapping and quantifying mammalian transcriptomes by RNA-Seq. *Nature Methods*, *5*, 621–628. http://dx.doi.org/10.1038/nmeth.1226.

Mpindi, J. P., Sara, H., Haapa-Paananen, S., Kilpinen, S., Pisto, T., Bucher, E., et al. (2011). Gti: A novel algorithm for identifying outlier gene expression profiles from integrated microarray datasets. *PLoS One*, *6*(2), e17259. http://dx.doi.org/10.1371/journal.pone.0017259.

Pawlikowska, I., Wu, G., Edmonson, M., Liu, Z., Gruber, T., Zhang, J., et al. (2014). The most informative spacing test effectively discovers biologically relevant outliers or multiple modes in expression. *Bioinformatics (Oxford, England)*, *30*(10), 1400–1408. http://dx.doi.org/10.1093/bioinformatics/btu039.

Pearson, W. R. (2013). An introduction to sequence similarity ("homology") searching. *Current Protocols in Bioinformatics*. http://dx.doi.org/10.1002/0471250953.bi0301s42, Editorial Board, Andreas D. Baxevanis et al., Chapter 3, Unit3.1.

Pinese, M., Scarlett, C. J., Kench, J. G., Colvin, E. K., Segara, D., Henshall, S. M., et al. (2009). Messina: A novel analysis tool to identify biologically relevant molecules in disease. *PLoS One*, *4*(4), e5337.

Pruitt, K. D., Tatusova, T., Klimke, W., & Maglott, D. R. (2009). NCBI reference sequences: Current status, policy and new initiatives. *Nucleic Acids Research*, *37*, D32–D36. http://dx.doi.org/10.1093/nar/gkn721.

Qu, H., & Fang, X. (2013). A brief review on the Human Encyclopedia of DNA Elements (ENCODE) project. *Genomics, Proteomics & Bioinformatics*, *11*, 135–141. http://dx.doi.org/10.1016/j.gpb.2013.05.001.

Quinlan, A. R., & Hall, I. M. (2010). BEDTools: A flexible suite of utilities for comparing genomic features. *Bioinformatics (Oxford, England)*, *26*, 841–842. http://dx.doi.org/10.1093/bioinformatics/btq033.

Ramskold, D., Wang, E. T., Burge, C. B., & Sandberg, R. (2009). An abundance of ubiquitously expressed genes revealed by tissue transcriptome sequence data. *PLoS Computational Biology*, *5*, e1000598. http://dx.doi.org/10.1371/journal.pcbi.1000598.

Rapaport, F., Khanin, R., Liang, Y., Pirun, M., Krek, A., Zumbo, P., et al. (2013). Comprehensive evaluation of differential gene expression analysis methods for RNA-seq data. *Genome Biology*, *14*, r95. http://dx.doi.org/10.1186/gb-2013-14-9-r95.

Risso, D., Schwartz, K., Sherlock, G., & Dudoit, S. (2011). GC-content normalization for RNA-Seq data. *BMC Bioinformatics*, *12*, 480. http://dx.doi.org/10.1186/1471-2105-12-480.

Roberts, A., Pimentel, H., Trapnell, C., & Pachter, L. (2011). Identification of novel transcripts in annotated genomes using RNA-Seq. *Bioinformatics (Oxford, England)*, *27*, 2325–2329. http://dx.doi.org/10.1093/bioinformatics/btr355.

Robinson, M. D., & Oshlack, A. (2010). A scaling normalization method for differential expression analysis of RNA-seq data. *Genome Biology*, *11*, r25. http://dx.doi.org/10.1186/gb-2010-11-3-r25.

Robinson, M. D., & Smyth, G. K. (2007). Moderated statistical tests for assessing differences in tag abundance. *Bioinformatics*, *23*, 2881–2887.

Robles, J. A., Qureshi, S. E., Stephen, S. J., Wilson, S. R., Burden, C. J., & Taylor, J. M. (2012). Efficient experimental design and analysis strategies for the detection of differential expression using RNA-sequencing. *BMC Genomics*, *13*, 484. http://dx.doi.org/10.1186/1471-2164-13-484.

Salwinski, L., Miller, C. S., Smith, A. J., Pettit, F. K., Bowie, J. U., & Eisenberg, D. (2004). The database of interacting proteins: 2004 update. *Nucleic Acids Research*, *32*, D449–D451. http://dx.doi.org/10.1093/nar/gkh086.

Schulz, M. H., Zerbino, D. R., Vingron, M., & Birney, E. (2012). Oases: Robust de novo RNA-seq assembly across the dynamic range of expression levels. *Bioinformatics (Oxford, England)*, *28*, 1086–1092. http://dx.doi.org/10.1093/bioinformatics/bts094.

Smyth, G. (2004). Linear models and empirical Bayes methods for assessing differential expression in microarray experiments. *Statistical Applications in Genetics and Molecular Biology*, *3*, 1–23.

Smyth, G. K. (2005). Limma: Linear Models for Microarray Data. Chapter 23 in Bioinformatics and Computational Biology Solutions Using R and Bioconductor. Publisher Springer. Eds. Robert Gentleman, Vincetn J. Carey, Wolfgang Huber, Rafael A. Irizarry, Sandrine Dudoit.

Srivastava, S., & Chen, L. (2010). A two-parameter generalized Poisson model to improve the analysis of RNA-seq data. *Nucleic Acids Research*, *38*(17), e170. http://dx.doi.org/10.1093/nar/gkq670.

Sun, Z., Asmann, Y. W., Nair, A., Zhang, Y., Wang, L., Kalari, K. R., et al. (2013). Impact of library preparation on downstream analysis and interpretation of RNA-Seq data: Comparison between Illumina PolyA and NuGEN Ovation protocol. *PLoS One*, *8*, e71745. http://dx.doi.org/10.1371/journal.pone.0071745.

Sun, J., Nishiyama, T., Shimizu, K., & Kadota, K. (2013). TCC: An R package for comparing tag count data with robust normalization strategies. *BMC Bioinformatics*, *14*, 219. http://dx.doi.org/10.1186/1471-2105-14-219.

Tarazona, S., García-Alcalde, F., Dopazo, J., Ferrer, A., & Conesa, A. (2011). Differential expression in RNA-seq: A matter of depth. *Genome Research*, *21*, 2213–2223. http://dx.doi.org/10.1101/gr.124321.111.

Tariq, M. A., Kim, H. J., Jejelowo, O., & Pourmand, N. (2011). Whole-transcriptome RNAseq analysis from minute amount of total RNA. *Nucleic Acids Research*, *39*, e120. http://dx.doi.org/10.1093/nar/gkr547.

Thierry-Mieg, D., & Thierry-Mieg, J. (2006). AceView: A comprehensive cDNA-supported gene and transcripts annotation. *Genome Biology*, 7(Suppl. 1), S12.1–S12.14. http://dx.doi.org/10.1186/gb-2006-7-s1-s12.

Tibshirani, R., & Hastie, T. (2007). Outlier sums for differential gene expression analysis. *Biostatistics*, *8*, 2–8.

Tomlins, S. A., Rhodes, D. R., Perner, S., Dhanasekaran, S. M., Mehra, R., Sun, X. W., et al. (2005). Recurrent fusion of TMPRSS2 and ETS transcription factor genes in prostate cancer. *Science*, *310*, 644–648.

Tong, P., Chen, Y., Su, X., & Coombes, K. R. (2013). SIBER: Systematic identification of bimodally expressed genes using RNAseq data. *Bioinformatics (Oxford, England)*, *29*, 605–613. http://dx.doi.org/10.1093/bioinformatics/bts713.

Trapnell, C., Hendrickson, D. G., Sauvageau, M., Goff, L., Rinn, J. L., & Pachter, L. (2013). Differential analysis of gene regulation at transcript resolution with RNA-seq. *Nature Biotechnology*, *31*, 46–53. http://dx.doi.org/10.1038/nbt.2450.

Trapnell, C., Pachter, L., & Salzberg, S. L. (2009). TopHat: Discovering splice junctions with RNA-Seq. *Bioinformatics (Oxford, England)*, *25*, 1105–1111. http://dx.doi.org/10.1093/bioinformatics/btp120.

Trapnell, C., Williams, B. A., Pertea, G., Mortazavi, A., Kwan, G., van Baren, M. J., et al. (2010). Transcript assembly and quantification by RNA-Seq reveals unannotated transcripts and isoform switching during cell differentiation. *Nature Biotechnology*, *28*, 511–515. http://dx.doi.org/10.1038/nbt.1621.

Van Dijk, E. L., Jaszczyszyn, Y., & Thermes, C. (2014). Library preparation methods for next-generation sequencing: Tone down the bias. *Experimental Cell Research*, *322*, 12–20. http://dx.doi.org/10.1016/j.yexcr.2014.01.008.

Wang, Y., & Rekaya, R. (2010). LSOSS: Detection of cancer outlier differential gene expression. *Biomarker Insights*, *5*, 69–78.

Wang, Y., Sun, G., Ji, Z., Xing, C., & Liang, Y. (2012). Weighted change-point method for detecting differential gene expression in breast cancer microarray data. *PLoS One*, *7*, e29860.

Wang, C., Taciroglu, A., Maetschke, S. R., Nelson, C. C., Ragan, M. A., & Davis, M. J. (2012). mCOPA: Analysis of heterogeneous features in cancer expression data. *Journal of Clinical Bioinformatics*, *2*, 22.

Wang, J., Wen, S., Symmans, W. F., Pusztai, L., & Coombes, K. R. (2009). The bimodality index: A criterion for discovering and ranking bimodal signatures from cancer gene expression profiling data. *Cancer Informatics*, 7, 199–216.
Wang, Y., Wu, C., Ji, Z., Wang, B., & Liang, Y. (2011). Non-parametric change-point method for differential gene expression detection. *PLoS One*, *6*(5), e20060. http://dx.doi.org/10.1371/journal.pone.0020060.
Wery, M., Descrimes, M., Thermes, C., Gautheret, D., & Morillon, A. (2013). Zinc-mediated RNA fragmentation allows robust transcript reassembly upon whole transcriptome RNA-Seq. *Methods*, *63*, 25–31. http://dx.doi.org/10.1016/j.ymeth.2013.03.009.
Wu, B. (2007). Cancer outlier differential gene expression detection. *Biostatistics*, *8*(3), 566–575.
Wu, T. D., & Nacu, S. (2010). Fast and SNP-tolerant detection of complex variants and splicing in short reads. *Bioinformatics (Oxford, England)*, *26*, 873–881. http://dx.doi.org/10.1093/bioinformatics/btq057.
Wu, P.-Y., Phan, J. H., & Wang, M. D. (2013). Assessing the impact of human genome annotation choice on RNA-seq expression estimates. *BMC Bioinformatics*, *14*, S8. http://dx.doi.org/10.1186/1471-2105-14-S11-S8.
Wu, H., Wang, C., & Wu, Z. (2013). A new shrinkage estimator for dispersion improves differential expression detection in RNA-seq data. *Biostatistics*, *14*, 232–243. http://dx.doi.org/10.1093/biostatistics/kxs033.
Wu, T. D., & Watanabe, C. K. (2005). GMAP: A genomic mapping and alignment program for mRNA and EST sequences. *Bioinformatics (Oxford, England)*, *21*, 1859–1875. http://dx.doi.org/10.1093/bioinformatics/bti310.
Xie, Y., Wu, G., Tang, J., Luo, R., Patterson, J., Liu, S., et al. (2014). SOAPdenovo-Trans: de novo transcriptome assembly with short RNA-Seq reads. *Bioinformatics*, *30*(12), 1660–1666.
Ye, Z., Chen, Z., Lan, X., Hara, S., Sunkel, B., Huang, T. H., et al. (2014). Computational analysis reveals a correlation of exon-skipping events with splicing, transcription and epigenetic factors. *Nucleic Acids Research*, *42*, 2856–2869. http://dx.doi.org/10.1093/nar/gkt1338.
Zhang, W., Chen, J., Yang, Y., Tang, Y., Shang, J., & Shen, B. (2011). A practical comparison of de novo genome assembly software tools for next-generation sequencing technologies. *PLoS One*, *6*, e17915. http://dx.doi.org/10.1371/journal.pone.0017915.

FURTHER READING

Alamancos, G. P., Agirre, E., & Eyras, E. (2014). Methods to study splicing from high-throughput RNA sequencing data. *Methods in Molecular Biology (Clifton, N.J.)*, *1126*, 357–397. http://dx.doi.org/10.1007/978-1-62703-980-2_26.
Alemu, E. Y., Carl, J. W., Bravo, H. C., & Hannenhalli, S. (2014). Determinants of expression variability. *Nucleic Acids Research*, *42*, 3503–3514. Retrieved from: http://nar.oxfordjournals.org/content/42/6/3503.
Anders, S., McCarthy, D. J., Chen, Y., Okoniewski, M., Smyth, G. K., Huber, W., et al. (2013). Count-based differential expression analysis of RNA sequencing data using R and bioconductor. *Nature Protocols*, *8*, 1765–1786. http://dx.doi.org/10.1038/nprot.2013.099.
Angiulli, F., & Pizzuti, C. (2002). Fast outlier detection in high dimensional spaces. *Principles of data mining and knowledge discovery. Lecture Notes in Computer Science Volume 2431.* (pp. 15–27). Springer: Heidelberg.
Au, K. F., Jiang, H., Lin, L., Xing, Y., & Wong, W. H. (2010). Detection of splice junctions from paired-end RNA-seq data by SpliceMap. *Nucleic Acids Research*, *38*, 4570–4578. http://dx.doi.org/10.1093/nar/gkq211.

Bicocca, V. T., Chang, B. H., Masouleh, B. K., Muschen, M., Loriaux, M. M., Druker, B. J., et al. (2012). Crosstalk between ROR1 and the pre-B cell receptor promotes survival of t (1; 19) acute lymphoblastic leukemia. *Cancer Cell, 22*, 656–667.
Carvalho, B. S., & Irizarry, R. A. (2010). A framework for oligonucleotide microarray preprocessing. *Bioinformatics, 26*, 2363–2367. http://dx.doi.org/10.1093/bioinformatics/btq431.
Casagrande, G., te Kronnie, G., & Basso, G. (2006). The effects of siRNA-mediated inhibition of E2A-PBX1 on EB-1 and Wnt16b expression in the 697 pre-B leukemia cell line. *Haematologica, 91*, 765–771.
Cerami, E., Gao, J., Dogrusoz, U., Gross, B. E., Sumer, S. O., Aksoy, B. A., et al. (2012). The cBio cancer genomics portal: An open platform for exploring multidimensional cancer genomics data. *Cancer Discovery, 2*, 401–404.
Cheadle, C., Vawter, M. P., Freed, W. J., & Becker, K. G. (2003). Analysis of microarray data using Z score transformation. *The Journal of Molecular Diagnostics, 5*, 73–81. http://dx.doi.org/10.1016/S1525-1578(10)60455-2.
Chen, G., Wang, C., & Shi, T. (2011). Overview of available methods for diverse RNA-Seq data analyses. *Science China. Life Sciences, 54*, 1121–1128. http://dx.doi.org/10.1007/s11427-011-4255-x.
Chung, N., Zhang, X. D., Kreamer, A., Locco, L., Kuan, P.-F., Bartz, S., et al. (2008). Median absolute deviation to improve hit selection for genome-scale RNAi screens. *Journal of Biomolecular Screening, 13*, 149–158. http://dx.doi.org/10.1177/1087057107312035.
Costa, V., Angelini, C., De Feis, I., & Ciccodicola, A. (2010). Uncovering the complexity of transcriptomes with RNA-Seq. *Journal of Biomedicine and Biotechnology*, 19 pages, Article ID 853916. http://dx.doi.org/10.1155/2010/853916.
Costa, V., Aprile, M., Esposito, R., & Ciccodicola, A. (2013). RNA-Seq and human complex diseases: Recent accomplishments and future perspectives. *European Journal of Human Genetics, 21*, 134–142. http://dx.doi.org/10.1038/ejhg.2012.129.
Courtney, E., Kornfeld, S., Janitz, K., & Janitz, M. (2010). Transcriptome profiling in neurodegenerative disease. *Journal of Neuroscience Methods, 193*, 189–202. http://dx.doi.org/10.1016/j.jneumeth.2010.08.018.
Dudoit, S., Yang, Y. H., Callow, M. J., & Speed, T. P. (2002). Statistical methods for identifying differentially expressed genes in replicated cDNA microarray experiments. *Statistica Sinica, 12*, 111–140.
Durinck, S., & Bullard J. GenomeGraphs: Plotting genomic information from Ensembl. R package version 1.24.0. http://www.bioconductor.org/packages/release/bioc/html/GenomeGraphs.html
Durinck, S., Moreau, Y., Kasprzyk, A., Davis, S., De Moor, B., Brazma, A., et al. (2005). BioMart and bioconductor: A powerful link between biological databases and microarray data analysis. *Bioinformatics, 21*, 3439–3440.
Ekins, S., Nikolsky, Y., Bugrim, A., Kirillov, E., & Nikolskaya, T. (2007). Pathway mapping tools for analysis of high content data. *Methods in Molecular Biology (Clifton, N.J.), 356*, 319–350.
Emerson, S. C., & Emerson, S. S. (2011). The importance of statistical theory in outlier detection. *UW Biostatistics Working Paper Series. Working Paper 381.* http://biostats.bepress.com/uwbiostat/paper381.
Esnaola, M., Puig, P., Gonzalez, D., Castelo, R., & Gonzalez, J. R. (2013). A flexible count data model to fit the wide diversity of expression profiles arising from extensively replicated RNA-seq experiments. *BMC Bioinformatics, 14*, 254. http://dx.doi.org/10.1186/1471-2105-14-254.
Finotello, F., Lavezzo, E., Barzon, L., Fontana, P., Si-Ammour, A., Toppo, S., et al. (2012). RNA sequencing data: Biases and normalization. *EMBnet.journal, 18*, 99. http://dx.doi.org/10.14806/ej.18.A.441.

Flicek, P., Amode, M. R., Barrell, D., Beal, K., Brent, S., Carvalho-Silva, D., et al. (2012). Ensembl 2012. *Nucleic Acids Research*, *40*, D84–D90.

Gentleman, R. C., Carey, V. J., Bates, D. M., Bolstad, B., Dettling, M., Dudoit, S., et al. (2004). Bioconductor: Open software development for computational biology and bioinformatics. *Genome Biology*, *5*, R80.

Gundem, G., & Lopez-Bigas, N. (2012). Sample level enrichment analysis unravels shared stress phenotypes among multiple cancer types. *Genome Medicine*, *4*, 28.

Haraksingh, R. R., & Snyder, M. P. (2013). Impacts of variation in the human genome on gene regulation. *Journal of Molecular Biology*, *425*, 3970–3977. http://dx.doi.org/10.1016/j.jmb.2013.07.015.

Hebenstreit, D., & Teichmann, S. A. (2011). Analysis and simulation of gene expression profiles in pure and mixed cell populations. *Physical Biology*, *8*. http://dx.doi.org/10.1088/1478-3975/8/3/035013.

Hitzemann, R., Bottomly, D., Darakjian, P., Walter, N., Iancu, O., Searles, R., et al. (2013). Genes, behavior and next-generation RNA sequencing. *Genes, Brain, and Behavior*, *12*(1), 1–12. http://dx.doi.org/10.1111/gbb.12007.

Huang da, W., Sherman, B. T., & Lempicki, R. A. (2009). Systematic and integrative analysis of large gene lists using DAVID bioinformatics resources. *Nature Protocols*, *4*, 44–57. http://dx.doi.org/10.1038/nprot.2008.211.

Irizarry, R. A., Hobbs, B., Collin, F., Beazer-Barclay, Y. D., Antonellis, K. J., Scherf, U., et al. (2003). Exploration, normalization, and summaries of high density oligonucleotide array probe level data. *Biostatistics*, *4*, 249–264. http://dx.doi.org/10.1093/biostatistics/4.2.249.

Johnson, W. E., Li, C., & Rabinovic, A. (2007). Adjusting batch effects in microarray expression data using empirical Bayes methods. *Biostatistics*, *8*, 118–127.

Kaur, H., Burden-Gulley, S. M., Phillips-Mason, P. J., Basilion, J. P., Sloan, A. E., & Brady-Kalnay, S. M. (2012). Protein tyrosine phosphatase mu regulates glioblastoma cell growth and survival in vivo. *Neuro-Oncology*, *14*, 561–573. http://dx.doi.org/10.1093/neuonc/nos066.

Kavanagh, T., Mills, J. D., Kim, W. S., Halliday, G. M., & Janitz, M. (2013). Pathway analysis of the human brain transcriptome in disease. *Journal of Molecular Neuroscience*, *51*, 28–36. http://dx.doi.org/10.1007/s12031-012-9940-0.

Kay, G. F., Barton, S. C., Surani, M. A., & Rastan, S. (1994). Imprinting and X chromosome counting mechanisms determine Xist expression in early mouse development. *Cell*, *77*, 639–650. http://dx.doi.org/10.1016/0092-8674(94)90049-3.

Klambauer, G., Unterthiner, T., & Hochreiter, S. (2013). DEXUS: Identifying differential expression in RNA-Seq studies with unknown conditions. *Nucleic Acids Research*, *41*, e198. http://dx.doi.org/10.1093/nar/gkt834.

Kou, Y., Lu, C.-T., & Chen, D. (2006). Spatial weighted outlier detection. In *Proceedings of the SIAM conference on data mining*.

Kvam, V. M., Liu, P., & Si, Y. (2012). A comparison of statistical methods for detecting differentially expressed genes from RNA-seq data. *American Journal of Botany*, *99*, 248–256. http://dx.doi.org/10.3732/ajb.1100340.

Lawrence, M., Huber, W., Pagès, H., Aboyoun, P., Carlson, M., Gentleman, R., et al. (2013). "Software for Computing and Annotating Genomic Ranges". *PLoS Computational Biology*. *9*. http://dx.doi.org/10.1371/journal.pcbi.1003118.

Lazar, C., Meganck, S., Taminau, J., Steenhoff, D., Coletta, A., Molter, C., et al. (2012). Batch effect removal methods for microarray gene expression data integration: A survey. *Briefings in Bioinformatics*, *14*(4), 469–490.

L'Ecuyer, P., Simard, R., Chen, E. J., & Kelton, W. D. (2002). An object-oriented random-number package with many long streams and substreams. *Operations Research*, *50*, 1073–1075.

Leek, J. T., Scharpf, R. B., Bravo, H. C., Simcha, D., Langmead, B., Johnson, W. E., et al. (2010). Tackling the widespread and critical impact of batch effects in high-throughput data. *Nature Reviews. Genetics*, *11*, 733–739.

Leek, J. T., & Storey, J. D. (2007). Capturing heterogeneity in gene expression studies by surrogate variable analysis. *PLoS Genetics*, *3*, 1724–1735.

Li, H., Ruan, J., & Durbin, R. (2008). Mapping short DNA sequencing reads and calling variants using mapping quality scores. *Genome Research*, *18*, 1851–1858. http://dx.doi.org/10.1101/gr.078212.108.

Lunter, G., & Goodson, M. (2011). Stampy: A statistical algorithm for sensitive and fast mapping of Illumina sequence reads. *Genome Research*, *21*, 936–939. http://dx.doi.org/10.1101/gr.111120.110.

Luo, W., Friedman, M. S., Shedden, K., Hankenson, K. D., & Woolf, P. J. (2009). GAGE: Generally applicable gene set enrichment for pathway analysis. *BMC Bioinformatics*, *10*, 161. http://dx.doi.org/10.1186/1471-2105-10-161.

Majewski, J., & Pastinen, T. (2011). The study of eQTL variations by RNA-seq: From SNPs to phenotypes. *Trends in Genetics*, *27*, 72–79. http://dx.doi.org/10.1016/j.tig.2010.10.006.

Maxson, J. E., Gotlib, J., Pollyea, D. A., Fleischman, A. G., Agarwal, A., Eide, C. A., et al. (2013). Oncogenic CSF3R mutations in chronic neutrophilic leukemia and atypical CML. *New England Journal of Medicine*, *368*, 1781–1790. http://dx.doi.org/10.1056/NEJMoa1214514.

McCarthy, D. J., Chen, Y., & Smyth, G. K. (2012). Differential expression analysis of multifactor RNA-Seq experiments with respect to biological variation. *Nucleic Acids Research*, *40*, 4288–4297. http://dx.doi.org/10.1093/nar/gks042.

McGettigan, P. A. (2013). Transcriptomics in the RNA-seq era. *Current Opinion in Chemical Biology*, *17*, 4–11. http://dx.doi.org/10.1016/j.cbpa.2012.12.008.

McHale, C. M., Zhang, L., Thomas, R., & Smith, M. T. (2013). Analysis of the transcriptome in molecular epidemiology studies. *Environmental and Molecular Mutagenesis*, *54*, 500–517. http://dx.doi.org/10.1002/em.21798.

Mostafavi, S., Battle, A., Zhu, X., Urban, A. E., Levinson, D., Montgomery, S. B., et al. (2013). Normalizing RNA-sequencing data by modeling hidden covariates with prior knowledge. *PLoS One*. *8*. http://dx.doi.org/10.1371/journal.pone.0068141.

Mutz, K.-O., Heilkenbrinker, A., Lönne, M., Walter, J.-G., & Stahl, F. (2013). Transcriptome analysis using next-generation sequencing. *Current Opinion in Biotechnology*, *24*, 22–30. http://dx.doi.org/10.1016/j.copbio.2012.09.004.

Oellrich, A., Sanger Mouse Genetics, P., & Smedley, D. (2014). Linking tissues to phenotypes using gene expression profiles. *Database*, *2014*, bau017.

Oh, S., Song, S., Dasgupta, N., & Grabowski, G. (2014). The analytical landscape of static and temporal dynamics in transcriptome data. *Frontiers in Genetics*, *5*, 35. http://dx.doi.org/10.3389/fgene.2014.00035.

Oshlack, A., Robinson, M. D., & Young, M. D. (2010). From RNA-seq reads to differential expression results. *Genome Biology*, *11*(12), 220. http://dx.doi.org/10.1186/gb-2010-11-12-220.

Piccolo, S. R., Withers, M. R., Francis, O. E., Bild, A. H., & Johnson, W. E. (2013). Multiplatform single-sample estimates of transcriptional activation. *Proceedings of the National Academy of Sciences*, *110*, 17778–17783. http://dx.doi.org/10.1073/pnas.1305823110.

Roberts, A., Schaeffer, L., & Pachter, L. (2013). Updating RNA-Seq analyses after re-annotation. *Bioinformatics*, *29*, 1631–1637. http://dx.doi.org/10.1093/bioinformatics/btt197.

Robinson, M. D., & Smyth, G. K. (2008). Small-sample estimation of negative binomial dispersion, with applications to SAGE data. *Biostatistics*, *9*, 321–332. Retrieved from: http://biostatistics.oxfordjournals.org/content/9/2/321.

Sevcikova, H., & Rossini, T. (2012). *rlecuyer: R interface to RNG with multiple streams.* Retrieved from: http://cran.r-project.org/package=rlecuyer; (2012).

Shiffler, R. E. (1988). Maximum Z scores and outliers. *The American Statistician, 42*, 79–80.

Shin, H., Shannon, C. P., Fishbane, N., Ruan, J., Zhou, M., Balshaw, R., et al. (2014). Variation in RNA-Seq transcriptome profiles of peripheral whole blood from healthy individuals with and without globin depletion. *PLoS One, 9*, e91041. http://dx.doi.org/10.1371/journal.pone.0091041.

Simon, R., Radmacher, M. D., & Dobbin, K. (2002). Design of studies using DNA microarrays. *Genetic Epidemiology, 23*, 21–36.

Tang, S., & Riva, A. (2013). PASTA: Splice junction identification from RNA-sequencing data. *BMC Bioinformatics, 14*, 116. http://dx.doi.org/10.1186/1471-2105-14-116.

Team, R. C. (2012). *R: A language and environment for statistical computing.* Vienna, Austria: R Foundation for Statistical Computing. Retrieved from: http://www.r-project.org/.

Tusher, V. G., Tibshirani, R., & Chu, G. (2001). Significance analysis of microarrays applied to the ionizing radiation response. *Proceedings of the National Academy of Sciences of the United States of America, 98*, 5116–5121.

Tyner, J. W., Deininger, M. W., Loriaux, M. M., Chang, B. H., Gotlib, J. R., Willis, S. G., et al. (2009). RNAi screen for rapid therapeutic target identification in leukemia patients. *Proceedings of the National Academy of Sciences of the United States of America, 106*, 8695–8700. http://dx.doi.org/10.1073/pnas.0903233106.

Vaske, C. J., Benz, S. C., Sanborn, J. Z., Earl, D., Szeto, C., Zhu, J., et al. (2010). Inference of patient-specific pathway activities from multi-dimensional cancer genomics data using PARADIGM. *Bioinformatics, 26*, i237–i245.

Wang, X., & Clark, A. G. (2014). Using next-generation RNA sequencing to identify imprinted genes. *Heredity.* http://dx.doi.org/10.1038/hdy.2014.18.

Wickham, H. (2007). Reshaping data with the reshape package. *Journal of Statistical Software, 21*, 1–20. Retrieved from: http://www.jstatsoft.org/v21/i12/.

Wickham, H. (2009). *ggplot2: Elegant graphics for data analysis.* New York: Springer.

Wu, G., Tang, J., Luo, R., Patterson, J., Liu, S., & Wang, J. (2014). SOAPdenovo-Trans: De novo transcriptome assembly with short RNA-Seq reads. *Bioinformatics (Oxford, England), 30*, 1660–1666. http://dx.doi.org/10.1093/bioinformatics/btu077.

Xu, J., & Wong, C. (2010). Hunting for robust gene signature from cancer profiling data: Sources of variability, different interpretations, and recent methodological developments. *Cancer Letters, 296*, 9–16.

Yang, Y. H., Dudoit, S., Luu, P., Lin, D. M., Peng, V., Ngai, J., et al. (2002). Normalization for cDNA microarray data: A robust composite method addressing single and multiple slide systematic variation. *Nucleic Acids Research, 30*, e15.

Yi, M., & Stephens, R. M. (2008). SLEPR: A sample-level enrichment-based pathway ranking method-seeking biological themes through pathway-level consistency. *PLoS One, 3*, e3288.

Yu, H. (2010). *Rmpi: Interface (wrapper) to MPI (message-passing interface).* Retrieved from: http://cran.r-project.org/package=Rmpi; (2010).

Zerbino, D. R., & Birney, E. (2008). Velvet: Algorithms for de novo short read assembly using de Bruijn graphs. *Genome Research, 18*, 821–829. http://dx.doi.org/10.1101/gr.074492.107.

Zhang, J., & Wang, H. (2006). Detecting outlying subspaces for high-dimensional data: The new task, algorithms, and performance. *Knowledge and Information Systems, 10*, 333–355.

CHAPTER THREE

Data Integration and Reproducibility for High-Throughput Transcriptomics

Michael Mooney*,[1], Shannon McWeeney*

*Division of Bioinformatics & Computational Biology, Department of Medical Informatics & Clinical Epidemiology, Oregon Health & Science University, Portland, Oregon, USA
[1]Corresponding author: e-mail address: mooneymi@ohsu.edu

Contents

Abstract

The rapid advances in high-throughput transcriptomics allow individual investigators to rapidly and comprehensively interrogate the transcriptome. This phenomenon has placed large volumes of gene expression data in public repositories presenting opportunities for secondary analysis, discovery, and *in silico* modeling. We focus here on guidelines for best practices for transcriptomics data integration and considerations for reproducibility. In addition, we discuss some considerations for multi-omic and cross-species comparisons.

1. OPPORTUNITIES FOR SECONDARY USE OF DATA AND META-ANLAYSIS IN TRANSCRIPTOMICS

Transcriptomic technologies have become a powerful tool for biologists and have provided a wealth of knowledge about the genetic activity

International Review of Neurobiology, Volume 116
ISSN 0074-7742
http://dx.doi.org/10.1016/B978-0-12-801105-8.00003-5

of cells under various conditions. The ability to detect and quantify virtually all transcripts within a tissue sample at a particular time point has allowed researchers to gain insight into the complex network of gene regulation that leads to specific phenotypes. In addition, profiling the transcriptome has also allowed researchers to identify factors that influence the regulation of gene expression, such as DNA variation and environmental exposures.

The enormous potential for new knowledge that is offered by transcriptomic studies has led to an explosion of gene expression data generated over the past decades. In the year 2000, 28 scholarly articles associated with the terms "transcriptome" or "transcriptomics" were indexed in PubMed. Since then, the number of studies has grown exponentially, and in 2013 alone there were 5039 papers published.

The large amount of data being produced across a wide variety of experimental conditions (animal models, tissue types, phenotypes, etc.) presents a great opportunity for secondary analysis of data, including meta-analyses and replication studies. However, the appropriate use of data from multiple studies (combining or comparing datasets) requires a thorough understanding of the differences among the technologies used to create the data.

1.1. Transcriptomics platforms

Next-generation sequencing technologies are quickly becoming a popular choice for transcriptomics experiments and are proving to have numerous advantages over older technologies (Hitzemann et al., 2013). However, over the past decade, the vast majority of transcriptomics studies have used other technologies, most notably the hybridization-based techniques (microarrays) and sequence-based approaches, such as serial analysis of gene expression (SAGE) and cap analysis of gene expression (CAGE).

Hybridization-based techniques use arrays of oligonucleotide probes, each targeted to a particular known transcript, to capture and measure the abundance of mRNA molecules present in a sample. RNA is isolated from the tissue sample of interest, reverse transcribed to produce complementary DNA (cDNA), and fluorescently labeled. The quantity of each RNA product is reported based on the level of fluorescence measured at the location on the array containing the corresponding probe. Typical microarrays contain only a few probes per gene, primarily located in the 3′-region of the transcript, and are typically not capable of distinguishing between alternatively spliced transcripts. However, specialized versions of microarrays, such as exon arrays and tiling arrays, allow for a higher resolution interrogation of the transcriptome.

Exon arrays contain probes in each exon of the transcript, including at exon/intron boundaries, and therefore increase the potential to detect alternatively spliced isoforms.

Tiling arrays, in contrast to the other microarray platforms mentioned above, do not require *a priori* knowledge about the local structure of the transcriptome. These arrays contain overlapping or closely spaced probes along continuous stretches of a sequenced genome. Because these arrays contain probes at much higher densities than traditional microarrays, they allow for detection of alternatively spliced transcripts as well as previously unknown transcripts.

While microarrays are an extremely powerful technique and have produced an enormous amount of new knowledge, they do have several limitations. First, compared to RNA-Seq, microarrays have a limited range of sensitivity on both the low and high ends. Cross-hybridization issues (i.e., nonspecific hybridization) make it impossible to distinguish low-abundance transcripts from random background noise. On the other hand, probe saturation can limit the ability to accurately quantify high-abundance transcripts.

Second, as mentioned above, the limited number of probes on traditional microarrays does not allow for the collection of information about alternative splicing or for the detection of novel transcripts. Exon and tiling arrays do provide some improvement in this area.

And finally, sequence variation within a probe's target region can have a significant impact on probe performance. Reduced hybridization or cross-hybridization due to DNA polymorphisms can produce misleading results. Masking (or removal) of affected probes can help reduce the impact of sequence variation but leads to loss of information and, of course, will not account for any unknown variants (i.e., this method is less effective for species with less information about genome variation) (Agarwal et al., 2010; Bottomly et al., 2011; Roy, Altermann, Park, & McNabb, 2011 May; Walter et al., 2007; Wang, Gerstein, & Snyder, 2009).

Sequence-based techniques do not use probes and therefore do not depend on prior knowledge of the transcriptome. High-throughput sequence-based methods are referred to as tag-based methods since small portions of transcripts, called "tags," are isolated, cloned into a vector, and then sequenced. These short tags are then mapped to reference genome for transcript identification. It should be noted that short tags often map to multiple locations in the genome, which can cause uncertainty in the detection and quantification of some transcripts, particularly those containing repetitive elements. Also, because only a small portion of the transcript is sequenced, it is often not possible to distinguish between splice variants.

There are a variety of closely related tag-based methods that differ based on the location and the size of the tag derived from the transcript. SAGE traditionally isolates a tag from the 3′-region of the transcript, while CAGE isolates a tag from the 5′-end. Long-SAGE and Super-SAGE produce longer tags, which are more likely to map uniquely to the genome and can increase the chances of detecting alternatively spliced transcripts (Anisimov, 2008; Harbers & Carninci, 2005; Horan, 2009; Kodzius et al., 2006; Matsumura, Krüger, Kahl, & Terauchi, 2008; Takahashi, Kato, Murata, & Carninci, 2012).

A number of repositories have been created to store and share the vast amount of data produced by these technologies (Table 3.1). The availability of these data not only makes secondary analyses possible but also aids the development of improved methodology for transcriptome studies.

Table 3.1 Transcriptome public repositories for secondary and meta-analysis

Repository	URL	References
Gene Expression Omnibus (GEO)	www.ncbi.nlm.nih.gov/geo/	Barrett et al. (2013)
ArrayExpress	www.ebi.ac.uk/arrayexpress/	Parkinson et al. (2011)
Expression Atlas	www.ebi.ac.uk/gxa/home/	Kapushesky et al. (2012)
Sequence Read Archive (SRA)	www.ncbi.nlm.nih.gov/sra/	Kodama, Shumway, Leinonen, and International Nucleotide Sequence Database Collaboration (2012), Leinonen, Sugawara, Shumway, and International Nucleotide Sequence Database Collaboration (2011)
RNA-Seq Atlas	medicalgenomics.org/rna_seq_atlas/	Krupp et al. (2012)
ReCount	bowtie-bio.sourceforge.net/recount/	Frazee, Langmead, and Leek (2011)

2. SELECTING THE UNIT OF COMPARISON

Given that different technologies interrogate the transcriptome at different resolutions, comparing the results across platforms is not a trivial exercise (Fig. 3.1). Depending on the data available, one must first choose the genomic or transcriptomic unit that will be used to make comparison across the datasets. For instance, in order to compare traditional microarray expression values to count data from RNA-Seq, both the expression values and counts must be mapped to common entities (e.g., genes or transcripts).

The choice of the unit of comparison will depend on the limitations of the datasets. Ideally, we would like to compare values that represent measurements of precisely the same entities (i.e., the exact same stretch of RNA). An example of this would be the counts of all reads that map to the corresponding sequence of a particular microarray probe. However, this would only be possible if there is enough sequence coverage (read depth) to get an accurate measurement at the precise location. In addition, there are known spatial biases present across the genome, which means that transcripts are not uniformly covered by RNA-Seq reads (Hansen, Brenner, & Dudoit, 2010; Mortazavi, Williams, McCue, Schaeffer, & Wold, 2008). Because of these issues, it is often more appropriate to produce summary measurements for larger entities, such as whole genes (Bottomly et al., 2011; Frazee et al., 2011). Typically, RNA-Seq reads are first mapped to known exons (discarding any ambiguously mapped reads), then the reads mapped to all exons associated with a particular gene are summed to provide a gene-level read count.

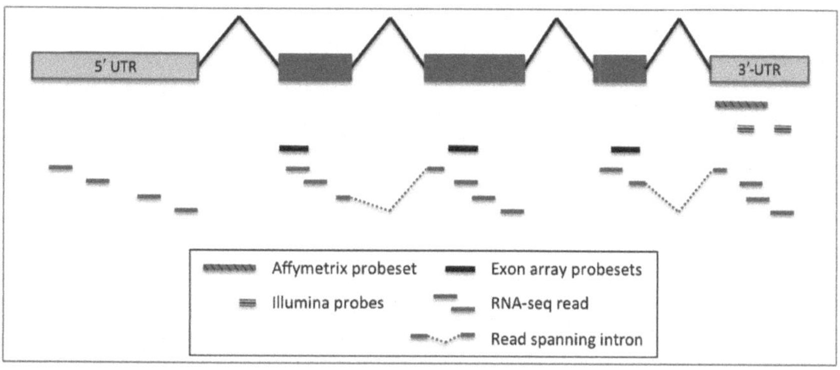

Figure 3.1 Visual comparison of how diverse technologies interrogate the transcriptome, providing different levels of resolution and influencing mapping and selection of the unit of analysis for cross-platform comparisons.

However, more precise mapping strategies, which map sequencing reads or microarray probes to specific transcript isoforms (a.k.a. splice variants), are also possible. Using transcript-level models (rather than gene-level models) allows for the detection of alternative exon usage (Laderas et al., 2011).

3. METRICS FOR AGREEMENT

A valid comparison of expression values produced under different experimental conditions can be performed only after appropriate normalization of the data has been done. This normalization procedure accounts for the various experimental conditions that can produce different base-line values across data sets, such as different starting concentrations of RNA or different sequencing depths.

Bullard and colleagues have suggested that the factor with the greatest impact on the ability to detect differential expression is the normalization procedure (Bullard, Purdom, Hansen, & Dudoit, 2010). Because the number of reads for a particular transcript will depend on both the concentration of the transcript in the sample and the length of the transcript, it is common practice to report a normalized read count as the reads per kilobase of exon model per million mapped reads (RPKM) (Mortazavi et al., 2008). However, this method can be affected by a small number of highly expressed transcripts. Bullard and colleagues suggest that a quantile-based normalization procedure is a more robust alternative (Bullard et al., 2010).

A comparison of seven different normalization methods applied to both real and simulated datasets found that most normalization procedures produced similar results. However, there were some exceptions. The results obtained using the total count normalization method, RPKM, and unnormalized raw counts appeared to cluster together and were somewhat different from the five other methods tested. The DESeq and Trimmed Mean of *M*-values (TMM) methods were recommended because of their ability to control the differential expression false-positive rate in the presence of high-count genes (Dillies et al., 2013).

Given the "analog" nature of hybridization-based technologies, and the variable performance of probes, normalization of microarray data can be a complex task. However, the accurate absolute expression levels produced by RNA-Seq data can be used to improve the quality and utility of microarray expression measures. Miller et al. (2014) found that filtering probes and scaling probe intensities using RNA-Seq expression values improved the biological reproducibility of microarray data from human brain tissues. This technique could significantly aide the secondary analysis of the large amount of publicly available microarray data.

After the normalization of datasets has been completed, a decision must be made about how to evaluate the agreement or concordance between the datasets. Agreement is often measured in three ways: transcript detection (i.e., the presence or absence of a particular transcript), correlation between normalized expression levels, and whether or not there is significant differential expression between two conditions of interest. Agreement in terms of differential expression should be further evaluated by examining the correlation of fold change between experiments, or at the very least comparing the direction of expression change.

For some research questions, the focus is on measures of absolute expression, or the ability to distinguish active from inactive transcripts within individual samples. Characterizing samples by their overall transcriptional state (i.e., the binary on/off pattern across all transcripts) also enables the comparison of samples across multiple experiments. The "bar-code" method has been used to estimate absolute expression values for microarray data by comparing probe-level expression values to a large reference database of microarray samples (McCall, Uppal, Jaffee, Zilliox, & Irizarry, 2011). This focuses on addressing the simple question of what is expressed and what is not expressed in a given sample. Piccolo and colleagues have proposed a related technique that does not require a reference database for comparison and is applicable to both hybridization-based technologies and RNA-seq (Piccolo, Withers, Francis, Bild, & Johnson, 2013). The algorithm, called Universal exPression Code (UPC), calculates "evidence codes" on a scale of 0–1, which indicate the likelihood that an expression value is a true deviation from a modeled background distribution.

The interpretation of these "evidence codes" is the same across platforms and therefore allows the integration of multiple datasets (particularly samples from the same subject across platforms). However, the increased sensitivity of RNA-seq compared to microarrays can lead to discrepancies regarding the classification of genes expressed at a low level. Therefore, care is still needed when combining datasets from these different technologies depending on the direction of discordance.

4. STUDIES ON REPRODUCIBILITY AND VALIDATION

Overall, both RNA-Seq and microarray technologies have been shown to generate highly reproducible results when standardized experimental protocols and data processing procedures are used.

A large-scaling RNA-Seq analysis was conducted by GEUVADIS (the Genetic European Variation in Disease, a European Medical Sequencing Consortium) to examine the sources of technical variation, and the feasibility of sharing and combining RNA-Seq data across laboratories ('t Hoen et al., 2013). Four hundred sixty-five samples were randomly distributed across seven laboratories, with five samples sequenced by all labs. Although they observed slight variations in GC content and insert size, overall the technical variation observed was small and the results consistent across labs.

Marioni and colleagues assessed technical variation of RNA-Seq data from liver and kidney samples by sequencing each sample seven times, using two different concentrations of cDNA. They evaluated technical variation by looking for lane effects, indicating systematic differences in the results for the same sample sequenced at the same concentration in different lanes. They consistently found very small numbers of genes (<0.5%) with significant lane effects (Marioni, Mason, Mane, Stephens, & Gilad, 2008).

Similar studies have been conducted to examine the reproducibility of microarray data. However, the conclusions drawn from these studies have been somewhat inconsistent. The MicroArray Quality Control (MAQC) project conducted a large-scale study of microarray technologies and found high levels of reproducibility both across platforms and within the same platform in terms of the ability to detect differential expression (MAQC Consortium et al., 2006). However, the methods used in this study have been criticized (Liang, 2007) and were inconsistent with other findings (Tan et al., 2003).

More recent advances in methodology and standardization of protocols, not to mention advancements in the technologies, have improved the reliability of microarray data (Draghici, Khatri, Eklund, & Szallasi, 2006; Larkin, Frank, Gavras, Sultana, & Quackenbush, 2005). However, secondary analysis and reproducibility of publicly archived high-throughput gene expression studies still have challenges given gaps in meta-data and study annotation needed to appropriately process and analyze the data (Rung & Brazma, 2013).

5. GUIDELINES FOR CROSS-PLATFORM STUDIES

Numerous studies have evaluated the comparability of RNA-Seq and hybridization-based technologies. Here, we will give an overview of the approaches used in these cross-platform studies and provide some guidelines for best practices (Fig. 3.2).

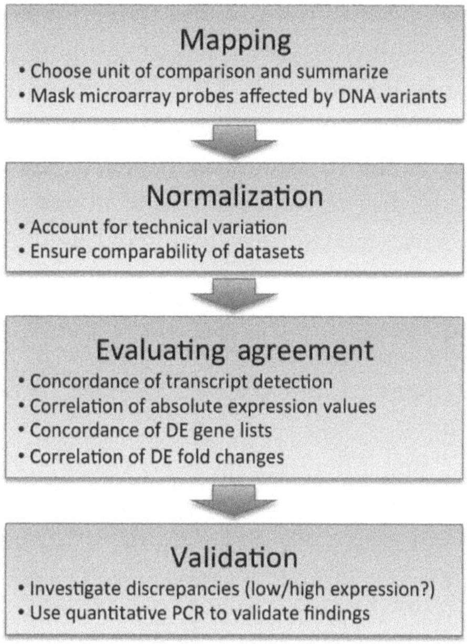

Figure 3.2 A Schematic of the Guidelines Framework for best practices for cross-platform studies. Each step highlights a key area for consideration during the planning of the study. This includes choosing the unit of comparison based on the primary research question for the datasets being analyzed; appropriately dealing with ambiguously mapped probes or sequencing reads, and accounting for the effects of DNA variation; selecting a robust normalization procedure; providing a thorough report of agreement; and validation/investigation of discrepancies via quantitative PCR.

Effective cross-platform comparisons depend on good quality data processing procedures that account for any biases inherent to the specific technologies. In addition, when comparing the detection of differential expression, appropriate test statistics must be used. Bullard et al. (2010) addressed these issues with a comparison of RNA-Seq normalization procedures and tests for differential expression. They recommend against using global normalization procedures and instead propose a quantile-based normalization procedure that matches the distribution of read counts per lane to a reference distribution derived from the median counts across lanes. They also warn against the use of the *t*-statistic for differential expression, since this statistic has low sensitivity for genes with low read counts (Bullard et al., 2010).

There is no consensus about the proper way to map and summarize expression values, although generally better concordance has been observed for gene-level summarization compared to transcript-level summarization, likely due to the uncertainty involved in distinguishing between splice variants of the same gene (Bullard et al., 2010; Marioni et al., 2008; Mortazavi et al., 2008).

An important component of summarizing expression values is dealing with sequencing reads that map equally well to multiple genome locations. Often researchers make the decision to discard all ambiguously mapped reads, but this can lead to inaccurate expression values for gene families with closely related sequences. An effective alternative is to assign ambiguously mapped probes to each matched location proportionally, based on the number of uniquely mapped probes assigned to each of the matched genes (Mortazavi et al., 2008).

Agreement regarding differential expression is the most common question asked in cross-platform gene expression studies. It is not sufficient to simply examine the overlap of the lists of differential expressed genes from each platform. Agreement should be assessed both in terms of the presence or absence of statistically significant differential expression, as well as the correlation between fold changes reported by each platform.

Any discrepancies between platforms should be investigated further to identify possible sources of the disagreement. For example, concordance may be lower for genes with low expression levels. The use of quantitative PCR can be used to validate findings (Guo et al., 2013; Li, Dai, Kang, & Zhou, 2014; Marioni et al., 2008; Nookaew et al., 2012; Soneson & Delorenzi, 2013; 't Hoen et al., 2013).

6. OTHER DATA INTEGRATION CONSIDERATIONS

We note that there are other key challenges in data integration related to transcriptomics beyond that of gene expression cross-platform issues, such as multi-omic integration and cross-species comparisons. As each of them warrants their own chapter, we provide a brief summary here (providing the reader with references for further information).

6.1. Multi-omic data integration

With an increasing focus on systems biology, experimental samples are being interrogated by multi-omic approaches (transcriptomics, proteomics, metabolomics, etc.) in order to gain insight into the dynamic relationship among diverse cellular components. It is thought that utilizing multiple data types may more accurately reflect the complex biology involved. To achieve this, similar guidelines such as those for the cross-platform analyses are needed. Processing and analyzing multiple omic data types require in-depth technical knowledge about how each individual data type was generated. In addition, understanding the resolution, temporal considerations, and accuracy for each technique is key to ensure appropriate comparisons across data types. Finally, multi-omic integrative statistical and computational methods are needed to provide a true systems perspective.

One of the areas of frequent interest for integration with the transcriptome is the proteome (Cox, Kislinger, & Emili, 2005; Waters, Pounds, & Thrall, 2006; Waters, Singhal, Webb-Robertson, Stephan, & Gephart, 2006). The correlation between mRNA and protein expression can be quite low for a wide range of reasons including differences in half-life, variability in expression level due to changes in cell cycle, and post-transcriptional modifications. This has led to a shift from single data type individual analysis to joint multi-omics approaches.

A wide range of integration approaches have been suggested including probabilistic networks (e.g., Hartemink, Gifford, Jaakkola, & Young, 2002; Troyanskaya, Dolinski, Owen, Altman, & Botstein, 2003), machine learning methods (e.g., Daemen et al., 2008; Zhang et al., 2006), and statistical models (e.g., Fagan, Culhane, & Higgins, 2007; Lê Cao, Rossouw, Robert-Granié, & Besse, 2008). The primary challenge for all of these methods is the extremely heterogeneous sources of data (Palsson & Zengler, 2010).

In a recent review by Haider and Pal (2013), they survey transcriptomic-proteomics integration approaches and classify them into eight classes (with the classification based on initial algorithm and final analysis goal). Gibbs, Gralinksi, Baric, and McWeeney (2014) proposed a classification system based on when the integration takes place from a network perspective. "Late" integration is when the integration step takes place after independent model building of each network. In this case, coexpression networks are individually constructed using transcript and peptide level data and then "later" integrated. This is in contrast to what they denote as "early" integration that involves joint or simultaneous integration of diverse omic data types (e.g., de Tayrac, Lê, Aubry, Mosser, & Husson, 2009; Tan et al., 2009; and others). With each of these classification systems, the goal is to make the underlying approaches, limitations, and assumptions accessible to researchers. It is critical to remember that selection of the appropriate integration method for a study is dependent upon the primary research hypothesis and the experimental design.

6.2. Cross-species comparisons

Comparing gene expression patterns across species is of great interest for understanding gene function, conservation, evolution, and for insights with respect to disease. Species-specific differences in the genome and corresponding transcriptome as well as gene duplication events after speciation make cross-species analyses challenging. A large number of approaches and databases allow cross-species comparisons of expression. We refer the readers to the Jay (2012) for an in-depth examination of cross-species integration methods. Here, we highlight a few of these approaches, highlighting consideration for analysis and interpretation.

Gene Ontology (GO) (www.geneontology.org) is one of the most well-known biomedical ontologies, providing annotations to over 600,000 gene products. To functionally describe gene products, annotations are made using GO describing the associated biological processes, cellular component in which the gene product is found, and molecular function. With regard to species comparisons, GO is species independent and enables retrieval and analysis of data from disparate database sources. Using data annotated with GO, it is possible to infer information about gene function in other species (e.g., Roslan et al., 2010) or compare candidate gene lists across species based on GO enrichment. For cross-species comparisons based on GO annotation or enrichment, it is critical to also evaluate the accompanying evidence codes that indicate how the annotation to a particular term is supported.

The database bgee (http://bgee.unil.ch/) compares expression patterns between species by leveraging ontologies for anatomy and development (Bastian et al., 2008). Homology relationships that allow anatomical comparisons are defined based on a modified ontology alignment. Expression is then mapped unto the aligned ontologies. Evaluation of expression comparisons should therefore be interpreted in the context of the ontology itself.

GeneWeaver.org and the tools within provide the ability to integrate functional genomics data across different species (Baker, Jay, Bubier, Langston, & Chesler, 2012). In GeneWeaver, it is critical to note that the definition of a "gene" actually refers to homologous gene clusters (i.e., an entire group of genes that share homology to each other). Utilizing both homology and semantic information, GeneWeaver is capable of more complex comparisons, such as examination of expression patterns in other species in human disease candidate genes, allowing it to be used in a wide range of studies (e.g., Bhandari et al., 2012; Chesler et al., 2012).

Xspecies (http://bioinformatics.math.chalmers.se/Xspecies/) provides an approach for cross-species meta-analysis of gene expression profiles that takes all orthologous and co-orthologous genes into account (Kristiansson et al., 2013). For each species, the expression data is analyzed individually resulting in a *p*-value for each measured gene's differential expression for the comparison of interest. Homology group information can be leveraged from public databases such as Homologene or inferred using *de novo* approaches. Gene-specific *p*-values are calculated in a method similar to Fisher's combined probability test that has been extended to take in-paralogous genes into account. It is important to note that this approach is heavily dependent upon the statistical analysis of the individual data sets. In addition, the approach assumes that the gene-specific *p*-values are independently and uniformly distributed under the null hypothesis. Therefore, careful examination of the original data sets, homology group definitions, and method assumptions should all be considered when utilizing this approach.

For both cross-platform and multi-omic integration, there is again a need for best practice guidelines for selection of data sets, processing, analysis, annotation, and integration of cross-species transcriptomics. We note that cross-platform issues can further complicate cross-species comparisons warranting careful study design, particularly for secondary analysis where metadata can be limited.

7. SUMMARY

The objective of integrating data from multiple transcriptome studies is to increase the reliability and generalizability of the results. This can now routinely be done given the vast amount of data available in public databases. However, within public transcriptomics data, there is tremendous heterogeneity due to platform, experimental conditions, and quality of annotation. In this chapter, we have developed an initial guideline for best practices for cross-platform transcriptomics integration. We note that the crux of this framework also applies to multi-omics integration and cross-species comparisons even though the underlying details of the components are distinct. In each case, it is key to carefully evaluate the datasets and identify the appropriate unit of comparison based on primary research question. A mapping strategy must be considered (across platform, data type, or species). The data sets will need to be carefully preprocessed to ensure comparisons can be made. Metrics will need to be identified *a priori* to evaluate agreement, concordance, and the integration itself. Finally, consideration should be made at the time the study is designed regarding how it will be validated. By following these best practices, we can increase the potential for reproducibility and meaningful data integration.

REFERENCES

Agarwal, A., Koppstein, D., Rozowsky, J., Sboner, A., Habegger, L., Hillier, L. W., et al. (2010). Comparison and calibration of transcriptome data from RNA-Seq and tiling arrays. *BMC Genomics*, *11*, 383. http://dx.doi.org/10.1186/1471-2164-11-383, PMID: 20565764.

Anisimov, S. V. (2008). Serial analysis of gene expression (SAGE): 13 years of application in research. *Current Pharmaceutical Biotechnology*, *9*(5), 338–350, Review. PMID: 18855686.

Baker, E. J., Jay, J. J., Bubier, J. A., Langston, M. A., & Chesler, E. J. (2012). GeneWeaver: A web-based system for integrative functional genomics. *Nucleic Acids Research*, *40*, D1067–D1076.

Barrett, T., Wilhite, S. E., Ledoux, P., Evangelista, C., Kim, I. F., Tomashevsky, M., et al. (2013). NCBI GEO: Archive for functional genomics data sets—Update. *Nucleic Acids Research*, *41*(Database issue), D991–D995. http://dx.doi.org/10.1093/nar/gks1193, Epub 2012 Nov 27. PMID: 23193258.

Bastian, F., Parmentier, G., Roux, J., Moretti, S., Laudet, V., & Robinson-Rechavi, M. (2008). Bgee: Integrating and comparing heterogeneous transcriptome data among species. Data integration in the life sciences. *Lecture Notes in Computer Science*, *5109*, 124–131.

Bhandari, P., Hill, J. S., Farris, S. P., Costin, B., Martin, I., Chan, C.-L., et al. (2012). Chloride intracellular channels modulate acute ethanol behaviors in Drosophila, Caenorhabditis elegans and mice. *Genes, Brain, and Behavior*, *11*, 387–397.

Bottomly, D., Walter, N. A., Hunter, J. E., Darakjian, P., Kawane, S., Buck, K. J., et al. (2011). Evaluating gene expression in C57BL/6J and DBA/2J mouse striatum using RNA-Seq and microarrays. *PLoS One*, *6*(3), e17820. http://dx.doi.org/10.1371/journal.pone.0017820, PMID: 21455293.

Bullard, J. H., Purdom, E., Hansen, K. D., & Dudoit, S. (2010). Evaluation of statistical methods for normalization and differential expression in mRNA-Seq experiments. *BMC Bioinformatics*, *11*, 94. http://dx.doi.org/10.1186/1471-2105-11-94, PMID: 20167110.

Chesler, E., Plitt, A., Fisher, D., Hurd, B., Lederle, L., Bubier, J., et al. (2012). Quantitative trait loci for sensitivity to ethanol intoxication in a c57bl/6j × 129s1/svimj inbred mouse cross. *Mammalian Genome*, *23*, 305–321.

Cox, B., Kislinger, T., & Emili, A. (2005). Integrating gene and protein expression data: Pattern analysis and profile mining. *Methods*, *35*, 303–314.

Daemen, A., Gevaert, O., De Bie, T., Debucquoy, A., Machiels, J. P., De Moor, B., et al. (2008). Integrating microarray and proteomics data to predict the response on cetuximab in patients with rectal cancer. *Pacific Symposium on Biocomputing*, *13*, 166–177.

de Tayrac, M., Lê, S., Aubry, M., Mosser, J., & Husson, F. (2009). Simultaneous analysis of distinct Omics data sets with integration of biological knowledge: Multiple factor analysis approach. *BMC Genomics*, *10*, 32. http://dx.doi.org/10.1186/1471-2164-10-32.

Dillies, M. A., Rau, A., Aubert, J., Hennequet-Antier, C., Jeanmougin, M., Servant, N., et al. (2013). A comprehensive evaluation of normalization methods for Illumina high-throughput RNA sequencing data analysis. *Briefings in Bioinformatics*, *14*(6), 671–683. http://dx.doi.org/10.1093/bib/bbs046, Epub 2012 Sep 17; PMID: 22988256.

Draghici, S., Khatri, P., Eklund, A. C., & Szallasi, Z. (2006). Reliability and reproducibility issues in DNA microarray measurements. *Trends in Genetics*, *22*(2), 101–109, Epub 2005 Dec 27. Review. PMID: 16380191.

Fagan, A., Culhane, A. C., & Higgins, D. G. (2007). A multivariate analysis approach to the integration of proteomic and gene expression data. *Proteomics*, 7, 2162–2171.

Frazee, A. C., Langmead, B., & Leek, J. T. (2011). ReCount: A multi-experiment resource of analysis-ready RNA-seq gene count datasets. *BMC Bioinformatics*, *12*, 449. http://dx.doi.org/10.1186/1471-2105-12-449, PMID: 22087737.

Gibbs, D. L., Gralinksi, L., Baric, R. S., & McWeeney, S. K. (2014). Multi-omic network signatures of disease. *Frontiers in Genetics*, *4*, 309. http://dx.doi.org/10.3389/fgene.2013.00309.

Guo, Y., Sheng, Q., Li, J., Ye, F., Samuels, D. C., & Shyr, Y. (2013). Large scale comparison of gene expression levels by microarrays and RNAseq using TCGA data. *PLoS One*, *8*(8), e71462. http://dx.doi.org/10.1371/journal.pone.0071462, eCollection 2013. PMID: 23977046.

Haider, S., & Pal, R. (2013). Integrated analysis of transcriptomic and proteomic data. *Current Genomics*, *14*, 91–110.

Hansen, K. D., Brenner, S. E., & Dudoit, S. (2010). Biases in Illumina transcriptome sequencing caused by random hexamer priming. *Nucleic Acids Research*, *38*(12), e131. http://dx.doi.org/10.1093/nar/gkq224, Epub 2010 Apr 14. PMID: 20395217.

Harbers, M., & Carninci, P. (2005). Tag-based approaches for transcriptome research and genome annotation. *Nature Methods*, *2*(7), 495–502, Review. PMID: 15973418.

Hartemink, A. J., Gifford, D. K., Jaakkola, T. S., & Young, R. A. (2002). Combining location and expression data for principled discovery of genetic regulatory network models. *Pacific Symposium on Biocomputing*, 7, 437–449.

Hitzemann, R., Bottomly, D., Darakjian, P., Walter, N., Iancu, O., Searles, R., et al. (2013). Genes, behavior and next-generation RNA sequencing. *Genes, Brain, and Behavior*, *12*(1), 1–12. http://dx.doi.org/10.1111/gbb.12007, Epub 2012 Dec 28. Review. PMID: 23194347.

Horan, M. P. (2009). Application of serial analysis of gene expression to the study of human genetic disease. *Human Genetics*, *126*(5), 605–614. http://dx.doi.org/10.1007/s00439-009-0719-5, Epub 2009 Jul 10. Review. PMID: 19590894.

Jay, J. (2012). Cross species integration of functional genomics experiments. *International Review of Neurobiology*, *104*, 1–24.

Kapushesky, M., Adamusiak, T., Burdett, T., Culhane, A., Farne, A., Filippov, A., et al. (2012). Gene Expression Atlas update—A value-added database of microarray and sequencing-based functional genomics experiments. *Nucleic Acids Research*, *40*(Database issue), D1077–D1081. http://dx.doi.org/10.1093/nar/gkr913, Epub 2011 Nov 7. PMID: 22064864.

Kodama, Y., Shumway, M., Leinonen, R., & International Nucleotide Sequence Database Collaboration (2012). The Sequence Read Archive: Explosive growth of sequencing data. *Nucleic Acids Research*, *40*(Database issue), D54–D56. http://dx.doi.org/10.1093/nar/gkr854, Epub 2011 Oct 18. PMID: 22009675.

Kodzius, R., Kojima, M., Nishiyori, H., Nakamura, M., Fukuda, S., Tagami, M., et al. (2006). CAGE: Cap analysis of gene expression. *Nature Methods*, *3*(3), 211–222, No abstract available. PMID: 16489339.

Kristiansson, E., Österlund, T., Gunnarsson, L., Arne, G., Larsson, D. G. J., & Nerman, O. (2013). A novel method for cross-species gene expression analysis. *BMC Bioinformatics*, *14*, 70. http://dx.doi.org/10.1186/1471-2105-14-70.

Krupp, M., Marquardt, J. U., Sahin, U., Galle, P. R., Castle, J., & Teufel, A. (2012). RNA-Seq Atlas—A reference database for gene expression profiling in normal tissue by next-generation sequencing. *Bioinformatics*, *28*(8), 1184–1185. http://dx.doi.org/10.1093/bioinformatics/bts084, Epub 2012 Feb 17. PMID: 22345621.

Laderas, T. G., Walter, N. A., Mooney, M., Vartanian, K., Darakjian, P., Buck, K., et al. (2011). Computational detection of alternative exon usage. *Frontiers in Neuroscience*, *5*, 69. http://dx.doi.org/10.3389/fnins.2011.00069, eCollection 2011. PMID: 21625610.

Larkin, J. E., Frank, B. C., Gavras, H., Sultana, R., & Quackenbush, J. (2005). Independence and reproducibility across microarray platforms. *Nature Methods*, *2*(5), 337–344, Epub 2005 Apr 21. PMID: 15846360.

Lê Cao, K.-A., Rossouw, D., Robert-Granié, C., & Besse, P. (2008). A sparse PLS for variable selection when integrating omics data. *Statistical Applications in Genetics and Molecular Biology*, 7, Article 35.

Leinonen, R., Sugawara, H., Shumway, M., & International Nucleotide Sequence Database Collaboration (2011). The sequence read archive. *Nucleic Acids Research*, *39*(Database issue), D19–D21, PMID: 21062823.

Li, W., Dai, C., Kang, S., & Zhou, X. J. (2014). Integrative analysis of many RNA-seq datasets to study alternative splicing. *Methods*, *67*(3), 313–324. http://dx.doi.org/10.1016/j.ymeth.2014.02.024, pii: S1046-2023(14)00068-1.

Liang, P. (2007). MAQC papers over the cracks. *Nature Biotechnology*, *25*(1), 27–28, author reply 28–29. PMID: 17211385.

MAQC Consortium, Shi, L., Reid, L. H., Jones, W. D., Shippy, R., Warrington, J. A., et al. (2006). The MicroArray Quality Control (MAQC) project shows inter- and intraplatform reproducibility of gene expression measurements. *Nature Biotechnology*, *24*(9), 1151–1161, PMID: 16964229.

Marioni, J. C., Mason, C. E., Mane, S. M., Stephens, M., & Gilad, Y. (2008). RNA-seq: An assessment of technical reproducibility and comparison with gene expression arrays. *Genome Research*, *18*(9), 1509–1517. http://dx.doi.org/10.1101/gr.079558.108, Epub 2008 Jun 11. PMID: 18550803.

Matsumura, H., Krüger, D. H., Kahl, G., & Terauchi, R. (2008). SuperSAGE: A modern platform for genome-wide quantitative transcript profiling. *Current Pharmaceutical Biotechnology*, *9*(5), 368–374, Review. PMID: 18855689.

McCall, M. N., Uppal, K., Jaffee, H. A., Zilliox, M. J., & Irizarry, R. A. (2011). The gene expression barcode: Leveraging public data repositories to begin cataloging the human and murine transcriptomes. *Nucleic Acids Research*, *39*, D1011–D1015.

Miller, J. A., Menon, V., Goldy, J., Kaykas, A., Lee, C. K., Smith, K. A., et al. (2014). Improving reliability and absolute quantification of human brain microarray data by filtering and scaling probes using RNA-Seq. *BMC Genomics*, *15*(1), 154. http://dx.doi.org/10.1186/1471-2164-15-154, PMID: 24564186.

Mortazavi, A., Williams, B. A., McCue, K., Schaeffer, L., & Wold, B. (2008). Mapping and quantifying mammalian transcriptomes by RNA-Seq. *Nature Methods*, *5*(7), 621–628. http://dx.doi.org/10.1038/nmeth.1226, Epub 2008 May 30. PMID: 18516045.

Nookaew, I., Papini, M., Pornputtapong, N., Scalcinati, G., Fagerberg, L., Uhlén, M., et al. (2012). A comprehensive comparison of RNA-Seq-based transcriptome analysis from reads to differential gene expression and cross-comparison with microarrays: A case study in Saccharomyces cerevisiae. *Nucleic Acids Research*, *40*(20), 10084–10097. http://dx.doi.org/10.1093/nar/gks804, Epub 2012 Sep 10. PMID: 22965124.

Palsson, B., & Zengler, K. (2010). The challenges of integrating multi-omic data sets. *Nature Chemical Biology*, *6*, 787–789.

Parkinson, H., Sarkans, U., Kolesnikov, N., Abeygunawardena, N., Burdett, T., Dylag, M., et al. (2011). ArrayExpress update—An archive of microarray and high-throughput sequencing-based functional genomics experiments. *Nucleic Acids Research*, *39*(Database issue), D1002–D1004. http://dx.doi.org/10.1093/nar/gkq1040, Epub 2010 Nov 10. PMID: 21071405.

Piccolo, S. R., Withers, M., Francis, O. E., Bild, A. H., & Johnson, W. E. (2013). Multiplatform single-sample estimates of transcriptional activation. *Proceedings of the National Academy of Sciences of the United States of America*, *110*, 17778–17783.

Roslan, R., Othman, R. M., Shah, Z. A., Kasim, S., Asmuni, H., Taliba, J., et al. (2010). Utilizing shared interacting domain patterns and Gene Ontology information to improve protein–protein interaction prediction. *Computers in Biology and Medicine*, *40*(6), 555–564.

Roy, N. C., Altermann, E., Park, Z. A., & McNabb, W. C. (2011). A comparison of analog and next-generation transcriptomic tools for mammalian studies. *Briefings in Functional Genomics*, *10*(3), 135–150. http://dx.doi.org/10.1093/bfgp/elr005, Epub 2011 Mar 9. Review. PMID: 21389008.

Rung, J., & Brazma, A. (2013). Reuse of public genome-wide gene expression data. *Nature Reviews Genetics*, *14*, 89–99. http://dx.doi.org/10.1038/nrg3394.

Soneson, C., & Delorenzi, M. (2013). A comparison of methods for differential expression analysis of RNA-seq data. *BMC Bioinformatics*, *14*, 91. http://dx.doi.org/10.1186/1471-2105-14-91, PMID: 23497356.

Takahashi, H., Kato, S., Murata, M., & Carninci, P. (2012). CAGE (cap analysis of gene expression): A protocol for the detection of promoter and transcriptional networks. *Methods in Molecular Biology*, *786*, 181–200. http://dx.doi.org/10.1007/978-1-61779-292-2_11, PMID: 21938627.

Tan, C. S., Salim, A., Ploner, A., Lehtiö, J., Chia, K. S., & Pawitan, Y. (2009). Correlating gene and protein expression data using correlated factor analysis. *BMC Bioinformatics*, *10*, 272.

Tan, P. K., Downey, T. J., Spitznagel, E. L., Jr., Xu, P., Fu, D., Dimitrov, D. S., et al. (2003). Evaluation of gene expression measurements from commercial microarray platforms. *Nucleic Acids Research*, *31*(19), 5676–5684, PMID: 14500831.

't Hoen, P. A., Friedländer, M. R., Almlöf, J., Sammeth, M., Pulyakhina, I., Anvar, S. Y., et al. (2013). Reproducibility of high-throughput mRNA and small RNA sequencing across laboratories. *Nature Biotechnology*, *31*(11), 1015–1022. http://dx.doi.org/10.1038/nbt.2702, Epub 2013 Sep 15. PMID: 24037425.

Troyanskaya, O. G., Dolinski, K., Owen, A. B., Altman, R. B., & Botstein, D. (2003). A Bayesian framework for combining heterogeneous data sources for gene function prediction (in Saccharomyces cerevisiae). *Proceedings of the National Academy of Sciences of the United States of America*, *100*, 8348–8353.

Walter, N. A., McWeeney, S. K., Peters, S. T., Belknap, J. K., Hitzemann, R., & Buck, K. J. (2007). SNPs matter: Impact on detection of differential expression. *Nature Methods*, *4*(9), 679–680, PMID: 17762873.

Wang, Z., Gerstein, M., & Snyder, M. (2009). RNA-Seq: A revolutionary tool for transcriptomics. *Nature Reviews Genetics*, *10*(1), 57–63. http://dx.doi.org/10.1038/nrg2484, Review. PMID: 19015660.

Waters, K. M., Pounds, J. G., & Thrall, B. D. (2006). Data merging for integrated microarray and proteomic analysis. *Briefings in Functional Genomics & Proteomics*, *5*, 261–272.

Waters, K. M., Singhal, M., Webb-Robertson, B. J. M., Stephan, E. G., & Gephart, J. M. (2006). Breaking the high-throughput bottleneck: New tools help biologists integrate complex datasets. *Scientific Computing*, *23*(4), 22–26.

Zhang, X., Lu, X., Shi, Q., Xu, X.-q., Leung, H.-c. E., Harris, L. N., et al. (2006). Recursive SVM feature selection and sample classification for mass-spectrometry and microarray data. *BMC Bioinformatics*, *7*, 197.

CHAPTER FOUR

Coexpression and Cosplicing Network Approaches for the Study of Mammalian Brain Transcriptomes

Ovidiu Dan Iancu*[,1], Alexandre Colville*, Priscila Darakjian*, Robert Hitzemann*[,†]

*Department of Behavioral Neuroscience, Oregon Health & Science University, Portland, Oregon, USA
[†]Research Service, Veterans Affairs Medical Center, Portland, Oregon, USA
[1]Corresponding author: e-mail address: iancuo@ohsu.edu

Contents

Abstract

Next-generation sequencing experiments have demonstrated great potential for transcriptome profiling. While transcriptome sequencing greatly increases the level of biological detail, system-level analysis of these high-dimensional datasets is becoming essential. We illustrate gene network approaches to the analysis of transcriptional data, with particular focus on the advantage of RNA-Seq technology compared to microarray platforms. We introduce a novel methodology for constructing cosplicing networks, based on distance measures combined with matrix correlations. We find that the cosplicing network is distinct and complementary to the coexpression network, although it shares the scale-free properties. In the cosplicing network, we find a set of novel hubs that have unique characteristics distinguishing them from coexpression hubs: they are heavily represented in neurobiological functional pathways and have strong overlap with markers of neurons and neuroglia, long-coding lengths, and high number of both exons and annotated transcripts. We also find that gene networks are plastic in the face of genetic and environmental pressures.

International Review of Neurobiology, Volume 116
ISSN 0074-7742
http://dx.doi.org/10.1016/B978-0-12-801105-8.00004-7

1. INTRODUCTION

Several different methodologies can be used to construct gene networks based on expression data (Allen, Xie, Chen, Girard, & Xiao, 2012; Jay et al., 2012). Correlation networks such as the Weighted Gene Coexpression Network Analysis (WGCNA) (Langfelder & Horvath, 2008) infer relationships between genes based on correlations between their expression levels, either across samples or across time points. Generally, coexpression networks are derived *de novo*, which implies that new data from a particular experiment and biological context are used to infer the topology of the network. Alternative approaches use previously curated databases such as protein–protein interactions to infer a network topology.

Coexpression analysis has established several principles regarding the organization of gene networks. First, it has been shown that coexpression structure follows a power-law distribution (Zhang & Horvath, 2005), which from a computational point of view implies that a few select genes (denoted as network hubs) have a relatively high number of connections with other genes, while the majority of genes have a low number of connections and are denoted as network "leaves." From a biological point of view, a scale-free structure implies that within a specific context (a pathway, tissue, or cell type), a few genes are key drivers of the biological activity. Given these observations, the emphasis on gene connectivity is driven to a large extent by the focus on detecting the highly connected hub genes that are potential targets for therapeutic manipulation. The relationship between a gene's network role and its biological significance has been validated experimentally in protein networks (Jeong, Mason, Barabasi, & Oltvai, 2001).

A second important general principle is "guilt by association": genes or transcripts that cluster together are assumed to share some similar function, likely context dependent. Importantly, this principle allows role inference for genes without current annotations. Additionally, in some configurations, transcriptome sequencing experiments allow the quantification of noncoding RNAs and inference of their roles by virtue of their association with the generally better annotated protein-coding genes.

The third principle regards the plasticity of biological networks and their responsiveness to environmental pressures. Even though biological interactions are dynamic, many network analysis approaches have been based on an implicitly static model. In contrast, differential network analysis leverages the dynamic nature of network interactions and aims to detect connections between genes or gene clusters (modules) that are most affected by changing environments, often under the assumption that these genes are most amenable to potential therapeutic intervention. Under this model, the quantity of interest is the change in the strength of the connections between genes, as opposed to the change in the expression levels of one or more genes.

This chapter reviews brain gene networks in the context of behavioral genotypes and mouse genetics. I will review construction, annotation, and analysis of gene networks, with several examples that illustrate each of the principles listed above. I will especially emphasize the advantages that RNA-Seq offers in comparison to microarray technologies.

2. CONSTRUCTION OF COEXPRESSION AND COSPLICING NETWORKS

The construction of a gene coexpression network starts typically from a gene expression experimental dataset, either microarray or RNA-Seq. The number of samples must be sufficient for reliable inference of correlation values; the exact number is dependent on the level of noise in specific experiments. We investigated the effect of using low number of samples in previous studies, finding that about 15–20 samples were at the lower limit for reliable inference of the network structure (Iancu, Darakjian, et al., 2012). Importantly, when comparing the stability of network structure for comparable microarray versus RNA-Seq data, we found that for similar number of samples, RNA sequencing data appear to give more reliable and stable estimates of network structure (Iancu, Kawane, et al., 2012).

The next step in network construction is the selection of the number of genes to include in the analysis. While recognizing that one of the main advantages of network approaches is the capacity to evaluate the transcriptome as a large system, several factors limit the inclusion of very large number of genes (>10,000). First, even with adequate computing power and with efficient implementation of network construction functions, computational issues still arise. The number of edges (connections between genes) grows with the square of the number of nodes/genes, so even assuming that computational needs grow linearly with the number of edges, a network with 10,000 genes will require twice as much computing power as a 7000 genes network. We have generally limited the size of our networks at 10,000 genes or less.

In addition to computational power, lack of sufficiently high expression and especially low variability in expression also limits the number of genes to be included in the network. As noted above, network edges are inferred based on correlations between expression levels. If a gene does not have sufficient variability in expression, then all its correlations with other genes will be low, regardless of its average expression level. Once again, RNA-Seq appears advantageous from this point of view: for roughly equivalent samples, the variability of sequencing data appears much higher than variability of microarray-derived expression data (Iancu, Kawane, et al., 2012). This higher variability is likely due to the fact that microarray inference of gene expression levels is subject to saturation at high levels, while at low levels transcriptome sequencing has been shown repeatedly to offer more reliable estimates (Marioni, Mason, Mane, Stephens, & Gilad, 2008; Wang, Gerstein, & Snyder, 2009). To facilitate robust network construction, often the coefficient of variability (CV) is computed for each gene, and only the top quantile(s) are retained for network construction. As an alternative, the network connectivity (sum of connection strengths) for each gene is computed, and only genes with sufficiently high connectivity are retained. When comparing gene connectivity in microarray platforms versus RNA-Seq, we observed higher values for the sequencing data, not unexpectedly given the higher variability.

Regardless of the source, one common approach in construction of the coexpression network is implemented in the WGCNA package (Langfelder & Horvath, 2008; Zhang & Horvath, 2005). Here, the first step is the computation of the correlation coefficient between transcript pairs to be included in the network. The correlation matrix is subsequently transformed into an adjacency matrix (A) using a power function. The connection strength a_{ij} between transcripts x_i and x_j then becomes $a_{ij} = |\mathrm{corr}(x_i, x_j)|^{\beta}$; β is selected in accordance to the scale-free topology criterion (Zhang & Horvath, 2005). In some case, the adjacency is further processed and the "topological overlap" (TO) between two transcripts is computed. In essence, the TO averages adjacency information over several network "neighbors" of the two transcripts (Ravasz, Somera, Mongru, Oltvai, & Barabasi, 2002; Zhang & Horvath, 2005). The TO between two transcripts i, j is computed as

$$w_{ij} = \frac{l_{ij} + a_{ij}}{\min\left\{k_i, k_j\right\} + 1 - a_{ij}},$$

where $l_{ij} = \sum_u a_{iu} a_{uj}$ represents the number of transcripts connected to both transcripts i, j, while u indexes all the transcripts in the network. Both the adjacency matrix A and the TO matrix are $N \times N$ square matrices, where

N is the number of genes selected for network construction. For each gene, the sum of its connection strengths to all other genes is denoted as the connectivity.

The next step in network analysis is the clustering of the adjacency or TO matrices, which results in the detection of gene modules, or groups of cotranscribed genes. While in principle any clustering procedure can be used, the specifics of transcriptional data have spurred the development of customized gene expression clustering procedures (Langfelder, Zhang, & Horvath, 2008). One potential limitation of clustering procedures, although one not limited to genomic data, regards the optimal number of clusters to be detected. In many applications, this is difficult to determine and in practice the level of granularity is ultimately dependent on the biological questions that are addressed. The number of modules detected in transcriptional data has ranged from as few as 6 (Iancu, Kawane, et al., 2012) to as many as 95 (Drnevich et al., 2012) or even hundreds (Vanderlinden et al., 2013). It is important to note that the number of modules detected does not necessarily reflect on the biological properties of the system under analysis, but rather the level of granularity chosen by the investigator.

To facilitate comparison of gene network properties across experiments, and to allow for a more rigorous evaluation of the gene modules, Langfelder, Luo, Oldham, and Horvath (2011) introduced a comprehensive procedure for the analysis of network modularity. This procedure addresses two related questions. First, how does a module differ from a random group of genes of the same size? A high-quality module will have much stronger edges than random groups of genes. Second, if the same genes are quantified in different experiments and different networks are constructed, are modules detected in one network still detectable (preserved) in the second network? To address these questions, three different measures of quality/preservation were defined. Density preservation implies that network hubs remain highly connected across the networks compared. Connectivity preservation implies that the pattern of connections between groups of genes remains unaltered. Separability measures whether genes assigned to modules/clusters are indeed more connected to each other than to genes outside the module. All network measurements can be expressed as Z scores; Z scores < 2 are taken to imply poor module quality. For complete mathematical definitions of these concepts, the reader is referred to Langfelder et al. (2011). In the following sections, we will offer specific examples of the level of module quality we detected in mammalian brain transcriptome data, as well as the level of preservation of modules across brain regions and species.

3. COSPLICING NETWORK CONSTRUCTION

The network construction steps outlined above are equally applicable to microarray or RNA-Seq data. However, RNA-Seq offers an important advantage over many microarray platforms: it is possible to infer the expression levels of each individual exon. The availability of exon-level data has allowed the construction of cosplicing networks. Two specific observations have led us to the development of cosplicing methodology. First, it is well established that a majority of mammalian genes have alternative isoforms, which are distinguished from different exon inclusion rates. The isoform diversity is particularly pronounced in the mammalian brain. Second, we observed in our data that in many cases, individual exons from different genes show high levels of correlation, even though the overall gene expression levels are not correlated. These observations suggested to us that the isoform production of different genes could be jointly regulated. To evaluate and quantify the extent of this phenomenon transcriptome wide, we devised a methodology for evaluating correlations in the isoform ratios across different genes. Previous descriptions of cosplicing networks (Chen & Zheng, 2009; Dai, Li, Liu, & Zhou, 2012) have revealed that in some cases individual exons from different genes are correlated when the overall gene expression levels appear unrelated. However, these approaches resulted in networks where the nodes were individual exons. While exon-level analysis has a high level of granularity, it lacks ease of integration and summarization, since annotation databases are focused on genes and not individual exons. Additionally, since in our framework the nodes in both the coexpression and cosplicing networks are the genes, a direct comparison of network properties is possible.

The essential feature of this new approach is a more complex but mathematically complete and rigorous representation of gene transcripts. Gene transcripts are represented as a list/vector of exon expression levels, which for RNA-Seq datasets are proportional to the exon-level read counts. Relative abundance of isoforms translates into relative exon inclusion rates and the goal of the cosplicing analysis is to detect correlations between the exon inclusion rates. We can illustrate this approach by example. Consider the case where two kinds of information are collected about the same set of N individuals: geographical location and DNA sequence information. Two sets of pairwise distances between samples/individuals can then be constructed: geographical distance and genetic distance. To see if these distances are related, Mantel tests (Mantel, 1967) can be performed, in essence computing the correlations between the $N(N-1)/2$ unique distances. A similar approach can be used to the evaluation of the gene expression vectors derived from the exon data. Each gene defines a different set of distances among samples; correlating these distances can offer an estimate of the relatedness between the gene's exon inclusion profiles. The advantage of adopting this representation of a gene rests in moving away from a single scalar value as a summary of gene expression. Scalar representation is unavoidable in the case of some microarray platforms, where each gene is probed by one or a few individual probes. Gene cosplicing networks are constructed using the WGCNA approach with the networks based on the Mantel correlations to derive network edge weights. The cosplicing network inference procedure is illustrated in Fig. 4.1.

In the case of count data such as generated by RNA-Seq, two commonly used distance measures are the Manhattan and Canberra distance measures:

$$d_A^M(i,j) = \sum_{e=1}^{p} \left|a_e^i - a_e^j\right|, \quad d_A^C(i,j) = \sum_{e=1}^{p} \frac{\left|a_e^i - a_e^j\right|}{\left|a_e^i + a_e^j\right|},$$

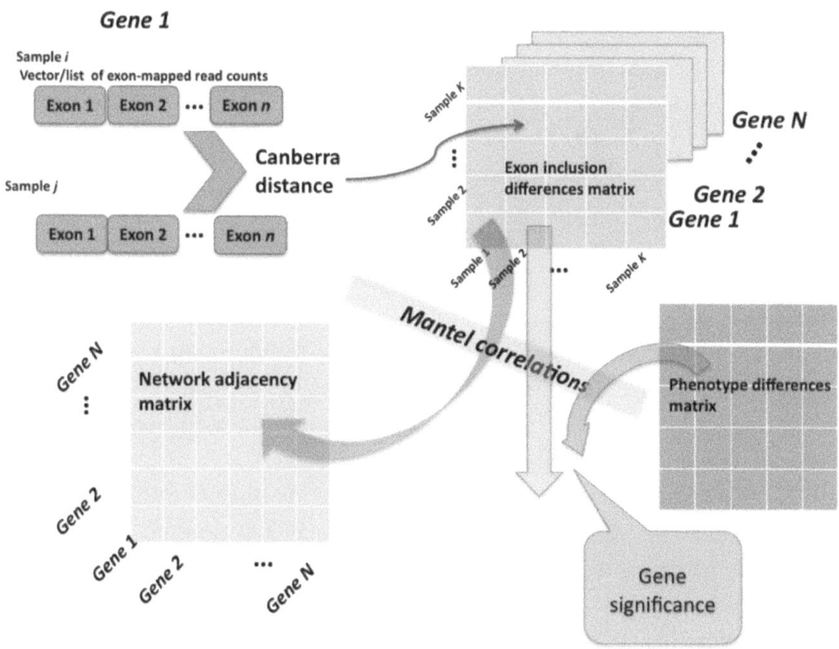

Figure 4.1 Illustration of cosplicing network construction.

where i and j are two samples, and e iterates over all exon counts a. The Manhattan metric has the disadvantage that it can be dominated by long exons with many reads. Therefore, we deemed more appropriate to use the Canberra metric, where each exon contributes a value between 0 and 1 to the total distance. On the other hand, the Canberra metric can potentially be skewed by the weighing equally exons with very low or noisy counts; this situation can be remedied by removing exons with low (bottom quartile) network connectivity, under the assumption that noisy exon data will be uncorrelated for different genes, resulting in low connectivity.

The network information derived from Mantel correlations appears distinct from networks based on Pearson or Spearman correlations and we argue is generated by including an accurate representation of cosplicing. Experimental support for coordinated gene cosplicing is extensive, particularly in the mammalian nervous system (Calarco, Zhen, & Blencowe, 2011; Fagnani et al., 2007); computational studies have also suggested that coordination between splicing and transcription events plays a pivotal role in gene regulation (Kosti, Radivojac, & Mandel-Gutfreund, 2012).

We recently derived both coexpression and cosplicing networks from the same mouse striatal dataset. We have found that both networks are scale free, with the cosplicing network displaying faster convergence as the power β is increased in accordance to the scale-free topology criterion. In terms of network properties, the cosplicing network displayed faster convergence to a scale-free structure as the power β is increased. As discussed in detail in Zhang and Horvath (2005) and Langfelder and Horvath (2008), the role of the soft threshold β is to emphasize the stronger correlations between genes and to minimize the effect of low and presumably less reliable correlations. We also contrasted other network properties such as connectivity, density, centralization, and heterogeneity as defined in Langfelder and Horvath (2008). Briefly, heterogeneity is the CV (standard deviation over mean) of the connectivity, while centrality is the difference between the maximum and average connectivity. Generally, we observed that the coexpression network was more connected and dense, while the cosplicing network was more centralized and heterogeneous.

To evaluate whether the two networks provide distinct information, we compared the two networks node connectivity distributions. We found that the connectivity values in the two networks were largely unrelated; in particular, we found that some hub nodes (top 10% connectivity) in one type of network were in the bottom 80% of connectivity ("leaf" nodes) in the other type of network—in other words, these hubs are highly connected in only one type of network. We illustrate and contrast these differences in connectivity patterns in Fig. 4.2A, where the cosplicing hubs with low coexpression connectivity are outlined in top left, while the coexpression hubs with low cosplicing connectivity are outlined in bottom right. We found that the coexpression hubs could be distinguished from the cosplicing hubs by several characteristics. In terms of variability, the cosplicing hubs had lower CV in terms of gene counts, but higher variability in terms of Canberra pairwise distances. Cosplicing hubs have higher average number of exons (Fig. 4.2B) and annotated transcripts, more protein domains, longer coding size, and lower GC content. In mouse embryonic stem cells, the measured mRNA decay rate is higher for the splicing hubs. We also overlayed the hubs with known markers of neurons and neuroglia (Cahoy et al., 2008), finding that cosplicing hubs were overrepresented among neurons, astrocytes, and oligodendrocytes (Fisher exact test p values 2×10^{-5}, 0.03, and 4×10^{-8}, respectively), while the coexpression hubs were underrepresented (p values 5×10^{-11}, 0.01, and 0.004, respectively).

In terms of GO annotations, the differences between the two hub categories were striking. The cosplicing hubs, nearly half the significant GO enrichment categories, corresponded to neural development, neural cell components, or synaptic transmission. In contrast, the coexpression hubs had no annotation in the above categories but were significantly enriched in terms generally associated with cellular energy exchange and metabolism.

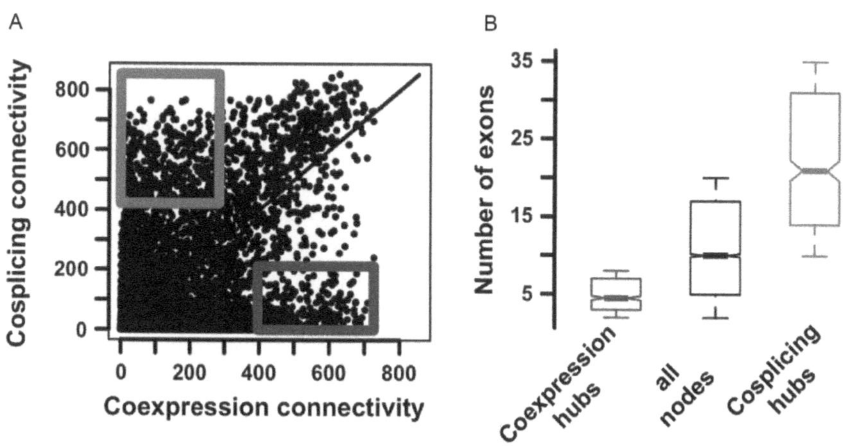

Figure 4.2 Distinct properties of coexpression and cosplicing hubs. We focus on hubs that are exclusive to either the coexpression (A—bottom right outline) or cosplicing (A—top left outline). These hubs differ in the number of exons per gene (B).

In summary, our approach is utilizing a distance measure approach, coupled with matrix correlations, to examine genome-wide alternative splicing coordination among genes. In this framework, the contribution of each exon is weighted equally, utilizing most of the data generated in the RNA-Seq experiment. The detection of strong network edges allows one to conclude that two genes coordinate their isoform production but does not directly indicate which isoform or isoforms contribute the most to this coordination. However, it is certainly possible to further examine exon–exon correlations and infer the most likely isoforms involved in the cosplicing. This additional analysis is facilitated by the fact that for most genes, there are relatively few and in a majority of cases only two highly expressed isoforms. Splicing is estimated to occur for 95% of all multiexon genes (Pan, Shai, Lee, Frey, & Blencowe, 2008), 86% of which have a minor isoform that accounts for >15% of total gene expression (Barrie, Smith, Sanford, & Sadee, 2012; Wang et al., 2008).

4. BIOLOGICAL ANNOTATION OF COEXPRESSION AND COSPLICING NETWORKS

One essential question in gene network analysis is the biological significance of the observed connectivity patterns: it is important to distinguish between randomly occurring structure in high-dimensional data and network structure clearly attributable to biological factors. To address this issue, we quantified functional and shared regulatory mechanisms that we predicted are likely to influence coexpression and cosplicing between gene pairs. Leveraging several public bioinformatics databases, we evaluated the following factors: known protein–protein interactions, shared transcription factor-binding sites (TFBSs), shared splicing factor-binding sites (SFBSs), shared microRNA-binding sites, shared chromosomal folding topological domains, spatial colocalization within the striatum, and membership in gene sets that are markers of cell types within the mouse brain.

We found that for genes with known protein–protein interactions (PPI), the edge weights were significantly higher than for randomly selected gene. In the consensus network, there were more cosplicing edges overlapping with known PPI. For gene pairs sharing TFBSs, we detected increased coexpression, but these common TFBSs did not affect cosplicing. The reverse was true for SFBSs: gene pairs sharing SFBSs had stronger cosplicing but coexpression was only mildly increased. For genes sharing microRNA-binding sites, we detected increased coexpression, but decreased cosplicing. This latter observation is more challenging to interpret biologically, but it is consistent with one putative microRNA activity, namely decreasing transcript stability (Valencia-Sanchez, Liu, Hannon, & Parker, 2006). Decreased transcript stability is likely to generate uncorrelated variability in the exon counts, decreasing the matrix correlations.

Comparison of the SFBS sharing by two genes required a special representation of the SFBSs present within the gene region. Binding sites can be present within both exonic and intronic sequences. To identify the SFBSs for all network genes, we used the SFMap database of predicted SFBSs for the mouse genome (Akerman, David-Eden, Pinter, & Mandel-Gutfreund, 2009; Paz, Akerman, Dror, Kosti, & Mandel-Gutfreund, 2010). We started by using biomaRt to retrieve the genomic coordinates of all genes in our networks. The SFMap database returned predicted binding sites of the following 21 splicing factors: SF2ASF, 9G8, SC35, Tra2alpha, Tra2beta, SRp20, SRp40, SRp55, hnRNPA1, hnRNPA2B1, hnRNPF, hnRNPH1, hnRNPM, hnRNPU, MBNL, NOVA1, PTB, CUG-BP, YB1, FOX1, and QK1. An accurate representation of the SFBS structure within a gene needs to take into account their both identity and multiplicity. We therefore constructed vectors of length 21: each entry in this vector simply records the number of the corresponding SFBSs within the gene sequence. Next, two different genes are compared by computing the distance measure between these vectors. We deemed two genes to have similar splicing regulatory regions if the distance was in the lower quantile of the collected distances between all pairs of genes.

Recent findings (Dixon et al., 2012) have revealed that chromosomal folding is a complex and highly regulated process that results in chromosomal regions in close spatial proximity (chromosomal spatial domains) that could nevertheless be well separated in terms of base-pair distance. Using the chromosomal domain boundaries detected in mouse cortex, we investigated whether the spatial folding of the chromosome affects coexpression and cosplicing, comparing gene pairs that are within the same topological domain versus gene pairs that belong to different domains but are within comparable base-pair distance. We found that genes sharing topological domains display increased coexpression but not cosplicing.

We previously found (Iancu et al., 2010; Iancu, Kawane, et al., 2012) that gene coexpression is strengthened for genes with spatially overlapping patterns of expression within the striatum. These results leveraged Allen Brain Atlas data on the spatial extent of gene expression, together with the extent of spatial overlap of different genes (Ng et al., 2007). We extended these results to cosplicing analysis, finding that gene cosplicing is strengthened for genes with spatial overlap, to an even larger extent than coexpression.

Biological functionality of gene modules is often inferred from the Gene Ontology (GO) annotation database. Here, the annotation of the genes within a module is evaluated for enrichment in specific GO categories, as compared to what is expected for random groups of genes of the same size. An important consideration in this analysis is the identity of the overall groups of genes considered. There are several choices: the whole transcriptome, the genes represented on a particular microarray or sequencing platform, or the genes included in network construction. We have opted for the third choice, contrasting the genes in a particular module against the genes selected for network construction. This choice was based on the following consideration: contrasting modules against the whole transcriptome revealed that each module was enriched in similar GO categories, namely categories under the general "neurological system process." Furthermore, contrasting the genes selected for network construction against the transcriptome revealed the same enrichment for "neurological system process" GO categories. This is not unexpected given that the genes selected for the network have relatively high expression and variability and our data originate from the mouse brain and therefore, as expected, expressed gene will be involved in neurological processes. We conclude that the general neurological function annotation applies to the whole network and by extension to each module of the network. Evaluation of individual modules against the network genes will then reveal module specific annotation above and beyond the general participation in neurological processes, which is common to all modules.

We conclude that membership in specific modules is generated by several interacting biological factors. First, there is spatial proximity between genes, either as chromosomal position on the same folding domain or as expression within the same brain subregions. Second, there is regulation by the same elements such as transcription factors, splicing factors, and microRNAs. Finally, there is participation in the same molecular mechanisms, as evidenced by shared GO annotations. For the genes that are poorly annotated, the principle of "guilt by association" allows inference of their functionality by virtue of their association or membership in the same module with well-annotated genes.

5. EFFECTS OF GENETIC SELECTION ON GENE NETWORKS

Differential network analysis has recently emerged as one of the most active research areas in computational biology; it has been suggested that: "[it] will become a standard mode of network analysis in the future, just as differential gene expression and protein phosphorylation studies are already pervasive in genomic and proteomic analysis" (Ideker & Krogan, 2012). To facilitate a better understanding of network plasticity, it might prove useful to offer an analogy in terms of evolutionary biology. A gene network or module could be considered equivalent in a general sense with an organ. Preservation/homology is important from a translational perspective: quantifying network effects on model organisms can potentially inform investigations in human populations.

For two species with a common ancestor, a homologous organ can be traced back to an ancestral structure and any differences accrued during speciation do not obscure the fact that both species have a physical structure fulfilling basically the same functionality. In contrast with physical organs where homology is more easily established, gene networks/modules are comprised of dispersed interactions between large number of genes and therefore it is necessary to computationally establish their homology or preservation across species or subpopulations. It is also important to clearly state the null hypothesis when establishing module homology/preservation: one tests the alternative hypothesis that two modules share some common structure versus the null hypothesis that two modules, even though composed of the same genes, do not share any commonality in terms of their interactions.

In contrast to module preservation, module disruption aims to detect significant changes in module structure that overlap with differences in external phenotypes such as behavioral measures. Here, homology of the modules is either assumed or formally established. However, homology does not preclude the emergence of more subtle differences in network structure that are not strong enough to completely abolish homology, but are still statistically significant. Here, we test the alternative hypothesis that there are significant changes in module structure versus the null hypothesis that network structure differences are small and not above the level expected from random variability in the data. It is important to note that two modules could be preserved and disrupted at the same time.

6. MODULE PRESERVATION ACROSS SUBPOPULATIONS AND SPECIES

Statistically, network preservation can be evaluated at several different levels (Langfelder et al., 2011). In addition to density, connectivity, and separability measures described above, tabulation-based metrics can also be used to evaluate whether modules independently detected in different networks share resemblance even though they are not identical in composition. We have examined the preservation of overall network structure between mouse populations of dramatically different genetic backgrounds (Iancu et al., 2010). We independently constructed networks from three different mouse populations of vastly different genetic backgrounds. Our populations were comprised of an B6 × DBA/2J (D2) F_2 intercross, a 4-way (HS4) cross between the B6, D2, BALB/cJ, and LP/J strains, and an eight-way cross between the same eight inbred strains used to create the used to create the collaborative cross (CC; Threadgill, Miller, Churchill, & de Villena, 2011). The eight inbred lines used for the CC are C57BL/6J (B6), A/J (A), 129S1/SvImJ (129), NOD/LtJ (NOD), NZO/HILtJ (NZO), CAST/EiJ (CAST), PWK/PhJ (PWK), and WSB/EiJ (WSB). Between any two of our three populations, there are no more than one inbred line in common (B6). Additionally, the HS-CC captures >90% of the available genetic diversity within *Mus musculus*, mainly because of the inclusion of three wild-derived strains CAST (*Mus musculus castaneus*), PWK (*M. m. musculus*), and WSB (*M. m. domesticus*).

In comparing module structure across these three mouse populations, we focused on tabulation-based preservation measures. Modules were detected independently in the three networks, and their overlaps in terms of gene membership were evaluated statistically. We observed that nearly all modules had one, or at most two to three counterpart modules in the other network, signifying very strong preservation. The results of the comparison between F_2 and HS-CC modules are presented in Fig. 4.3.

Evaluation of module preservation was also performed across species. Using RNA-seq data from mouse (striatum), macaque (ventral–medial prefrontal cortex), and human (prefrontal cortex), we constructed both coexpression and cosplicing networks. Here, we used a slightly different methodology for comparison of module preservation, focusing on density, connectivity, and separability. We also detect modules only in the HS-CC mouse data, and these modules were tested for preservation in a mouse

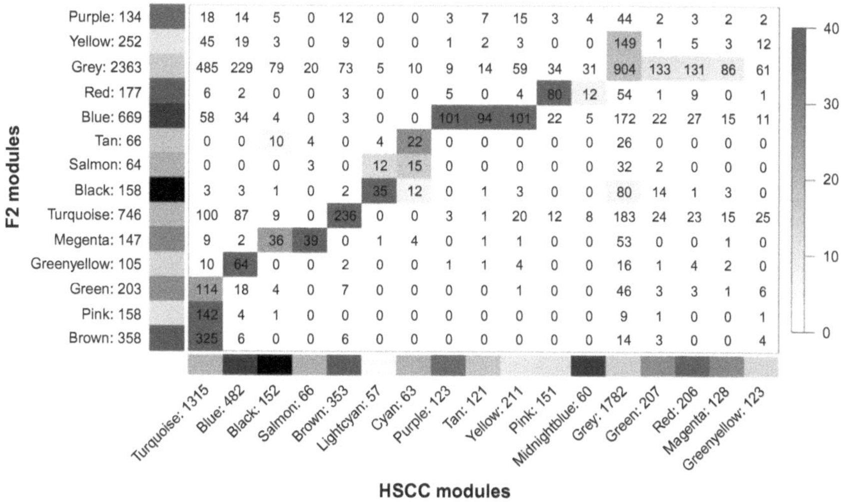

Figure 4.3 Gene membership overlap between F_2 and HS-CC modules. Module identity and gene numbers note on sides. Intensity of red color (dark gray in the print version) corresponds to the $-\log_{10}(p)$ overlap evaluated using the Fisher exact test.

population of different genetic background (HS-Npt), and also in macaque and human cortical RNA-Seq data; preservation measures are presented in Fig. 4.4. Network measurements were expressed as Z scores; Z scores <2 were taken to imply poor module preservation, while higher values were taken to signify moderate (>2) or high (>10) module preservation.

The results presented in Fig. 4.4 indicate that cosplicing modules are more preserved across species than coexpression modules (compare Fig. 4.4A and C versus B and D, respectively). Connectivity and separability were more preserved than density measures. Intuitively, we can think of connectivity as the patterns or the blueprints of organization of gene interactions. Conversely, density measures quantify the actual strength of the particular patterns. There results therefore indicate that nearly all mouse modules have preserved connectivity in the human and primate data; however, the strength (density) of these patterns varies more widely across species.

7. MODULE DISRUPTION RELATED TO BEHAVIORAL CHANGES

In a more current study (Iancu, Oberbeck, Darakjian, Kawane, et al., 2013), we addresses the relationship between gene network topology and

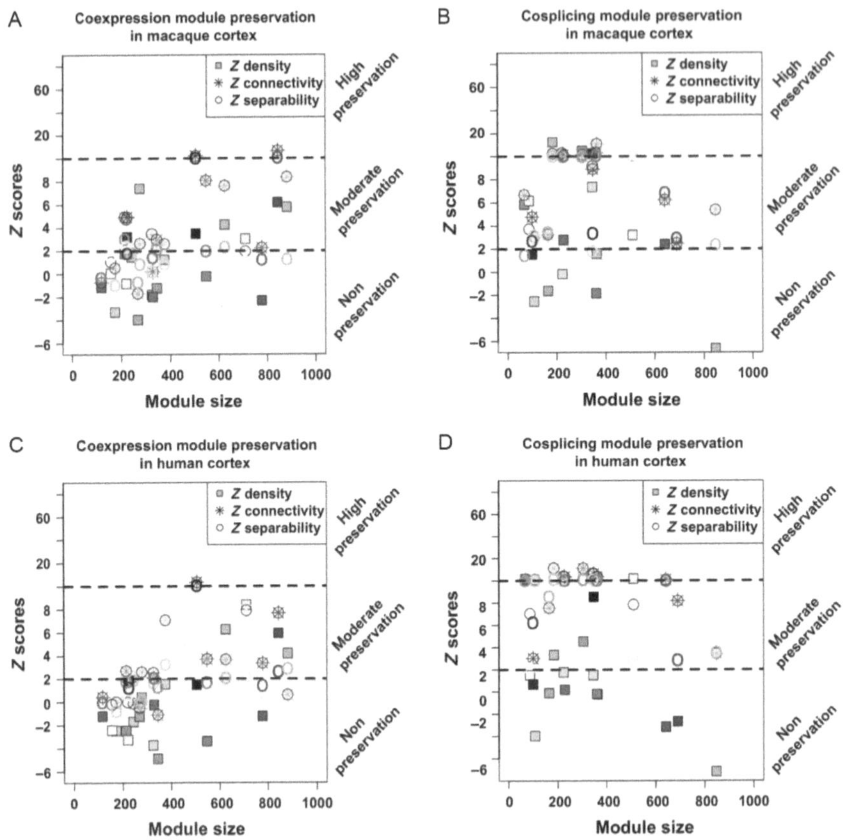

Figure 4.4 Gene module preservation across genetic backgrounds and species. We compare coexpression (A, C) and cosplicing (B, D) module preservation of HS-CC modules compared to HS-Npt mice striatum (A, B), macaque cortical tissue (C, D), and human cortical tissue (E, F). We observe generally better preservation of cosplicing modules compared to coexpression modules.

excessive ethanol consumption as reflected in blood ethanol level (BEC). Mice from a genetically diverse population used were selectively bred for high BEC achieved after 2–4 h access to 20% alcohol solution early in the circadian dark period, when the animals consume the most liquid (Crabbe et al., 2009). Animals often will drink to intoxication, and at the end of the alcohol access period, the BEC is measured and used as a selection phenotype. After 15–20 generations of selected breeding, selected populations differ significantly in BEC from their unselected ancestors. We compared striatal gene network structure between the selected populations and the unselected ancestors. Importantly, because two

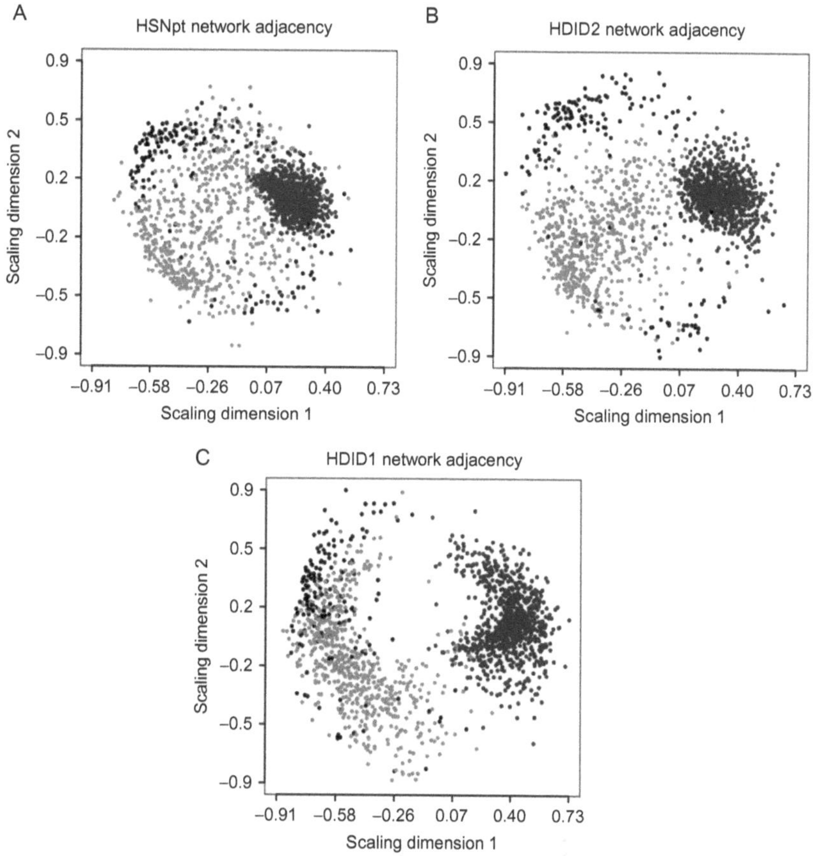

Figure 4.5 Multidimensional scaling plots of the coexpression networks in the unselected ancestors and selected group networks. For visual clarity, only the four modules most consistently affected by selection ("black," "magenta" (gray in the print version), "dark-red" (dark gray in the print version), and "green" (light gray in the print version)) are depicted. Each dot represents a transcript, with colors corresponding to module assignments. The distances between points correspond to network adjacency. The figure illustrates (1) the modularity of the networks, with similar colors clustered together and (2) the effect of selection on the network structure. The "dark-red" (dark gray in the print version) module appears more dispersed, while the "magenta" (gray in the print version) module appears more compacted in the selection networks.

selections were performed independently, we could evaluate reproducibility of the network analysis results.

To facilitate comparison across networks from selected and unselected groups, we first constructed a "consensus" network using all available samples. Module assignment was detected in this network and used for subsequent comparisons. To evaluate statistical significance of observed changes in network structure, we constructed a set of $N=1000$ networks composed random mixtures of samples from the two categories. Next, we compared the changes between selected/unselected groups versus changes that could occur between networks inferred between random groups of samples. Significant changes were quantified as disruption Z scores. At the whole module level, we observed that some modules became more connected, while another set of modules displayed lower connectivity. These changes can be visualized in Fig. 4.5 as increases or decreases in the compactness of the module representation.

The module representation in Fig. 4.5 is well suited to illustrate both module preservation and module disruption. For each of the four modules depicted, the shape and distribution are clearly similar in all three networks, illustrating preservation. At the same time, the modules are not identical in the three networks and the more subtle differences in dispersion/compactness illustrate module disruption. Importantly, the number of generations in the two selections is predictive of the extent of module disruption.

8. SUMMARY AND FUTURE DIRECTIONS

In this chapter, we presented an overview of the network methodologies that can be applied to transcriptional sequencing data. Using several example datasets, we illustrate several network analysis concepts, including scale-free structure, modularity, and the inference of functional roles for unannotated genes, and most importantly differential network analysis. Our results illustrate that genetic factors influence the network topology in a predictable and reproducible manner (Iancu, Oberbeck, Darakjian, Kawane, et al., 2013, Iancu, Oberbeck, Darakjian, Metten, et al., 2013). While in our work we focused on effects of genetic differences on network topology, others (Mulligan et al., 2011) have considered the effects of alcohol consumption on network structure.

The advent of RNA-Seq technology is likely to provide even more impetus to network approaches. As we have demonstrated, RNA-Seq data offer better estimation of the gene expression levels and their variability, greatly improving the accuracy of network measurements (Iancu, Kawane, et al., 2012). Additionally, exon-level data allow the construction and analysis of gene cosplicing networks.

Network analysis can potentially be extended in several additional directions. One particularly promising potential area of inquiry is cross-regional network analysis. Recent results have revealed that gene expression levels can be correlated between distinct tissues (Dobrin et al., 2009). Extending network analysis to several brain regions is potentially of high interest given the complexity of behavioral phenotypes and their likely recruitment of several brain regions. A second direction with particular promise is the integration of gene network with other modalities of system-level evaluation of brain function. In this respect, recent advances in rodent MRI brain imaging hold the promise of overlaying molecular/genetic mechanisms with structural and functional brain imaging results.

REFERENCES

Akerman, M., David-Eden, H., Pinter, R. Y., & Mandel-Gutfreund, Y. (2009). A computational approach for genome-wide mapping of splicing factor binding sites. *Genome Biology*, *10*(3), R30.

Allen, J. D., Xie, Y., Chen, M., Girard, L., & Xiao, G. (2012). Comparing statistical methods for constructing large scale gene networks. *PLoS One*, 7(1), e29348.

Barrie, E. S., Smith, R. M., Sanford, J. C., & Sadee, W. (2012). mRNA transcript diversity creates new opportunities for pharmacological intervention. *Molecular Pharmacology*, *81*(5), 620–630.

Cahoy, J. D., Emery, B., Kaushal, A., Foo, L. C., Zamanian, J. L., Christopherson, K. S., et al. (2008). A transcriptome database for astrocytes, neurons, and oligodendrocytes: A new resource for understanding brain development and function. *Journal of Neuroscience*, *28*(1), 264–278.

Calarco, J. A., Zhen, M., & Blencowe, B. J. (2011). Networking in a global world: Establishing functional connections between neural splicing regulators and their target transcripts. *RNA*, *17*(5), 775–791.

Chen, L., & Zheng, S. (2009). Studying alternative splicing regulatory networks through partial correlation analysis. *Genome Biology*, *10*(1), R3.

Crabbe, J. C., Metten, P., Rhodes, J. S., Yu, C. H., Brown, L. L., Phillips, T. J., et al. (2009). A line of mice selected for high blood ethanol concentrations shows drinking in the dark to intoxication. *Biological Psychiatry*, *65*(8), 662–670.

Dai, C., Li, W., Liu, J., & Zhou, X. J. (2012). Integrating many co-splicing networks to reconstruct splicing regulatory modules. *BMC Systems Biology*, *6*(Suppl. 1), S17.

Dixon, J. R., Selvaraj, S., Yue, F., Kim, A., Li, Y., Shen, Y., et al. (2012). Topological domains in mammalian genomes identified by analysis of chromatin interactions. *Nature*, *485*(7398), 376–380.

Dobrin, R., Zhu, J., Molony, C., Argman, C., Parrish, M. L., Carlson, S., et al. (2009). Multi-tissue co-expression networks reveal unexpected subnetworks associated with disease. *Genome Biology*, *10*(5), R55.

Drnevich, J., Replogle, K. L., Lovell, P., Hahn, T. P., Johnson, F., Mast, T. G., et al. (2012). Impact of experience-dependent and -independent factors on gene expression in songbird brain. *Proceedings of the National Academy of Sciences of the United States of America*, *109*(Suppl. 2), 17245–17252.

Fagnani, M., Barash, Y., Ip, J. Y., Misquitta, C., Pan, Q., Saltzman, A. L., et al. (2007). Functional coordination of alternative splicing in the mammalian central nervous system. *Genome Biology*, *8*(6), R108.

Iancu, O. D., Darakjian, P., Malmanger, B., Walter, N. A., McWeeney, S., & Hitzemann, R. (2012). Gene networks and haloperidol-induced catalepsy. *Genes, Brain, and Behavior*, *11*(1), 29–37.

Iancu, O. D., Darakjian, P., Walter, N. A., Malmanger, B., Oberbeck, D., Belknap, J., et al. (2010). Genetic diversity and striatal gene networks: Focus on the heterogeneous stock-collaborative cross (HS-CC) mouse. *BMC Genomics*, *11*, 585.

Iancu, O. D., Kawane, S., Bottomly, D., Searles, R., Hitzemann, R., & McWeeney, S. (2012). Utilizing RNA-Seq data for de novo co-expression network inference. *Bioinformatics*, *28*(12), 1592–1597.

Iancu, O. D., Oberbeck, D., Darakjian, P., Kawane, S., Erk, J., McWeeney, S., et al. (2013). Differential network analysis reveals genetic effects on catalepsy modules. *PLoS One*, *8*(3), e58951.

Iancu, O. D., Oberbeck, D., Darakjian, P., Metten, P., McWeeney, S., Crabbe, J. C., et al. (2013). Selection for drinking in the dark alters brain gene co-expression networks. *Alcoholism, Clinical and Experimental Research*, *37*(8), 1295–1303.

Ideker, T., & Krogan, N. J. (2012). Differential network biology. *Molecular Systems Biology*, *8*, 565.

Jay, J. J., Eblen, J. D., Zhang, Y., Benson, M., Perkins, A. D., Saxton, A. M., et al. (2012). A systematic comparison of genome-scale clustering algorithms. *BMC Bioinformatics*, *13*(Suppl. 10), S7.

Jeong, H., Mason, S. P., Barabasi, A. L., & Oltvai, Z. N. (2001). Lethality and centrality in protein networks. *Nature*, *411*(6833), 41–42.

Kosti, I., Radivojac, P., & Mandel-Gutfreund, Y. (2012). An integrated regulatory network reveals pervasive cross-regulation among transcription and splicing factors. *PLoS Computational Biology*, *8*(7), e1002603.

Langfelder, P., & Horvath, S. (2008). WGCNA: An R package for weighted correlation network analysis. *BMC Bioinformatics*, *9*, 559.

Langfelder, P., Luo, R., Oldham, M. C., & Horvath, S. (2011). Is my network module preserved and reproducible? *PLoS Computational Biology*, 7(1), e1001057.

Langfelder, P., Zhang, B., & Horvath, S. (2008). Defining clusters from a hierarchical cluster tree: The Dynamic Tree Cut package for R. *Bioinformatics*, *24*(5), 719–720.

Mantel, N. (1967). The detection of disease clustering and a generalized regression approach. *Cancer Research*, *27*(2), 209–220.

Marioni, J. C., Mason, C. E., Mane, S. M., Stephens, M., & Gilad, Y. (2008). RNA-seq: An assessment of technical reproducibility and comparison with gene expression arrays. *Genome Research*, *18*(9), 1509–1517.

Mulligan, M. K., Rhodes, J. S., Crabbe, J. C., Mayfield, R. D., Adron Harris, R., & Ponomarev, I. (2011). Molecular profiles of drinking alcohol to intoxication in C57BL/6J mice. *Alcoholism, Clinical and Experimental Research*, *35*(4), 659–670.

Ng, L., Pathak, S. D., Kuan, C., Lau, C., Dong, H., Sodt, A., et al. (2007). Neuroinformatics for genome-wide 3D gene expression mapping in the mouse brain. *IEEE/ACM Transactions on Computational Biology and Bioinformatics*, *4*(3), 382–393.

Pan, Q., Shai, O., Lee, L. J., Frey, B. J., & Blencowe, B. J. (2008). Deep surveying of alternative splicing complexity in the human transcriptome by high-throughput sequencing. *Nature Genetics*, *40*(12), 1413–1415.

Paz, I., Akerman, M., Dror, I., Kosti, I., & Mandel-Gutfreund, Y. (2010). SFmap: A web server for motif analysis and prediction of splicing factor binding sites. *Nucleic Acids Research*, *38*(Web Server issue), W281–W285.

Ravasz, E., Somera, A. L., Mongru, D. A., Oltvai, Z. N., & Barabasi, A. L. (2002). Hierarchical organization of modularity in metabolic networks. *Science*, *297*(5586), 1551–1555.

Threadgill, D. W., Miller, D. R., Churchill, G. A., & de Villena, F. P. (2011). The collaborative cross: A recombinant inbred mouse population for the systems genetic era. *ILAR Journal*, *52*(1), 24–31.

Valencia-Sanchez, M. A., Liu, J., Hannon, G. J., & Parker, R. (2006). Control of translation and mRNA degradation by miRNAs and siRNAs. *Genes & Development, 20*(5), 515–524.

Vanderlinden, L. A., Saba, L. M., Kechris, K., Miles, M. F., Hoffman, P. L., & Tabakoff, B. (2013). Whole brain and brain regional co-expression network interactions associated with predisposition to alcohol consumption. *PLoS One, 8*(7), e68878.

Wang, Z., Gerstein, M., & Snyder, M. (2009). RNA-Seq: A revolutionary tool for transcriptomics. *Nature Reviews. Genetics, 10*(1), 57–63.

Wang, E. T., Sandberg, R., Luo, S., Khrebtukova, I., Zhang, L., Mayr, C., et al. (2008). Alternative isoform regulation in human tissue transcriptomes. *Nature, 456*(7221), 470–476.

Zhang, B., & Horvath, S. (2005). A general framework for weighted gene co-expression network analysis. *Statistical Applications in Genetics and Molecular Biology, 4,* Article 17.

CHAPTER FIVE

Splicing in the Human Brain

Ammar Zaghlool[*,†], **Adam Ameur**[*,†], **Lucia Cavelier**[*,‡], **Lars Feuk**[*,†,1]
[*]Department of Immunology, Genetics and Pathology, Uppsala University, Uppsala, Sweden
[†]Science for Life Laboratory, Uppsala University, Uppsala, Sweden
[‡]Uppsala University Hospital, Uppsala, Sweden
[1]Corresponding author: e-mail address: lars.feuk@igp.uu.se

Contents

Abstract

It has become increasingly clear over the past decade that RNA has important functions in human cells beyond its role as an intermediate translator of DNA to protein. It is now known that RNA plays highly specific roles in pathways involved in regulatory, structural, and catalytic functions. The complexity of RNA production and regulation has become evident with the advent of high-throughput methods to study the transcriptome. Deep sequencing has revealed an enormous diversity of RNA types and transcript isoforms in human cells. The transcriptome of the human brain is particularly interesting as it contains more expressed genes than other tissues and also displays an extreme diversity of transcript isoforms, indicating that highly complex regulatory pathways are present in the brain. Several of these regulatory proteins are now identified, including RNA-binding proteins that are neuron specific. RNA-binding proteins also play important roles in regulating the splicing process and the temporal and spatial isoform production. While significant progress has been made in understanding the human transcriptome, many questions still remain regarding the basic mechanisms of splicing and subcellular localization of RNA. A long-standing question is to what extent the splicing of pre-mRNA is cotranscriptional and posttranscriptional, respectively. Recent data, including studies of the human brain, indicate that splicing is primarily cotranscriptional in human cells. This

International Review of Neurobiology, Volume 116
ISSN 0074-7742
http://dx.doi.org/10.1016/B978-0-12-801105-8.00005-9

chapter describes the current understanding of splicing and splicing regulation in the human brain and discusses the recent global sequence-based analyses of transcription and splicing.

1. Pre-mRNA SPLICING IN HUMAN CELLS

Pre-mRNA splicing is one of the fundamental processes in the intrinsic and regulated gene expression in eukaryotes. It is a highly precise process that involves the removal of noncoding intronic sequences from the premature RNA transcript (pre-mRNA) to produce the mature form of protein-coding messenger RNA (mRNA). Several signals exist within introns that are critical for the splicing process including a 5′ splice site, a branch site, and a 3′ splice site. The splice sites are located at the 5′ and 3′ ends of introns and contain almost invariant sequence: GU at the 5′ splice site and AG at the 3′ splice site. The branch site is often located anywhere between 21 and 34 nucleotides upstream of the 3′ end of introns and typically contains the consensus yUn<u>A</u>y, where the underlined A is the branch point, n is any nucleotide, and the lowercase pyrimidines are not as conserved as the uppercase A and U. Pre-mRNA splicing is a sequential process catalyzed by a macromolecular machine called the spliceosome, a large complex of small nuclear ribonucleoprotein units (snRNPs) and a large number of non-snRNP factors (Fig. 5.1).

2. ALTERNATIVE Pre-mRNA SPLICING

Pre-mRNA splicing can give rise to different RNA transcripts from a single gene by alternative splicing, a process where different combinations of exons are joined together to create a diverse pool of mRNA transcripts. Alternative splicing events can be classified into five major categories: (a) exon skipping, representing the most common mode of alternative splicing; (b) intron retention; (c) alternative 5′ splice site usage; (d) alternative 3′ splice site usage; and (e) mutually exclusive exons (Fig. 5.2).

Alternative splicing is more common in multicellular than unicellular eukaryotes. For example in *Saccharomyces cerevisiae*, alternative splicing is very rare (Ares, Grate, & Pauling, 1999; Howe, Kane, & Ares, 2003) as it contains only 250–300 introns and pre-mRNA splicing affects only 3% of the genes. In other unicellular eukaryotes like trypanosomes, most genes are even intronless (Liang, Haritan, Uliel, & Michaeli, 2003). In contrast, alternative

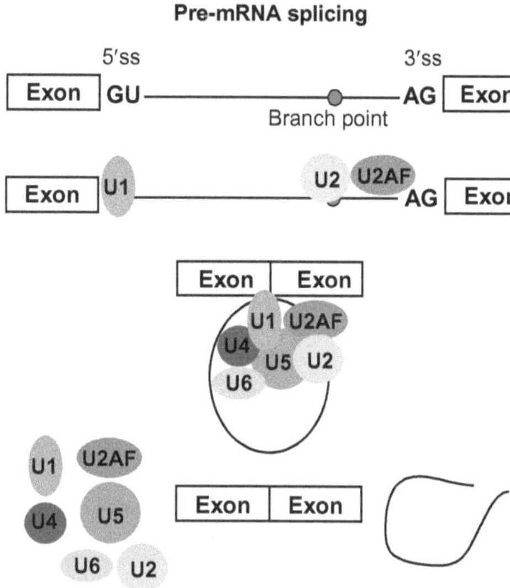

Figure 5.1 Pre-mRNA splicing. The spliceosomal factors are sequentially recruited to the 5′ splice site, 3′ splice site, and to the branch point. Once the complete spliceosome is assembled over the intron, the exons are joined together and the intron lariat is released.

splicing in metazoans is prevalent and represents an important component of gene expression regulation. Several examples from *Drosophila melanogaster* highlight the biological importance of alternative splicing in controlling gene expression. Sex determination in *D. melanogaster* is controlled by alternative splicing that ultimately gives rise to sex-specific expression of two splicing variants of the Doublesex transcription factor (Salz, 2011; Salz & Erickson, 2010). The *Drosophila's DSCAM* gene exemplifies the extreme capability of alternative splicing in expanding the coding capacity of metazoan genes through encoding more than 38,000 distinctive mRNA isoforms essential for nervous system development (Park & Graveley, 2007; Schmucker et al., 2000). Remarkably, the number of *DSCAM* isoforms exceeds the total number of genes encoded in *D. melanogaster* genome.

In recent years, alternative splicing has emerged as one of the most essential mechanisms underlying the complex phenotypic and functional diversity displayed by the mammalian cells. In the human genome, the number of protein-coding genes is estimated to be around 21,000. These genes are able to code for more than 120,000 proteins (Flicek et al., 2013; Harrow et al., 2012; Modrek & Lee, 2002). The poor correlation between the number of genes and proteome diversity suggests an important association between

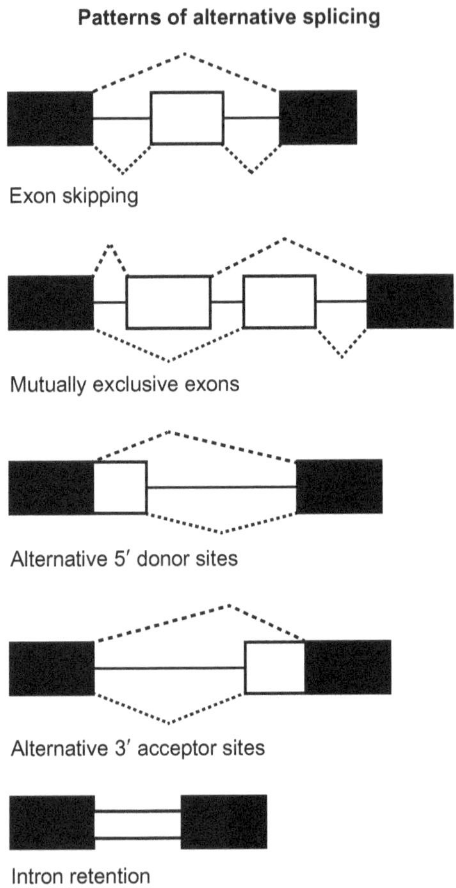

Figure 5.2 Major patterns of alternative splicing. In this figure, exons are represented by boxes and introns are represented by lines. Constitutive exons are shown in black and alternatively spliced regions are white. Dashed lines indicate splicing options.

alternative splicing and organismal complexity. Moreover, recent research suggests that alternative splicing may be the driving force of evolutionary changes differentiating primates and humans from other species (Merkin, Russell, Chen, & Burge, 2012). It is estimated that up to 95% of multiexon genes in human are subjected to alternative splicing, thereby greatly expanding the functional diversity of the human proteome and adding additional layers of regulation of gene expression (Black, 2000; Irimia & Blencowe, 2012; Nilsen & Graveley, 2010). This enormous diversity is achieved by a tight regulation of alternative splicing outcomes. Alternative splicing of a gene may create mRNA isoforms that differ in their coding capacity, transcript stability, and which can even display distinctive protein properties (Kelemen et al., 2013; Stamm et al., 2005). In fact, particular mRNA isoforms might be specific to a sex, developmental stage, cell or tissue type, or environmental conditions (Ellis et al., 2012; Grabowski & Black, 2001; Yap & Makeyev, 2013). The production of multiple mRNA transcripts from a single gene often occurs through the inclusion or exclusion of different combinations of exons. In a typical mRNA transcript, some exons are constitutively spliced, i.e., they are always included in the transcript. Other exons are alternatively spliced; therefore, they are sometimes included and sometimes excluded. Although the pre-mRNA splicing machinery depends mainly on splice sites for defining intron borders, the inclusion or exclusion of an alternatively spliced exon relies on a complex network of regulatory mechanisms that primarily involve specialized RNA-binding proteins called splicing factors that recognize *cis*-acting sequence elements located in the vicinity of the regulated splice sites (Black, 2003; Chen & Manley, 2009).

Cis-acting elements located in both exonic and intronic regions can either promote exon inclusion (splicing enhancers) or exon skipping (splicing silencers). These elements are classified as exonic splicing enhancers (ESEs) or exonic splicing silencers (ESSs) if they function from exonic locations, and as intronic splicing enhancers (ISEs) or intronic splicing silencers (ISSs) if they function from intronic locations (Fig. 5.3). ESEs are the best-characterized sequence elements, serving as binding sites for a family of

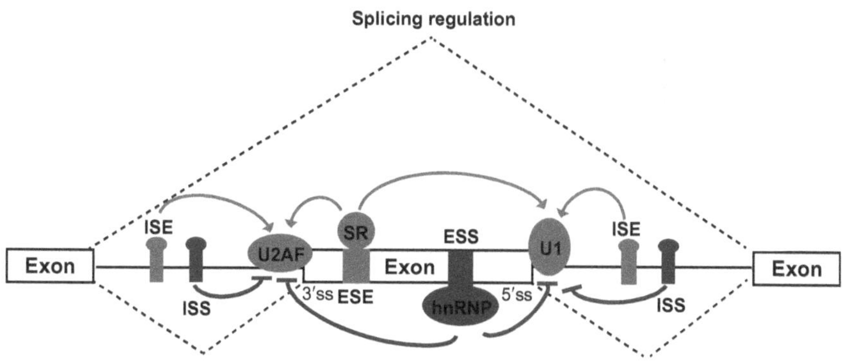

Figure 5.3 Splicing regulation by the *cis*-acting elements and *trans*-acting splicing factors. Two alternative splicing pathways with the middle exon either included or excluded. Splicing enhancing signals are shown in green (light gray in the print version), while splicing inhibitory signals are shown in red (dark gray in the print version). In this model, ESSs or ISSs are bound by hnRNPs and repress exon inclusion by inhibiting the recruitment of the splicing machinery (e.g., U2AF and U1). ESEs or ISEs serve as binding sites for SR proteins that influence the inclusion of exons by recruiting the splicing machinery.

proteins called SR (Ser-Arg) proteins. SR proteins influence the inclusion of exons by recruiting the splicing machinery to the adjacent exon/intron borders (Chen & Manley, 2009; Graveley, 2000). ESS and ISS repress exon inclusion by recruiting members of the heterogeneous nuclear ribonucleoprotein family (hnRNP), such as hnRNPA1 (Del Gatto-Konczak, Olive, Gesnel, & Breathnach, 1999). *Cis*-elements may also influence splice site selection by forming secondary structures that prevent the accessibility of certain regulatory proteins to the transcript (Grover et al., 1999). Furthermore, the concentration ratios between negative and positive splicing factors in a given cell influence splice site selection (Zhu, Mayeda, & Krainer, 2001).

3. TISSUE-SPECIFIC ALTERNATIVE SPLICING

Alternative splicing also acts as an important mechanism in defining tissue specificity by regulating the tissue-specific expression of transcript isoforms. Tissue-specific alternative splicing is usually controlled by the combinatorial effect of ubiquitous and tissue-specific expression of splicing regulatory proteins (splicing activators and splicing suppressors) that interact with *cis*-acting elements to influence splicing at nearby splice sites (Black, 2003; Matlin, Clark, & Smith, 2005). There are several known examples of tissue-specific regulation of alternative splicing. Prominent examples include the tissue-specific alternative splicing of *MTMR1* in muscles (Buj-Bello et al., 2002), the brain-specific isoform of the renin gene (Lee-Kirsch, Gaudet, Cardoso, & Lindpaintner, 1999), cancer-associated splicing of *CD44* (Naor, Nedvetzki, Golan, Melnik, & Faitelson, 2002), and the tissue-specific pattern of alternative splicing of *TCF7L2* (Prokunina-Olsson et al., 2009). Initial genome-wide analyses of tissue-specific alternative splicing have identified hundreds of tissue-specific splice isoforms and revealed high levels of alternative splicing events in the human brain, liver, and testis (Xu, Modrek & Lee, 2002; Yeo, Holste, Kreiman, & Burge, 2004). In addition, recent high-throughput studies indicated that, among the tissues analyzed, around 50% of alternative splicing isoforms exhibit differential expression between tissues (Wang et al., 2008). These observations strongly suggest that alternative splicing is subjected to tissue-specific regulation.

4. ALTERNATIVE SPLICING IN THE BRAIN

The human brain is an immensely complex system, containing more than a trillion neurons, which are accurately connected to one another via the synapses. Neurons have an enormous biochemical complexity that builds up the physical and biological basis of the humans' superior cognitive abilities. The outstanding complexity of the human brain depends on flexible and precise regulation of its underlying transcriptome. Remarkably, the human brain exhibits high levels of gene expression. It has been estimated that around 50% of known protein-coding genes are expressed across the human brain (International Human Genome Sequencing Consortium, 2004), and the expression patterns of these genes showed distinctive profiles compared to other tissues in a study that included 45 different human tissues (Roth et al., 2006). Moreover, using exon arrays (de la Grange, Gratadou, Delord, Dutertre, & Auboeuf, 2010) and RNA-seq (Ramskold, Wang, Burge, & Sandberg, 2009), several studies have suggested that the human brain, together with the testes and kidney, has a higher gene expression levels and transcriptome complexity than other tissues. Intriguingly, a recent study from the Allen Human Brain Atlas explored the adult human brain transcriptome in unprecedented anatomical details and demonstrated significant variation of transcriptional landscape across different brain regions. They further investigated transcripts of 700 genes, previously shown to be differentially expressed in the postsynaptic density, and found that 31% of these transcripts exhibit highly specialized regional expression (Hawrylycz et al., 2012). Another important recent study analyzed transcriptional dynamics during human brain development from 57 postmortem human brains ranging in age from 5.7 weeks postconception to 82 years and found that over 80% of the genes examined are differentially regulated across brain regions and/or over time (Kang et al., 2011).

Alternative splicing is a critical mechanism throughout the body, but probably most important in the brain. The high level and complexity of gene expression of the human brain transcriptome is accompanied with high levels of alternative splicing. In fact, the brain contains the highest number of alternative splicing events and highest occurrence of tissue-specific alternative splicing compared to other tissues (de la Grange et al., 2010; Xu, Modrek, & Lee, 2002). These observations suggest that alternative splicing can serve as a major mechanism contributing to the enormous levels of molecular and cellular diversity observed in the brain. Indeed, alternative splicing has been implicated in several neurodevelopmental and neurological processes such as the development of neurons and the formation of functional synapses (Norris & Calarco, 2012).

The human brain contains billions of neurons that communicate through formation of highly specific and complex patterns of synaptic connections.

The identity, development, and maintenance of these synaptic connections are likely to be mediated through specific interaction between highly polymorphic pre- and postsynaptic cell adhesion molecules. The genome size is not sufficient to achieve such diversity; nevertheless, alternative splicing may represent the most versatile way to amplify the number of cell adhesion molecules to account for the complexity of synaptic connections. One of the outstanding examples that support the functional importance of alternative splicing in neuronal interactions is the extensive splicing of *Neurexin* genes. Neurexins are a family of neural proteins that function as cell adhesion molecules during synaptogenesis and interneuronal signaling. Although there are only three *Neurexin* genes in human genome, designated *NRXN1*, *NRXN2*, and *NRXN3*, each of these genes contains multiple cassette exons and one alternative promoter (Ullrich, Ushkaryov, & Sudhof, 1995). Recent estimates indicate that thousands of Neurexin isoforms are generated and regulated by alternative splicing (Treutlein, Gokce, Quake, & Sudhof, 2014). Particular presynaptic alternative Neurexin isoforms bind preferentially to a postsynaptic Neuroligin receptor, which also undergoes alternative splicing, although to a lesser extent (Scheiffele, Fan, Choih, Fetter, & Serafini, 2000). Accordingly, it has been proposed that the differential binding affinity between the alternative isoforms of Neurexins and Neurologins may lead to unique neuron identity that tunes the trans-synaptic signaling properties (Boucard, Chubykin, Comoletti, Taylor, & Sudhof, 2005; Chih, Gollan, & Scheiffele, 2006; Graf, Kang, Hauner, & Craig, 2006). For instance, the binding of presynaptic β-Neurexins to the postsynaptic Neuroligins is sufficient to trigger synaptic function (Scheiffele et al., 2000). Notably, the interaction occurs only if exon 20 of β-neurexins is excluded from the transcript (Ichtchenko et al., 1995). Although the precise function of the exon 20 containing transcript is not well understood, it has been speculated that the differential inclusion or exclusion of this exon may control the formation of a functional synapse (Graveley, 2001). Moreover, a recent study showed that constitutive inclusion of exon 20 decreased the postsynaptic AMPA receptor levels and enhanced its endocytosis (Aoto, Martinelli, Malenka, Tabuchi, & Sudhof, 2013). From these observations, the authors suggested that alternative splicing of exon 20 controls the postsynaptic AMPA receptor trafficking and further supported the role of alternative splicing in trans-synaptic regulation of synaptic strength and long-term plasticity.

Alternative splicing can also influence neural activity through expanding the diversity of responses to neural signals. For example, calcium signaling has been previously reported to play a role in neuronal activity modulation and in learning and memory formation (West et al., 2001). It was also reported that calcium signaling regulates gene expression at the transcription and alternative splicing levels (Alonso & Garcia-Sancho, 2011; Xie, 2008). There are several examples of altered alternative splicing patterns in neurons in response to variation in the intracellular levels of calcium. Increased internal levels of calcium cause the exclusion of exon 5 and exon 21 of NMDA receptor type 1 in hippocampal neurons (Han, Yeo, An, Burge, & Grabowski, 2005; Zhu et al., 2001). The inclusion or exclusion of these exons influences the localization and trafficking of the NMDA receptor to regulates synaptic strength and plasticity (Lee et al., 2007).

5. BRAIN-SPECIFIC SPLICING REGULATION

Alternative splicing requires highly accurate and orchestrated regulation especially in the nervous system, where subtle changes in splicing outputs may lead to profound effects on properties of the many different types of neurons (Lipscombe, 2005). Alternative splicing is primarily regulated by combinatorial interaction of ubiquitously expressed splicing factors such as hnRNPs and ER proteins (Black, 2003; Chen & Manley, 2009; Graveley, 2000). Therefore, tissue-specific splicing events are, at least partially, achieved by the differential expression of these factors. During the last years, several tissue-specific splicing factors have been identified especially in the brain, including NOVA1, NOVA2, nPTB, Fox1, and Fox2 (Boutz et al., 2007; Dredge, Stefani, Engelhard, & Darnell, 2005; Fogel et al., 2012; Li, Lee, & Black, 2007; Ule et al., 2005; Zhang et al., 2008). Because they are exclusively expressed in the nervous system and control the choice of mRNAs expressed and thereby influence RNA metabolism, processing, localization, and expression, these factors are considered as key contributors to spatial–temporal control of neuronal RNA functions (Doyle & Kiebler, 2011; Kusek et al., 2012; Vessey et al., 2012). Therefore, extended understanding of their functions, sites of action, and regulation may contribute to better understanding for the functional and physiological complexity of the nervous system. In fact, recent technologies that employ cross-linking and immunoprecipitation followed by high-throughput sequencing (HITS-CLIP) have provided valuable insights on the function of some tissue-specific splicing factors, among the most studied are NOVA and PTB/nPTB (Licatalosi et al., 2008; Xue et al., 2009).

NOVAs are a family of RNA-binding proteins that were discovered as a target antigen for the autoantibodies in patients with paraneoplastic opsoclonus-myoclonus ataxia syndrome, a rare neurologic disease that causes defects in the motor systems (Buckanovich, Posner, & Darnell, 1993; Luque et al., 1991; Yang, Yin, & Darnell, 1998). In human, the NOVA family consists of two proteins designated NOVA1 and NOVA2. The amino acid sequences of the two proteins are highly similar and contain three KH-type RNA-binding domains (Buckanovich & Darnell, 1997; Yang et al., 1998). NOVA1 and NOVA2 were among the first known tissue-specific alternative splicing regulators. They bind to sequences in nascent transcripts containing UCAY motifs (Dredge & Darnell, 2003). The expression of *NOVA1* and *NOVA2* is highly tissue specific. While NOVA2 appears to be primarily expressed in the neocortex, NOVA1 is exclusively expressed in the subcortical regions and in the postmitotic neurons of the central nervous system (Buckanovich et al., 1993; Racca et al., 2010; Yang et al., 1998; Yano, Hayakawa-Yano, Mele, & Darnell, 2010). The first insights into NOVA1 function came from analysis of NOVA knockout mice; in those mice, NOVA1 was shown to be essential for neuronal viability through regulating alternative splicing (Jensen et al., 2000). Of note, the binding position of NOVA within the nascent transcript is a key determinant for splicing outcomes. In other words, NOVA binding within regions located upstream of alternative exons causes their exclusion, while binding to downstream sites induced alternative exon inclusion (Ule et al., 2006). Subsequent genome-wide studies have supported the regulatory role of NOVA proteins in various alternative splicing events and identified a NOVA regulatory network active in transcripts associated with functions in the synapse, in formation of neuromuscular junctions, in motor neuron function, in Reelin signaling, and in neural migration (Park & Curran, 2010; Ruggiu et al., 2009; Ule et al., 2005; Yano et al., 2010). Recently, HITS-CLIP analysis and advanced computational approaches led to the identification of ~700 alternative exons regulated by NOVA in mouse brain (Licatalosi et al., 2008; Zhang et al., 2010). In line with the previous findings, these studies demonstrated that NOVA2 null mice showed major splicing defects particularly in the neocortex, where NOVA2 is exclusively expressed. Intriguingly, a more recent study demonstrated that NOVA-dependent splicing regulation induced dramatic changes in the abundance of several synaptic proteins implicated in epilepsy, likely through inclusion of cryptic exons and nonsense-mediated decay (NMD) (Eom et al., 2013).

nPTB (also called PTB2) is another well-studied neuronal RNA-binding protein that contains four RNA-binding motifs. PTB, an nPTB paralog, is among the first identified alternative splicing regulators (Garcia-Blanco, Jamison, & Sharp, 1989). Generally, PTB and nPTB function as splicing silencers by binding to the polypyrimidine tract near the 3′ end of introns. The binding of PTB to this region interferes with the assembly of the spliceosome, which ultimately leads to the exclusion of downstream exons (Sharma, Falick, & Black, 2005). In spite of their high similarity in polypeptide sequence and RNA-binding specificity, PTB and nPTB are expressed in a mutually exclusive pattern in different brain parts (Boutz et al., 2007). While PTB is expressed in neuronal precursors, nPTB is specifically expressed in differentiated neurons (Boutz et al., 2007; Makeyev, Zhang, Carrasco, & Maniatis, 2007). This fascinating pattern of reciprocal expression is achieved by a cross-regulatory mechanism where, in neuronal precursors, PTB induces skipping of exon 10 in nPTB, thereby producing a transcript degraded by NMD (Spellman, Llorian, & Smith, 2007). On the other hand, differentiated neurons express miR-124, which was reported to mediate PTB downregulation, and consequently relieves the expression of nPTB (Makeyev et al., 2007). These observations suggest that these proteins have different physiological functions during neuronal differentiation (Boutz et al., 2007). In fact, upregulation of nPTB and downregulation of PTB are suggested to explain 25% of neuron-specific alternative splicing events (Boutz et al., 2007). Taken together, these observations provide a strong evidence for the fundamental role of alternative splicing regulation in the function and development of the brain and nervous system.

6. TRANSCRIPTION-COUPLED REGULATION OF ALTERNATIVE SPLICING

Alternative splicing is not only regulated through *trans*-acting splicing factors but also by processes linked to the transcription machinery. Recently, several lines of research have established that splicing is physically and functionally coupled with transcription, and that this coupling may influence alternative splicing regulation and other downstream RNA processing mechanisms (Kornblihtt et al., 2013; Moore & Proudfoot, 2009; Shukla & Oberdoerffer, 2012). This implies that RNA processing factors are recruited to the emerging RNA transcript during transcription and that splicing occurs cotranscriptionally. Two models have been suggested to explain the coupling between the transcription unit and the splicing machinery (Kornblihtt, 2006). The first model is called recruitment coupling where the CTD of RNA pol II plays a major role in cotranscriptional coupling between RNA biogenesis and processing (Munoz, de la Mata, & Kornblihtt, 2010). In this model, the CTD of RNA pol II offers a flexible landing pad for various transcription and splicing factors and facilitates their recruitment to the emerging nascent RNA transcript (Phatnani & Greenleaf, 2006). This recruitment may influence alternative splicing regulation by the ability of different transcription factors to recruit distinct splicing factors (Cramer et al., 1999). The second model is called the kinetic coupling (Carrillo Oesterreich, Bieberstein, & Neugebauer, 2011). This model proposes that RNA pol II-mediated elongation rate influences the regulation of alternative splicing and splicing outcomes by affecting the period during which splice signals are exposed to the splice factors in the growing nascent RNA transcript (Perales & Bentley, 2009). For example, if the RNA pol II elongation rate is high due to a strong upstream promoter or an open chromatin structure, it increases the possibility that weak 3′ splice sites around cassette exons will be outcompeted by an already transcribed strong 3′ splice site downstream. In comparison, low RNA pol II processivity will provide enough time for the splicing machinery to recognize any weak 3′ splice sites leading to the inclusion of a cassette exon (de la Mata et al., 2003). There are several lines of evidence that support this model (de la Mata, Lafaille, & Kornblihtt, 2010; Dutertre et al., 2010; Schmidt et al., 2011). For example, it has been demonstrated that slow transcription of the fibronectin gene (FN1) leads to the inclusion of the fibronectin extra domain 1 (ED1) exon which is preceded by a weak 3′ splice site. However, when transcription elongation rate was higher, this exon was excluded (Kadener, Fededa, Rosbash, & Kornblihtt, 2002).

More recent global studies have revealed a pausing of RNA pol II near the 3′ end of intron-containing genes. It is suggested that the pausing represents a check point to allow the splicing machinery to cope with transcription (Alexander, Innocente, Barrass, & Beggs, 2010). There is also increasing evidence that chromatin status and histone modifications also play a key role in alternative splicing regulation (Allo et al., 2010; Luco, Allo, Schor, Kornblihtt, & Misteli, 2011). The outcome of alternative splicing might also be affected by other posttranscriptional mechanisms such as RNA editing, mRNA decay, and microRNA binding (Graveley, 2009; Hughes, 2006; Luco & Misteli, 2011).

7. COTRANSCRIPTIONAL AND POSTTRANSCRIPTIONAL SPLICING

Transcription and pre-mRNA splicing are performed by different molecular complexes and were initially considered as two independent and separate processes. However, over the past decade, there has been an accumulation of evidence indicating that transcription and splicing are tightly coupled processes, and that splicing frequently occurs cotranscriptionally. Cotranscriptional splicing means that the splicing machinery is active, while the nascent RNA molecule is still being produced by the elongating RNA pol II, prior to the initiation of the polyadenylation process. Cotranscriptional splicing therefore occurs while the pre-mRNA is still bound to the chromatin. An illustration of cotranscriptional splicing is depicted in Fig. 5.4. Initial support for cotranscriptional splicing came from experiments using electron microscopy of *D. melanogaster* embryonic transcription units, visualizing lariat intron formation and associated ribonucleoprotein splicing complexes on transcripts still attached to the DNA (Beyer & Osheim, 1988). Additional evidence came from studies using immunofluorescent microscopy, providing proof of spliceosome assembly at sites of active transcription (Misteli, Caceres, & Spector,

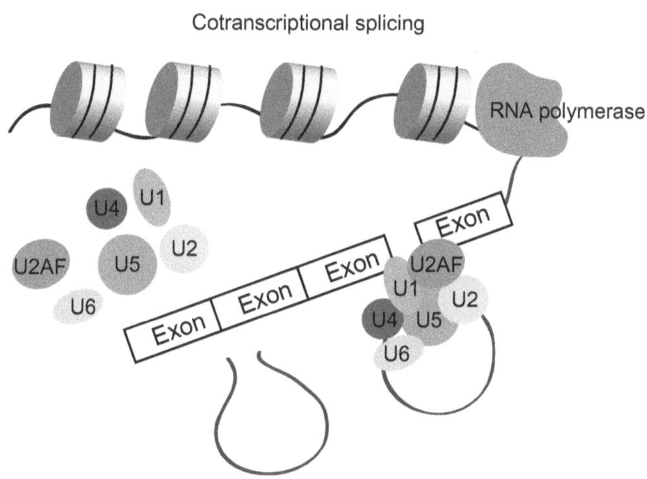

Figure 5.4 Cotranscriptional splicing. As the newly synthesized RNA emerges from RNA pol II, the splicing factors are recruited to the nascent transcript and the introns are excised prior to completion of the transcription process.

1997; Neugebauer & Roth, 1997). RNA-FISH targeted to exon–exon junctions also showed partially spliced mRNA located at their DNA template, indicating that pre-mRNA splicing occurs prior to transcript release (Zhang, Taneja, Singer, & Green, 1994). Further studies identified spliced mRNA, spliceosome components, and splicing factors associated with chromatin at transcriptionally active genes (Kotovic, Lockshon, Boric, & Neugebauer, 2003; Lacadie & Rosbash, 2005; Listerman, Sapra, & Neugebauer, 2006; Pandya-Jones & Black, 2009). The current consensus is that splicing primarily occurs cotranscriptionally, where introns are removed prior to completion of transcription, as opposed to posttranscriptional splicing, where introns are removed from a fully transcribed template. However, the data show evidence for both processes occurring in human cells, and although several recent studies support cotranscriptional splicing as a general phenomenon (Ameur et al., 2011; Khodor et al., 2011; Tilgner et al., 2012), there is also clear evidence that splicing can occur posttranscriptionally (Bauren & Wieslander, 1994; Brody et al., 2011; Wetterberg, Bauren, & Wieslander, 1996), and that cotranscriptional splicing rates can vary within individual genes (Tilgner et al., 2012). There are also some clear examples of how posttranscriptional splicing can provide an efficient way to regulate nuclear export and translation of transcripts (Boothby, Zipper, van der Weele, & Wolniak, 2013; Denis et al., 2005).

Analyses of splicing patterns in chromatin-associated and chromatin-free RNA in the nucleoplasm of human cells reveal that introns adjacent to constitutive exons are mostly cotranscriptionally spliced and that these introns are typically removed in a 5′–3′ order. Meanwhile, introns close to the 3′ end of transcripts are more likely to be posttranscriptionally excised (Pandya-Jones & Black, 2009; Tilgner et al., 2012). Internal introns harboring alternative exons are also removed during transcription, but with variable efficiencies (de la Mata et al., 2010; Pandya-Jones & Black, 2009). These observations were recently validated using single-molecule imaging of transcriptionally coupled and uncoupled splicing (Vargas et al., 2011). These data, in addition to the fact that cotranscriptional recruitment of SR proteins increases splicing efficiency (Das et al., 2007), suggest a regulatory potential of cotranscriptional splicing. The kinetic coupling between transcription and splicing has been shown by several studies. Analysis of chromatin-associated nascent transcripts shows that RNA pol II pauses at terminal exons (Carrillo Oesterreich et al., 2011). Terminal exons are typically short, and the authors therefore suggest that polymerase pausing occurs to allow sufficient time for intron removal before transcript release. Another study shows a general RNA pol II pausing at the 3′ end of introns in reporter genes, coinciding with splicing factor recruitment (Alexander et al., 2010). All these indirect observations indicate that cotranscriptional splicing represents the rule rather than the exception. As shown below, this assumption is now supported by direct estimates of global analyses of cotranscriptional splicing.

8. GLOBAL ANALYSIS OF Pre-mRNA SPLICING

With the ability to perform global studies of the human transcriptome, it has become clear that there is an enormous diversity of alternative transcripts generated by alternative splicing. The first global studies using arrays designed to detect splicing events gave the first insight into this diversity (Calarco, Saltzman, Ip, & Blencowe, 2007; Pan et al., 2004). The introduction of RNA-seq, representing an even more unbiased view of the RNA in the cell, has further expanded the repertoire of alternative splice variants and as well as alternative starting and end sites of transcripts. At the same time, deep sequencing studies using targeted enrichment of specific transcripts have shown that the transcript diversity detected in a typical RNA-seq experiment is far from complete and that additional alternative splice variants are present at very low levels and can only be detected using extremely deep sequencing (Halvardson, Zaghlool, & Feuk, 2013; Mercer et al., 2012). However, the goal of a global RNA-seq experiment is to reconstruct the best possible representation of the transcripts present in the sequenced sample, i.e., to estimate both the qualitative (which transcripts and isoforms) and quantitative (expression levels) presence of RNA molecules, normalized in a way that samples can be compared. The relative importance of gene expression levels and different transcript isoforms to both function and total transcript abundance in cells has long been a topic of discussion. It has been shown that differential splicing and differential expression may be used for regulation of different biological processes (Dittmar et al., 2012) and that genes regulated by alternative splicing may often be different from genes regulated by expression level differences (Pan et al., 2004). However, recent data from ENCODE indicate that variability of gene expression contributes more than variability of splicing ratios to the variability in transcript abundance across cell lines (Djebali et al., 2012; Gonzalez-Porta, Calvo, Sammeth, & Guigo, 2012). It is not yet clear whether the same is true for different cell types in the human brain. It is important to consider that there are many regulatory processes in the cell exerting their effect at the level of RNA, which will affect the mRNA transcripts that are eventually used for translation. In general, the difficulty of obtaining high-quality brain tissue, and the limited possibilities to perform *in vivo* studies in brain, means that we have a relatively limited understanding of the RNA processing regulation in brain as compared to other tissues. At the same time, the brain seems to have more regulation at the level of RNA, with more genes expressed, more splice isoforms, and more RNA editing, compared with other human tissues. In order to describe the transcriptome, it is therefore important to consider how the RNA to be sequenced is extracted. Furthermore, the computational analysis of the resulting RNA-seq data for detection of transcript isoforms will also have an effect on the final results. Overall, the main experimental steps that will influence the outcome of RNA-seq analyses are the way the samples are prepared, how the RNA is sequenced, and how the resulting data are analyzed. Here, we will go into further detail to describe different steps in the preparation of RNA and analysis of the data.

9. THE INFLUENCE OF RNA EXTRACTION METHODS ON TRANSCRIPTOME ANALYSIS

Preparation of RNA for downstream analysis is a crucial decision that will have important implications for the results of any analysis performed, whether that is based on PCR, targeted sequencing, or global RNA sequencing. The RNA preparation should be guided by what hypothesis to test or what question to address. Since RNA goes through multiple processing steps, a cell contains a mixture of RNA molecules at different levels of maturation. This implies that analysis of, e.g., splicing and alternative transcripts will give different results depending on what RNA population is investigated. To attain a higher-resolution view of specific pools of RNA within the cell, a number of different protocols have been developed to fractionate the cell or to extract RNA associated with certain subcellular structures. The most common strategy for RNA-seq is to isolate the polyadenylated fraction of RNA, representing mature transcripts that have gone through capping and splicing. Most of the polyadenylated RNA molecules will be present in the cytosol, but some will remain in the nucleus awaiting transport. Normally, poly-A is selected by capture with synthesized poly-T oligonucleotides in a column or by magnetic beads. If there is very limited material, e.g., single cell sequencing, the poly-A RNA is instead amplified using a poly-T primer (Picelli et al., 2014). This is the most commonly used approach as it focuses on mRNA that is mature and therefore regarded as representative of the transcripts that will be used for translation, i.e., better reflecting the protein production. However, it is clear that the situation is more complex as many mRNAs are degraded or inhibited from being translated (Valencia-Sanchez, Liu, Hannon, & Parker, 2006), and the correlation between mRNA and protein is therefore far from perfect (Guo et al., 2008). Sequencing of poly-A RNA also simplifies analysis of splicing events and alternative transcripts as the intronic sequences have been removed.

Since not all RNA molecules in a cell are polyadenylated, an alternative approach that is commonly used is to sequence total RNA. With this approach, there is less bias in terms of which RNA molecules are included, except that most protocols, either at the stage of extraction or at the stage of library preparation, will lead to exclusion of small RNAs. The extraction of total RNA will also require an additional step to remove ribosomal RNA prior to RNA-seq, something that is not necessary in poly-A based RNA-seq. Two major reasons make total RNA sequencing less attractive for RNA-seq. First, there are significant amounts of pre-mRNA and nascent RNA molecules present in the cell, covering a large fraction of the genome. This means that a large fraction of sequence reads will map to intronic regions, making the approach less attractive for expression analysis (Ameur et al., 2011). It is also challenging to identify splice junctions in total RNA-seq data, as only a minor fraction of sequence reads will originate from mature transcripts. Total RNA-seq is therefore excellent for providing a global view of all the RNA present in the cell and works well for analysis of nascent transcripts but is less suitable for analysis of differential expression levels and alternative transcript detection.

In addition to the standard protocols of poly-A RNA-seq and total RNA-seq, several protocols have been developed to gain additional insight into specific RNA populations in the cell. One commonly used approach is to divide the nuclear and cytosolic RNA fractions and sequence them separately. The nuclear RNA will then primarily represent nascent transcripts and the cytosolic fraction will primarily contain the mRNA. The cytosolic fraction will also contain most of the rRNA, making it necessary to perform either rRNA depletion or poly-A selection of the cytosolic fraction to study mRNA. One challenge of using poly-A sequencing that is circumvented using cytosolic RNA-seq is that poly-A selection is not perfect, and there may exist a background of pre-mRNA leading to noise in the analysis. This background is especially pronounced in poly-A RNA-seq data from brain samples (Zaghlool et al., 2013). The reason this problem is more apparent in the brain is that many neuronal genes are very long, with long introns, and have very high levels of expression in the brain. The ratio of total RNA to mRNA for these transcripts is therefore very high, and even a low total level of background in poly-A selection may lead to significant amounts of sequence reads mapping to introns in the final results. An extreme example of this is shown in Fig. 5.5, which also shows how extraction of cytosolic RNA diminishes this problem. Another factor that differentiates poly-A selected and cytosolic RNA is that some introns seem to be spliced only after polyadenylation is completed (Bhatt et al., 2012). These introns will then be included in poly-A selected RNA, but will not be included in the cytosolic RNA fraction. Separating the nuclear fractions is excellent for studies of splicing dynamics and provides a view of which transcripts that are fully processed and exported from the nucleus (Bhatt et al., 2012; Tilgner et al., 2012).

Further refinement of subcellular populations of RNA may be performed by isolation of specific protein complexes, with subsequent extraction of RNA molecules associated with them. An approach to study RNA that is undergoing transcription or maturation is to extract RNA associated with the chromatin. Once the splicing and polyadenylation processes are completed, the mRNA will be released from the chromatin for export out of the nucleus. By studying both the chromatin-bound and nonchromatin-bound nuclear RNA, it is possible to gain further insight into the RNA processing and how it is linked to nuclear export (Djebali et al., 2012). The analysis of chromatin-bound RNA is also an excellent approach for investigation of the splicing process, as was shown in the ENCODE project (Pandya-Jones et al., 2013; Tilgner et al., 2012).

Once the RNA is exported to the cytosol, there are still several processes that will regulate which RNA transcripts that are eventually used for translation, including regulation by microRNAs, NMD, and differential ribosome recruitment. In order to facilitate the analysis of the subset of transcripts undergoing translation, methods have been developed to isolate the ribosome-associated RNA (Masek, Valasek, & Pospisek, 2011). By comparing the ribosome-associated RNA with the remaining cytosolic RNA, it is been determined that splicing has important effects also for the transcripts that are actually translated. Using an approach called Frac-seq (for subcellular fractionation and high-throughput RNA sequencing), Sterne-Weiler and colleagues partitioned the mRNA isoforms between the cytosolic and the polyribosome-associated fractions and were able to show that 30% of alternative splicing events exhibit isoform-specific polyribosome associations (Sterne-Weiler et al., 2013).

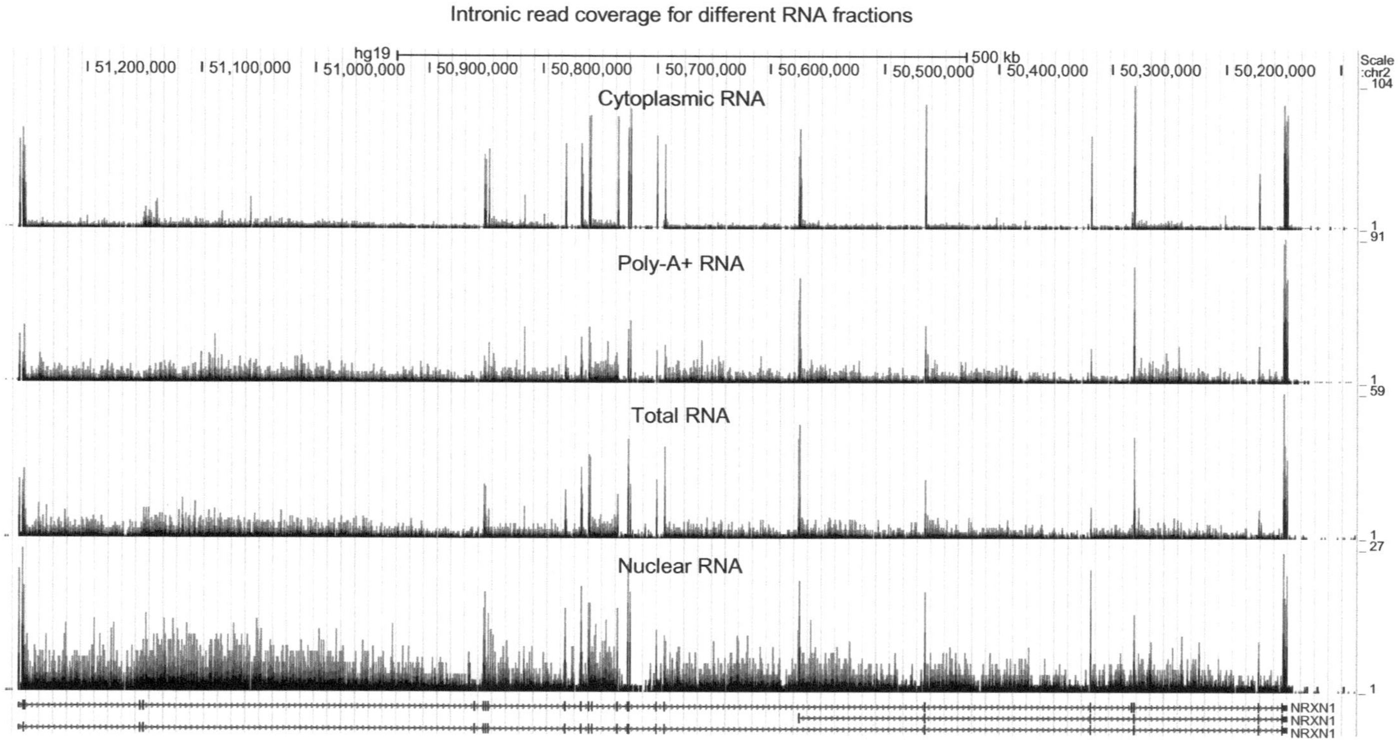

Figure 5.5 Intronic read coverage for different RNA fractions. The figure shows RNA-seq coverage for fetal frontal cortex across the gene *NRXN1* for four different RNA fractions (cytoplasmic RNA, poly-A+ RNA, total RNA, and nuclear RNA) viewed in the UCSC genome browser. Long neuronal genes tend to yield a high coverage in introns in conventional RNA-seq data. The figure shows that fractions except the cytoplasmic RNA show a high coverage across the entire transcript, including the introns (Zaghlool et al., 2013).

While the techniques described above will work also in tissue samples, including brain, there are additional strategies that can be used only in model systems and cell lines. These include, for example, protocols where modified nucleotides are introduced to perform time course studies of RNA transcription rates (Marras, Gold, Kramer, Smith, & Tyagi, 2004; Singh & Padgett, 2009), RNA splicing rates (Singh & Padgett, 2009), and RNA turnover (Munchel, Shultzaberger, Takizawa, & Weis, 2011). One way to study nascent transcripts is called native elongating transcript sequencing which relies on the stability of the DNA–RNA–RNAP ternary complex. By expressing a FLAG-tagged RNA polymerase II in yeast, the authors used an antibody to pull down the polymerase with attached nascent RNA molecules, which could then be sequenced (Churchman & Weissman, 2011). Another approach limited to cell lines is global run-on sequencing combined with a next-generation sequencing readout, which is a way to specifically sequence actively transcribed genes in a genome (Core, Waterfall, & Lis, 2008).

10. COMPUTATIONAL METHODS TO STUDY SPLICING DYNAMICS

Analyses of data from RNA-seq experiments enable us to investigate RNA processing and splicing on a global scale. A typical RNA-seq experiment generates several million reads of lengths around 100–200 bp, and these reads represent the diversity of RNA molecules present in the sample. The majority of computational tools developed for RNA-seq data are focused on analyzing the read densities over exons, aiming to quantify expression levels of genes and splice isoforms. However, instead of analyzing read densities over the introns, it is possible to also obtain information on the levels of nascent transcription and cotranscriptional splicing from the same experiment. The foundation for such analyses of intronic RNAs is the observation that nascent transcripts in combination with cotranscriptional splicing give rise to a 5′–3′ slope across each intron, generating a "saw-tooth"-like pattern over the full-length transcript (Ameur et al., 2011; Fig. 5.6). These 5′–3′ slopes are more apparent for longer introns, since they have more time to accumulate nascent transcripts at the 5′ ends of introns before they are spliced out. Since many of the highly expressed genes in the human brain contain a large number of exons separated by very long introns, this type of analysis is particularly suitable for studying nascent transcript formation in samples originating from the brain. Consistent with this, the percentage

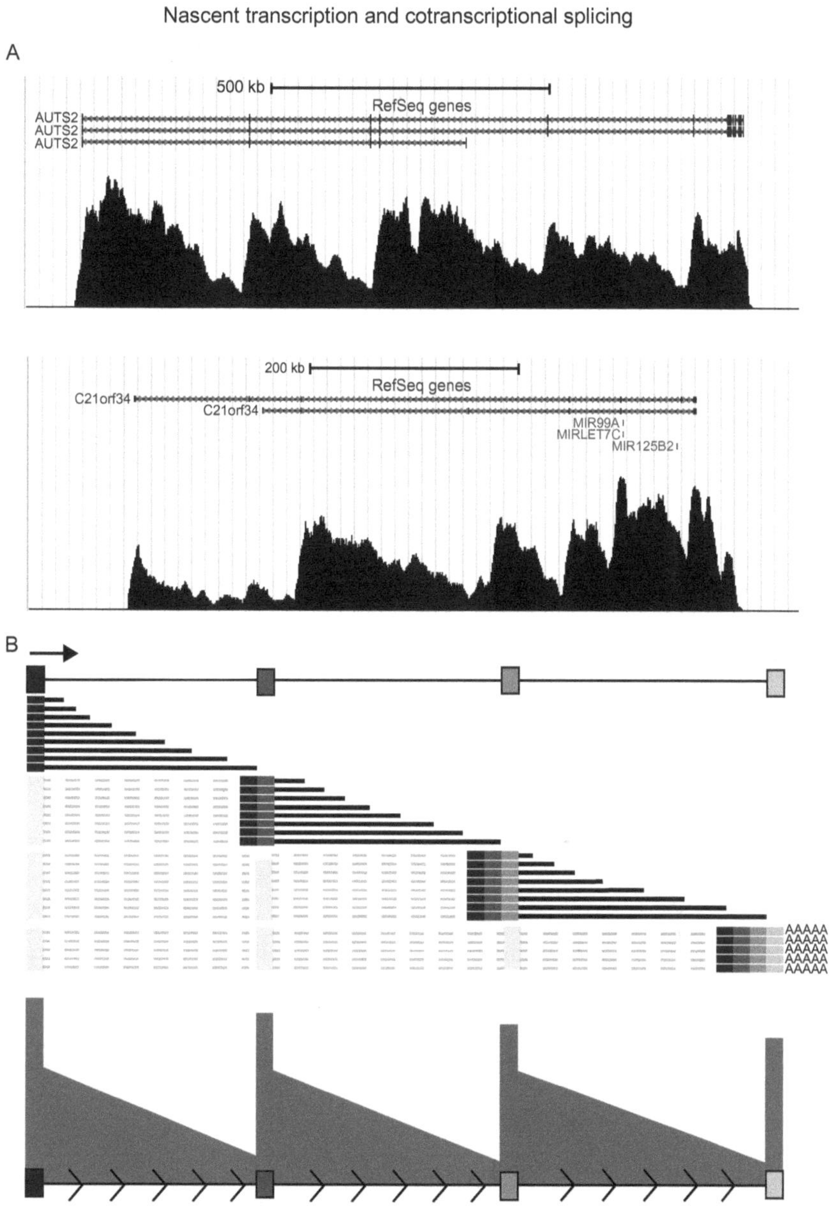

Figure 5.6 Nascent transcription and cotranscriptional splicing. (A) RNA-seq coverage pattern for *AUTS2* (top) and *C21orf34*, a noncoding RNA gene (bottom), viewed in the University of California, Santa Cruz (UCSC) Genome Browser. The RNA-seq signals have been smoothed using window averaging. For both protein-coding genes and long non-coding RNA genes, there is an apparent "saw-tooth" pattern with higher RNA-seq read signals toward the 5′ end of each intron. (B) Model for cotranscriptional splicing. The

(*Continued*)

of intronic RNA-seq reads is higher in brain samples compared to liver (Ameur et al., 2011). It is important to note that this does not necessarily mean that brain tissue contains a higher proportion of nascent transcription and cotranscriptional splicing as compared to other tissues. Rather, it is just a reflection that the brain is actively expressing genes with long introns, which makes these processes easier to study by RNA-seq in that particular tissue.

Recently, a number of computational methods have been developed for studying cotranscriptional splicing rates by analysis of RNA-seq data (Brugiolo, Herzel, & Neugebauer, 2013). What is common between all these methods is that they are making use of intronic read coverage to quantify the levels of cotranscriptional splicing. The cotranscriptional splicing rates can be calculated for each individual exon in the genome, either by calculating a score based on the coverage in intronic regions surrounding each exon (Ameur et al., 2011) or by also taking into account the reads that are mapping to exon junctions and borders (Tilgner et al., 2012). The estimation of the frequency of cotranscriptional splicing in human cells determined by these global computational methods is similar in several independent studies using different analytical approaches, all arriving at a cotranscriptional splicing frequency of 0.75–0.85 (Brugiolo et al., 2013). This frequency is further supported by analysis in other eukaryotes (Carrillo Oesterreich, Preibisch, & Neugebauer, 2010). Interestingly, several of these studies indicate that constitutive splicing is more cotranscriptional than alternative splicing.

Potentially, the intronic reads from total RNA-seq experiments can be used also for other purposes than just for the study of cotranscriptional splicing. Methods are being developed to also investigate other transcriptional events by exploiting the information that is contained in the intronic RNA-seq data, including rates of pre-mRNA synthesis, intron degradation, and mRNA decay (Gray et al., 2014). These new computational methods could make it possible to study the rates of ongoing transcription and splicing

Figure 5.6—Cont'd total RNA-seq data give rise to a typical saw-tooth pattern across genes that are actively transcribed. The gradient of RNA across the introns can be explained by a large number of nascent transcripts in various stages of completion. The pattern is repeated for each intron because the nascent transcript is spliced very rapidly after the polymerase completes transcribing each intron. The sequence read coverage is comparatively higher for exons, as the RNA-seq is measuring both the pool of nascent transcripts and the pool of mature polyadenylated RNA. *Figure adapted from Ameur et al. (2011).*

in relation to other RNA processing dynamics by a single RNA-seq experiment.

Similar computational strategies can also be used to study RNA-binding proteins that bind to intronic sequences and regulate splicing. By CLIP-seq experiments, the binding profiles of FUS and TDP-43, two RNA-binding proteins linked to amyotrophic lateral sclerosis and frontotemporal lobar degeneration, were recently investigated in mouse and human brain (Lagier-Tourenne et al., 2012; Rogelj et al., 2012). In both cases, a saw-tooth pattern was observed over FUS-regulated genes, indicating that FUS is binding to intronic sequences within the newly formed transcript and remains bound until cotranscriptional splicing of the intron occurs. TDP-43, on the other hand, was mainly bound to other targets and did not show any clear evidence of cotranscriptional deposition.

As new techniques are constantly developed, we will certainly see more applications that can be targeted also to increase our understanding of the complex brain transcriptome. The revolution in single cell biology will make it possible to obtain cells from patients' brains as a side product from other medical procedures where thin needle biopsies are utilized. The advances in long-read sequencing will also enhance the ability to sequence across intron–exon boundaries and will provide better insight into alternative transcript variants and splicing mechanisms. We are at the start of very exciting times in research of the human brain transcriptome science.

REFERENCES

Alexander, R. D., Innocente, S. A., Barrass, J. D., & Beggs, J. D. (2010). Splicing-dependent RNA polymerase pausing in yeast. *Molecular Cell*, *40*(4), 582–593.

Allo, M., Schor, I. E., Munoz, M. J., de la Mata, M., Agirre, E., Valcarcel, J., et al. (2010). Chromatin and alternative splicing. *Cold Spring Harbor Symposia on Quantitative Biology*, *75*, 103–111.

Alonso, M. T., & Garcia-Sancho, J. (2011). Nuclear Ca(2+) signalling. *Cell Calcium*, *49*(5), 280–289.

Ameur, A., Zaghlool, A., Halvardson, J., Wetterbom, A., Gyllensten, U., Cavelier, L., et al. (2011). Total RNA sequencing reveals nascent transcription and widespread co-transcriptional splicing in the human brain. *Nature Structural & Molecular Biology*, *18*(12), 1435–1440.

Aoto, J., Martinelli, D. C., Malenka, R. C., Tabuchi, K., & Sudhof, T. C. (2013). Presynaptic neurexin-3 alternative splicing trans-synaptically controls postsynaptic AMPA receptor trafficking. *Cell*, *154*(1), 75–88.

Ares, M., Jr., Grate, L., & Pauling, M. H. (1999). A handful of intron-containing genes produces the lion's share of yeast mRNA. *RNA*, *5*(9), 1138–1139.

Bauren, G., & Wieslander, L. (1994). Splicing of Balbiani ring 1 gene pre-mRNA occurs simultaneously with transcription. *Cell*, *76*(1), 183–192.

Beyer, A. L., & Osheim, Y. N. (1988). Splice site selection, rate of splicing, and alternative splicing on nascent transcripts. *Genes & Development*, *2*(6), 754–765.

Bhatt, D. M., Pandya-Jones, A., Tong, A. J., Barozzi, I., Lissner, M. M., Natoli, G., et al. (2012). Transcript dynamics of proinflammatory genes revealed by sequence analysis of subcellular RNA fractions. *Cell*, *150*(2), 279–290.

Black, D. L. (2000). Protein diversity from alternative splicing: A challenge for bioinformatics and post-genome biology. *Cell*, *103*(3), 367–370.

Black, D. L. (2003). Mechanisms of alternative pre-messenger RNA splicing. *Annual Review of Biochemistry*, *72*, 291–336.

Boothby, T. C., Zipper, R. S., van der Weele, C. M., & Wolniak, S. M. (2013). Removal of retained introns regulates translation in the rapidly developing gametophyte of Marsilea vestita. *Developmental Cell*, *24*(5), 517–529.

Boucard, A. A., Chubykin, A. A., Comoletti, D., Taylor, P., & Sudhof, T. C. (2005). A splice code for trans-synaptic cell adhesion mediated by binding of neuroligin 1 to alpha- and beta-neurexins. *Neuron*, *48*(2), 229–236.

Boutz, P. L., Stoilov, P., Li, Q., Lin, C. H., Chawla, G., Ostrow, K., et al. (2007). A post-transcriptional regulatory switch in polypyrimidine tract-binding proteins reprograms alternative splicing in developing neurons. *Genes & Development*, *21*(13), 1636–1652.

Brody, Y., Neufeld, N., Bieberstein, N., Causse, S. Z., Bohnlein, E. M., Neugebauer, K. M., et al. (2011). The in vivo kinetics of RNA polymerase II elongation during co-transcriptional splicing. *PLoS Biology*, *9*(1), e1000573.

Brugiolo, M., Herzel, L., & Neugebauer, K. M. (2013). Counting on co-transcriptional splicing. *F1000Prime Reports*, *5*, 9.

Buckanovich, R. J., & Darnell, R. B. (1997). The neuronal RNA binding protein Nova-1 recognizes specific RNA targets in vitro and in vivo. *Molecular and Cellular Biology*, *17*(6), 3194–3201.

Buckanovich, R. J., Posner, J. B., & Darnell, R. B. (1993). Nova, the paraneoplastic Ri antigen, is homologous to an RNA-binding protein and is specifically expressed in the developing motor system. *Neuron*, *11*(4), 657–672.

Buj-Bello, A., Furling, D., Tronchere, H., Laporte, J., Lerouge, T., Butler-Browne, G. S., et al. (2002). Muscle-specific alternative splicing of myotubularin-related 1 gene is impaired in DM1 muscle cells. *Human Molecular Genetics*, *11*(19), 2297–2307.

Calarco, J. A., Saltzman, A. L., Ip, J. Y., & Blencowe, B. J. (2007). Technologies for the global discovery and analysis of alternative splicing. *Advances in Experimental Medicine and Biology*, *623*, 64–84.

Carrillo Oesterreich, F., Bieberstein, N., & Neugebauer, K. M. (2011). Pause locally, splice globally. *Trends in Cell Biology*, *21*(6), 328–335.

Carrillo Oesterreich, F., Preibisch, S., & Neugebauer, K. M. (2010). Global analysis of nascent RNA reveals transcriptional pausing in terminal exons. *Molecular Cell*, *40*(4), 571–581.

Chen, M., & Manley, J. L. (2009). Mechanisms of alternative splicing regulation: Insights from molecular and genomics approaches. *Nature Reviews. Molecular Cell Biology*, *10*(11), 741–754.

Chih, B., Gollan, L., & Scheiffele, P. (2006). Alternative splicing controls selective trans-synaptic interactions of the neuroligin-neurexin complex. *Neuron*, *51*(2), 171–178.

Churchman, L. S., & Weissman, J. S. (2011). Nascent transcript sequencing visualizes transcription at nucleotide resolution. *Nature*, *469*(7330), 368–373.

Core, L. J., Waterfall, J. J., & Lis, J. T. (2008). Nascent RNA sequencing reveals widespread pausing and divergent initiation at human promoters. *Science*, *322*(5909), 1845–1848.

Cramer, P., Caceres, J. F., Cazalla, D., Kadener, S., Muro, A. F., Baralle, F. E., et al. (1999). Coupling of transcription with alternative splicing: RNA pol II promoters modulate SF2/ASF and 9G8 effects on an exonic splicing enhancer. *Molecular Cell*, *4*(2), 251–258.

Das, R., Yu, J., Zhang, Z., Gygi, M. P., Krainer, A. R., Gygi, S. P., et al. (2007). SR proteins function in coupling RNAP II transcription to pre-mRNA splicing. *Molecular Cell, 26*(6), 867–881.

de la Grange, P., Gratadou, L., Delord, M., Dutertre, M., & Auboeuf, D. (2010). Splicing factor and exon profiling across human tissues. *Nucleic Acids Research, 38*(9), 2825–2838.

de la Mata, M., Alonso, C. R., Kadener, S., Fededa, J. P., Blaustein, M., Pelisch, F., et al. (2003). A slow RNA polymerase II affects alternative splicing in vivo. *Molecular Cell, 12*(2), 525–532.

de la Mata, M., Lafaille, C., & Kornblihtt, A. R. (2010). First come, first served revisited: Factors affecting the same alternative splicing event have different effects on the relative rates of intron removal. *RNA, 16*(5), 904–912.

Del Gatto-Konczak, F., Olive, M., Gesnel, M. C., & Breathnach, R. (1999). hnRNP A1 recruited to an exon in vivo can function as an exon splicing silencer. *Molecular and Cellular Biology, 19*(1), 251–260.

Denis, M. M., Tolley, N. D., Bunting, M., Schwertz, H., Jiang, H., Lindemann, S., et al. (2005). Escaping the nuclear confines: Signal-dependent pre-mRNA splicing in anucleate platelets. *Cell, 122*(3), 379–391.

Dittmar, K. A., Jiang, P., Park, J. W., Amirikian, K., Wan, J., Shen, S., et al. (2012). Genome-wide determination of a broad ESRP-regulated posttranscriptional network by high-throughput sequencing. *Molecular and Cellular Biology, 32*(8), 1468–1482.

Djebali, S., Davis, C. A., Merkel, A., Dobin, A., Lassmann, T., Mortazavi, A., et al. (2012). Landscape of transcription in human cells. *Nature, 489*(7414), 101–108.

Doyle, M., & Kiebler, M. A. (2011). Mechanisms of dendritic mRNA transport and its role in synaptic tagging. *EMBO Journal, 30*(17), 3540–3552.

Dredge, B. K., & Darnell, R. B. (2003). Nova regulates GABA(A) receptor gamma2 alternative splicing via a distal downstream UCAU-rich intronic splicing enhancer. *Molecular and Cellular Biology, 23*(13), 4687–4700.

Dredge, B. K., Stefani, G., Engelhard, C. C., & Darnell, R. B. (2005). Nova autoregulation reveals dual functions in neuronal splicing. *EMBO Journal, 24*(8), 1608–1620.

Dutertre, M., Sanchez, G., De Cian, M. C., Barbier, J., Dardenne, E., Gratadou, L., et al. (2010). Cotranscriptional exon skipping in the genotoxic stress response. *Nature Structural & Molecular Biology, 17*(11), 1358–1366.

Ellis, J. D., Barrios-Rodiles, M., Colak, R., Irimia, M., Kim, T., Calarco, J. A., et al. (2012). Tissue-specific alternative splicing remodels protein-protein interaction networks. *Molecular Cell, 46*(6), 884–892.

Eom, T., Zhang, C., Wang, H., Lay, K., Fak, J., Noebels, J. L., et al. (2013). NOVA-dependent regulation of cryptic NMD exons controls synaptic protein levels after seizure. *Elife, 2*, e00178.

Flicek, P., Ahmed, I., Amode, M. R., Barrell, D., Beal, K., Brent, S., et al. (2013). Ensembl 2013. *Nucleic Acids Research, 41*(Database issue), D48–D55.

Fogel, B. L., Wexler, E., Wahnich, A., Friedrich, T., Vijayendran, C., Gao, F., et al. (2012). RBFOX1 regulates both splicing and transcriptional networks in human neuronal development. *Human Molecular Genetics, 21*(19), 4171–4186.

Garcia-Blanco, M. A., Jamison, S. F., & Sharp, P. A. (1989). Identification and purification of a 62,000-dalton protein that binds specifically to the polypyrimidine tract of introns. *Genes & Development, 3*(12A), 1874–1886.

Gonzalez-Porta, M., Calvo, M., Sammeth, M., & Guigo, R. (2012). Estimation of alternative splicing variability in human populations. *Genome Research, 22*(3), 528–538.

Grabowski, P. J., & Black, D. L. (2001). Alternative RNA splicing in the nervous system. *Progress in Neurobiology, 65*(3), 289–308.

Graf, E. R., Kang, Y., Hauner, A. M., & Craig, A. M. (2006). Structure function and splice site analysis of the synaptogenic activity of the neurexin-1 beta LNS domain. *Journal of Neuroscience, 26*(16), 4256–4265.

Graveley, B. R. (2000). Sorting out the complexity of SR protein functions. *RNA, 6*(9), 1197–1211.

Graveley, B. R. (2001). Alternative splicing: Increasing diversity in the proteomic world. *Trends in Genetics, 17*(2), 100–107.

Graveley, B. R. (2009). Alternative splicing: Regulation without regulators. *Nature Structural & Molecular Biology, 16*(1), 13–15.

Gray, J. M., Harmin, D. A., Boswell, S. A., Cloonan, N., Mullen, T. E., Ling, J. J., et al. (2014). SnapShot-Seq: A method for extracting genome-wide, in vivo mRNA dynamics from a single total RNA sample. *PLoS One, 9*(2), e89673.

Grover, A., Houlden, H., Baker, M., Adamson, J., Lewis, J., Prihar, G., et al. (1999). 5′ splice site mutations in tau associated with the inherited dementia FTDP-17 affect a stem-loop structure that regulates alternative splicing of exon 10. *The Journal of Biological Chemistry, 274*(21), 15134–15143.

Guo, Y., Xiao, P., Lei, S., Deng, F., Xiao, G. G., Liu, Y., et al. (2008). How is mRNA expression predictive for protein expression? A correlation study on human circulating monocytes. *Acta Biochimica et Biophysica Sinica, 40*(5), 426–436.

Halvardson, J., Zaghlool, A., & Feuk, L. (2013). Exome RNA sequencing reveals rare and novel alternative transcripts. *Nucleic Acids Research, 41*(1), e6.

Han, K., Yeo, G., An, P., Burge, C. B., & Grabowski, P. J. (2005). A combinatorial code for splicing silencing: UAGG and GGGG motifs. *PLoS Biology, 3*(5), e158.

Harrow, J., Frankish, A., Gonzalez, J. M., Tapanari, E., Diekhans, M., Kokocinski, F., et al. (2012). GENCODE: The reference human genome annotation for The ENCODE Project. *Genome Research, 22*(9), 1760–1774.

Hawrylycz, M. J., Lein, E. S., Guillozet-Bongaarts, A. L., Shen, E. H., Ng, L., Miller, J. A., et al. (2012). An anatomically comprehensive atlas of the adult human brain transcriptome. *Nature, 489*(7416), 391–399.

Howe, K. J., Kane, C. M., & Ares, M., Jr. (2003). Perturbation of transcription elongation influences the fidelity of internal exon inclusion in *Saccharomyces cerevisiae*. *RNA, 9*(8), 993–1006.

Hughes, T. A. (2006). Regulation of gene expression by alternative untranslated regions. *Trends in Genetics, 22*(3), 119–122.

Ichtchenko, K., Hata, Y., Nguyen, T., Ullrich, B., Missler, M., Moomaw, C., et al. (1995). Neuroligin 1: A splice site-specific ligand for beta-neurexins. *Cell, 81*(3), 435–443.

International Human Genome Sequencing Consortium. (2004). Finishing the euchromatic sequence of the human genome. *Nature, 431*(7011), 931–945.

Irimia, M., & Blencowe, B. J. (2012). Alternative splicing: Decoding an expansive regulatory layer. *Current Opinion in Cell Biology, 24*(3), 323–332.

Jensen, K. B., Dredge, B. K., Stefani, G., Zhong, R., Buckanovich, R. J., Okano, H. J., et al. (2000). Nova-1 regulates neuron-specific alternative splicing and is essential for neuronal viability. *Neuron, 25*(2), 359–371.

Kadener, S., Fededa, J. P., Rosbash, M., & Kornblihtt, A. R. (2002). Regulation of alternative splicing by a transcriptional enhancer through RNA pol II elongation. *Proceedings of the National Academy of Sciences of the United States of America, 99*(12), 8185–8190.

Kang, H. J., Kawasawa, Y. I., Cheng, F., Zhu, Y., Xu, X., Li, M., et al. (2011). Spatio-temporal transcriptome of the human brain. *Nature, 478*(7370), 483–489.

Kelemen, O., Convertini, P., Zhang, Z., Wen, Y., Shen, M., Falaleeva, M., et al. (2013). Function of alternative splicing. *Gene, 514*(1), 1–30.

Khodor, Y. L., Rodriguez, J., Abruzzi, K. C., Tang, C. H., Marr, M. T., 2nd., & Rosbash, M. (2011). Nascent-seq indicates widespread cotranscriptional pre-mRNA splicing in Drosophila. *Genes & Development, 25*(23), 2502–2512.

Kornblihtt, A. R. (2006). Chromatin, transcript elongation and alternative splicing. *Nature Structural & Molecular Biology, 13*(1), 5–7.

Kornblihtt, A. R., Schor, I. E., Allo, M., Dujardin, G., Petrillo, E., & Munoz, M. J. (2013). Alternative splicing: A pivotal step between eukaryotic transcription and translation. *Nature Reviews. Molecular Cell Biology, 14*(3), 153–165.

Kotovic, K. M., Lockshon, D., Boric, L., & Neugebauer, K. M. (2003). Cotranscriptional recruitment of the U1 snRNP to intron-containing genes in yeast. *Molecular and Cellular Biology, 23*(16), 5768–5779.

Kusek, G., Campbell, M., Doyle, F., Tenenbaum, S. A., Kiebler, M., & Temple, S. (2012). Asymmetric segregation of the double-stranded RNA binding protein Staufen2 during mammalian neural stem cell divisions promotes lineage progression. *Cell Stem Cell, 11*(4), 505–516.

Lacadie, S. A., & Rosbash, M. (2005). Cotranscriptional spliceosome assembly dynamics and the role of U1 snRNA:5′ss base pairing in yeast. *Molecular Cell, 19*(1), 65–75.

Lagier-Tourenne, C., Polymenidou, M., Hutt, K. R., Vu, A. Q., Baughn, M., Huelga, S. C., et al. (2012). Divergent roles of ALS-linked proteins FUS/TLS and TDP-43 intersect in processing long pre-mRNAs. *Nature Neuroscience, 15*(11), 1488–1497.

Lee, J. A., Xing, Y., Nguyen, D., Xie, J., Lee, C. J., & Black, D. L. (2007). Depolarization and CaM kinase IV modulate NMDA receptor splicing through two essential RNA elements. *PLoS Biology, 5*(2), e40.

Lee-Kirsch, M. A., Gaudet, F., Cardoso, M. C., & Lindpaintner, K. (1999). Distinct renin isoforms generated by tissue-specific transcription initiation and alternative splicing. *Circulation Research, 84*(2), 240–246.

Li, Q., Lee, J. A., & Black, D. L. (2007). Neuronal regulation of alternative pre-mRNA splicing. *Nature Reviews. Neuroscience, 8*(11), 819–831.

Liang, X. H., Haritan, A., Uliel, S., & Michaeli, S. (2003). trans and cis splicing in trypanosomatids: Mechanism, factors, and regulation. *Eukaryotic Cell, 2*(5), 830–840.

Licatalosi, D. D., Mele, A., Fak, J. J., Ule, J., Kayikci, M., Chi, S. W., et al. (2008). HITS-CLIP yields genome-wide insights into brain alternative RNA processing. *Nature, 456*(7221), 464–469.

Lipscombe, D. (2005). Neuronal proteins custom designed by alternative splicing. *Current Opinion in Neurobiology, 15*(3), 358–363.

Listerman, I., Sapra, A. K., & Neugebauer, K. M. (2006). Cotranscriptional coupling of splicing factor recruitment and precursor messenger RNA splicing in mammalian cells. *Nature Structural & Molecular Biology, 13*(9), 815–822.

Luco, R. F., Allo, M., Schor, I. E., Kornblihtt, A. R., & Misteli, T. (2011). Epigenetics in alternative pre-mRNA splicing. *Cell, 144*(1), 16–26.

Luco, R. F., & Misteli, T. (2011). More than a splicing code: Integrating the role of RNA, chromatin and non-coding RNA in alternative splicing regulation. *Current Opinion in Genetics & Development, 21*(4), 366–372.

Luque, F. A., Furneaux, H. M., Ferziger, R., Rosenblum, M. K., Wray, S. H., Schold, S. C., Jr., et al. (1991). Anti-Ri: An antibody associated with paraneoplastic opsoclonus and breast cancer. *Annals of Neurology, 29*(3), 241–251.

Makeyev, E. V., Zhang, J., Carrasco, M. A., & Maniatis, T. (2007). The MicroRNA miR-124 promotes neuronal differentiation by triggering brain-specific alternative pre-mRNA splicing. *Molecular Cell, 27*(3), 435–448.

Marras, S. A., Gold, B., Kramer, F. R., Smith, I., & Tyagi, S. (2004). Real-time measurement of in vitro transcription. *Nucleic Acids Research, 32*(9), e72.

Masek, T., Valasek, L., & Pospisek, M. (2011). Polysome analysis and RNA purification from sucrose gradients. *Methods in Molecular Biology*, *703*, 293–309.

Matlin, A. J., Clark, F., & Smith, C. W. (2005). Understanding alternative splicing: Towards a cellular code. *Nature Reviews. Molecular Cell Biology*, *6*(5), 386–398.

Mercer, T. R., Gerhardt, D. J., Dinger, M. E., Crawford, J., Trapnell, C., Jeddeloh, J. A., et al. (2012). Targeted RNA sequencing reveals the deep complexity of the human transcriptome. *Nature Biotechnology*, *30*(1), 99–104.

Merkin, J., Russell, C., Chen, P., & Burge, C. B. (2012). Evolutionary dynamics of gene and isoform regulation in Mammalian tissues. *Science*, *338*(6114), 1593–1599.

Misteli, T., Caceres, J. F., & Spector, D. L. (1997). The dynamics of a pre-mRNA splicing factor in living cells. *Nature*, *387*(6632), 523–527.

Modrek, B., & Lee, C. (2002). A genomic view of alternative splicing. *Nature Genetics*, *30*(1), 13–19.

Moore, M. J., & Proudfoot, N. J. (2009). Pre-mRNA processing reaches back to transcription and ahead to translation. *Cell*, *136*(4), 688–700.

Munchel, S. E., Shultzaberger, R. K., Takizawa, N., & Weis, K. (2011). Dynamic profiling of mRNA turnover reveals gene-specific and system-wide regulation of mRNA decay. *Molecular Biology of the Cell*, *22*(15), 2787–2795.

Munoz, M. J., de la Mata, M., & Kornblihtt, A. R. (2010). The carboxy terminal domain of RNA polymerase II and alternative splicing. *Trends in Biochemical Sciences*, *35*(9), 497–504.

Naor, D., Nedvetzki, S., Golan, I., Melnik, L., & Faitelson, Y. (2002). CD44 in cancer. *Critical Reviews in Clinical Laboratory Sciences*, *39*(6), 527–579.

Neugebauer, K. M., & Roth, M. B. (1997). Distribution of pre-mRNA splicing factors at sites of RNA polymerase II transcription. *Genes & Development*, *11*(9), 1148–1159.

Nilsen, T. W., & Graveley, B. R. (2010). Expansion of the eukaryotic proteome by alternative splicing. *Nature*, *463*(7280), 457–463.

Norris, A. D., & Calarco, J. A. (2012). Emerging roles of alternative pre-mRNA splicing regulation in neuronal development and function. *Frontiers in Neuroscience*, *6*, 122.

Pan, Q., Shai, O., Misquitta, C., Zhang, W., Saltzman, A. L., Mohammad, N., et al. (2004). Revealing global regulatory features of mammalian alternative splicing using a quantitative microarray platform. *Molecular Cell*, *16*(6), 929–941.

Pandya-Jones, A., Bhatt, D. M., Lin, C. H., Tong, A. J., Smale, S. T., & Black, D. L. (2013). Splicing kinetics and transcript release from the chromatin compartment limit the rate of Lipid A-induced gene expression. *RNA*, *19*(6), 811–827.

Pandya-Jones, A., & Black, D. L. (2009). Co-transcriptional splicing of constitutive and alternative exons. *RNA*, *15*(10), 1896–1908.

Park, T. J., & Curran, T. (2010). Alternative splicing disabled by Nova2. *Neuron*, *66*(6), 811–813.

Park, J. W., & Graveley, B. R. (2007). Complex alternative splicing. *Advances in Experimental Medicine and Biology*, *623*, 50–63.

Perales, R., & Bentley, D. (2009). "Cotranscriptionality": The transcription elongation complex as a nexus for nuclear transactions. *Molecular Cell*, *36*(2), 178–191.

Phatnani, H. P., & Greenleaf, A. L. (2006). Phosphorylation and functions of the RNA polymerase II CTD. *Genes & Development*, *20*(21), 2922–2936.

Picelli, S., Faridani, O. R., Bjorklund, A. K., Winberg, G., Sagasser, S., & Sandberg, R. (2014). Full-length RNA-seq from single cells using Smart-seq2. *Nature Protocols*, *9*(1), 171–181.

Prokunina-Olsson, L., Welch, C., Hansson, O., Adhikari, N., Scott, L. J., Usher, N., et al. (2009). Tissue-specific alternative splicing of TCF7L2. *Human Molecular Genetics*, *18*(20), 3795–3804.

Racca, C., Gardiol, A., Eom, T., Ule, J., Triller, A., & Darnell, R. B. (2010). The neuronal splicing factor Nova co-localizes with target RNAs in the dendrite. *Frontiers in Neural Circuits*, *4*, 5.

Ramskold, D., Wang, E. T., Burge, C. B., & Sandberg, R. (2009). An abundance of ubiquitously expressed genes revealed by tissue transcriptome sequence data. *PLoS Computational Biology*, *5*(12), e1000598.

Rogelj, B., Easton, L. E., Bogu, G. K., Stanton, L. W., Rot, G., Curk, T., et al. (2012). Widespread binding of FUS along nascent RNA regulates alternative splicing in the brain. *Scientific Reports*, *2*, 603.

Roth, R. B., Hevezi, P., Lee, J., Willhite, D., Lechner, S. M., Foster, A. C., et al. (2006). Gene expression analyses reveal molecular relationships among 20 regions of the human CNS. *Neurogenetics*, 7(2), 67–80.

Ruggiu, M., Herbst, R., Kim, N., Jevsek, M., Fak, J. J., Mann, M. A., et al. (2009). Rescuing Z+ agrin splicing in Nova null mice restores synapse formation and unmasks a physiologic defect in motor neuron firing. *Proceedings of the National Academy of Sciences of the United States of America*, *106*(9), 3513–3518.

Salz, H. K. (2011). Sex determination in insects: A binary decision based on alternative splicing. *Current Opinion in Genetics & Development*, *21*(4), 395–400.

Salz, H. K., & Erickson, J. W. (2010). Sex determination in Drosophila: The view from the top. *Fly (Austin)*, *4*(1), 60–70.

Scheiffele, P., Fan, J., Choih, J., Fetter, R., & Serafini, T. (2000). Neuroligin expressed in nonneuronal cells triggers presynaptic development in contacting axons. *Cell*, *101*(6), 657–669.

Schmidt, U., Basyuk, E., Robert, M. C., Yoshida, M., Villemin, J. P., Auboeuf, D., et al. (2011). Real-time imaging of cotranscriptional splicing reveals a kinetic model that reduces noise: Implications for alternative splicing regulation. *The Journal of Cell Biology*, *193*(5), 819–829.

Schmucker, D., Clemens, J. C., Shu, H., Worby, C. A., Xiao, J., Muda, M., et al. (2000). Drosophila Dscam is an axon guidance receptor exhibiting extraordinary molecular diversity. *Cell*, *101*(6), 671–684.

Sharma, S., Falick, A. M., & Black, D. L. (2005). Polypyrimidine tract binding protein blocks the 5′ splice site-dependent assembly of U2AF and the prespliceosomal E complex. *Molecular Cell*, *19*(4), 485–496.

Shukla, S., & Oberdoerffer, S. (2012). Co-transcriptional regulation of alternative pre-mRNA splicing. *Biochimica et Biophysica Acta*, *1819*(7), 673–683.

Singh, J., & Padgett, R. A. (2009). Rates of in situ transcription and splicing in large human genes. *Nature Structural & Molecular Biology*, *16*(11), 1128–1133.

Spellman, R., Llorian, M., & Smith, C. W. (2007). Crossregulation and functional redundancy between the splicing regulator PTB and its paralogs nPTB and ROD1. *Molecular Cell*, *27*(3), 420–434.

Stamm, S., Ben-Ari, S., Rafalska, I., Tang, Y., Zhang, Z., Toiber, D., et al. (2005). Function of alternative splicing. *Gene*, *344*, 1–20.

Sterne-Weiler, T., Martinez-Nunez, R. T., Howard, J. M., Cvitovik, I., Katzman, S., Tariq, M. A., et al. (2013). Frac-seq reveals isoform-specific recruitment to polyribosomes. *Genome Research*, *23*(10), 1615–1623.

Tilgner, H., Knowles, D. G., Johnson, R., Davis, C. A., Chakrabortty, S., Djebali, S., et al. (2012). Deep sequencing of subcellular RNA fractions shows splicing to be predominantly co-transcriptional in the human genome but inefficient for lncRNAs. *Genome Research*, *22*(9), 1616–1625.

Treutlein, B., Gokce, O., Quake, S. R., & Sudhof, T. C. (2014). Cartography of neurexin alternative splicing mapped by single-molecule long-read mRNA sequencing.

Proceedings of the National Academy of Sciences of the United States of America, 111(13), E1291–E1299.

Ule, J., Stefani, G., Mele, A., Ruggiu, M., Wang, X., Taneri, B., et al. (2006). An RNA map predicting Nova-dependent splicing regulation. *Nature, 444*(7119), 580–586.

Ule, J., Ule, A., Spencer, J., Williams, A., Hu, J. S., Cline, M., et al. (2005). Nova regulates brain-specific splicing to shape the synapse. *Nature Genetics, 37*(8), 844–852.

Ullrich, B., Ushkaryov, Y. A., & Sudhof, T. C. (1995). Cartography of neurexins: More than 1000 isoforms generated by alternative splicing and expressed in distinct subsets of neurons. *Neuron, 14*(3), 497–507.

Valencia-Sanchez, M. A., Liu, J., Hannon, G. J., & Parker, R. (2006). Control of translation and mRNA degradation by miRNAs and siRNAs. *Genes & Development, 20*(5), 515–524.

Vargas, D. Y., Shah, K., Batish, M., Levandoski, M., Sinha, S., Marras, S. A., et al. (2011). Single-molecule imaging of transcriptionally coupled and uncoupled splicing. *Cell, 147*(5), 1054–1065.

Vessey, J. P., Amadei, G., Burns, S. E., Kiebler, M. A., Kaplan, D. R., & Miller, F. D. (2012). An asymmetrically localized Staufen2-dependent RNA complex regulates maintenance of mammalian neural stem cells. *Cell Stem Cell, 11*(4), 517–528.

Wang, E. T., Sandberg, R., Luo, S., Khrebtukova, I., Zhang, L., Mayr, C., et al. (2008). Alternative isoform regulation in human tissue transcriptomes. *Nature, 456*(7221), 470–476.

West, A. E., Chen, W. G., Dalva, M. B., Dolmetsch, R. E., Kornhauser, J. M., Shaywitz, A. J., et al. (2001). Calcium regulation of neuronal gene expression. *Proceedings of the National Academy of Sciences of the United States of America, 98*(20), 11024–11031.

Wetterberg, I., Bauren, G., & Wieslander, L. (1996). The intranuclear site of excision of each intron in Balbiani ring 3 pre-mRNA is influenced by the time remaining to transcription termination and different excision efficiencies for the various introns. *RNA, 2*(7), 641–651.

Xie, J. (2008). Control of alternative pre-mRNA splicing by Ca(++) signals. *Biochimica et Biophysica Acta, 1779*(8), 438–452.

Xu, Q., Modrek, B., & Lee, C. (2002). Genome-wide detection of tissue-specific alternative splicing in the human transcriptome. *Nucleic Acids Research, 30*(17), 3754–3766.

Xue, Y., Zhou, Y., Wu, T., Zhu, T., Ji, X., Kwon, Y. S., et al. (2009). Genome-wide analysis of PTB-RNA interactions reveals a strategy used by the general splicing repressor to modulate exon inclusion or skipping. *Molecular Cell, 36*(6), 996–1006.

Yang, Y. Y., Yin, G. L., & Darnell, R. B. (1998). The neuronal RNA-binding protein Nova-2 is implicated as the autoantigen targeted in POMA patients with dementia. *Proceedings of the National Academy of Sciences of the United States of America, 95*(22), 13254–13259.

Yano, M., Hayakawa-Yano, Y., Mele, A., & Darnell, R. B. (2010). Nova2 regulates neuronal migration through an RNA switch in disabled-1 signaling. *Neuron, 66*(6), 848–858.

Yap, K., & Makeyev, E. V. (2013). Regulation of gene expression in mammalian nervous system through alternative pre-mRNA splicing coupled with RNA quality control mechanisms. *Molecular and Cellular Neurosciences, 56*, 420–428.

Yeo, G., Holste, D., Kreiman, G., & Burge, C. B. (2004). Variation in alternative splicing across human tissues. *Genome Biology, 5*(10), R74.

Zaghlool, A., Ameur, A., Nyberg, L., Halvardson, J., Grabherr, M., Cavelier, L., et al. (2013). Efficient cellular fractionation improves RNA sequencing analysis of mature and nascent transcripts from human tissues. *BMC Biotechnology, 13*, 99.

Zhang, C., Frias, M. A., Mele, A., Ruggiu, M., Eom, T., Marney, C. B., et al. (2010). Integrative modeling defines the Nova splicing-regulatory network and its combinatorial controls. *Science, 329*(5990), 439–443.

Zhang, G., Taneja, K. L., Singer, R. H., & Green, M. R. (1994). Localization of pre-mRNA splicing in mammalian nuclei. *Nature*, *372*(6508), 809–812.
Zhang, C., Zhang, Z., Castle, J., Sun, S., Johnson, J., Krainer, A. R., et al. (2008). Defining the regulatory network of the tissue-specific splicing factors Fox-1 and Fox-2. *Genes & Development*, *22*(18), 2550–2563.
Zhu, J., Mayeda, A., & Krainer, A. R. (2001). Exon identity established through differential antagonism between exonic splicing silencer-bound hnRNP A1 and enhancer-bound SR proteins. *Molecular Cell*, *8*(6), 1351–1361.

CHAPTER SIX

Understanding Complex Transcriptome Dynamics in Schizophrenia and Other Neurological Diseases Using RNA Sequencing

Xi Wang*, Murray J. Cairns*,†,1
*School of Biomedical Sciences and Pharmacy, Faculty of Health and Medicine, The University of Newcastle, Callaghan, New South Wales, Australia
†The Schizophrenia Research Institute, Sydney, Australia
[1]Corresponding author: e-mail address: murray.cairns@newcastle.edu.au

Contents

Abstract

How the human brain develops and adapts with its trillions of functionally integrated synapses remains one of the greatest mysteries of life. With tremendous advances in neuroscience, genetics, and molecular biology, we are beginning to appreciate the scope of this complexity and define some of the parameters of the systems that make it possible. These same tools are also leading to advances in our understanding of the pathophysiology of neurocognitive and neuropsychiatric disorders. Like the substrate for these problems, the etiology is usually complex—involving an array of genetic and environmental influences. To resolve these influences and derive better

International Review of Neurobiology, Volume 116
ISSN 0074-7742
http://dx.doi.org/10.1016/B978-0-12-801105-8.00006-0

interventions, we need to reveal every aspect of this complexity and model their interactions and define the systems and their regulatory structure. This is particularly important at the tissue-specific molecular interface between the underlying genetic and environmental influence defined by the transcriptome. Recent advances in transcriptome analysis facilitated by RNA sequencing (RNA-Seq) can provide unprecedented insight into the functional genomics of neurological disorders. In this review, we outline the advantages of this approach and highlight some early application of this technology in the investigation of the neuropathology of schizophrenia. Recent progress of RNA-Seq studies in schizophrenia has shown that there is extraordinary transcriptome dynamics with significant levels of alternative splicing. These studies only scratch the surface of this complexity and therefore future studies with greater depth and samples size will be vital to fully explore transcriptional diversity and its underlying influences in schizophrenia and provide the basis for new biomarkers and improved treatments.

1. INTRODUCTION

Complex polygenic neurobehavioral syndromes such as schizophrenia, bipolar, and autism spectrum disorders are associated with multiple genetic variants in combination with numerous environment risk factors. While significant progress has been made toward identifying the small number of more penetrant rare variants in family studies and a large number of less penetrant common variants through genome-wide association studies (McCarroll & Hyman, 2013; Mowry & Gratten, 2013; Owen, 2012), we are far from entirely understanding the underlying genetics of these disorders (Kim, Zerwas, Trace, & Sullivan, 2011; Tiwari, Zai, Muller, & Kennedy, 2010). Through epidemiology, twin studies, and animal models, we are also making progress toward understanding many of the environmental risks factors and their relationship with epigenetic marks that transform the fate of neural lineages and their developing tissues. These combined influences culminate in the transcriptional signal that provides the basis for cell growth, homeostasis, differentiation, and more specialized functions. Transcriptional dysregulation is also the daily reality driving many pathophysiological processes. Therefore to come to terms with this complex, dynamic, and highly heterogeneous signal and to identify programmatic alterations, we need to explore the rich texture of transcription using the most sophisticated high-resolution genomics. In only a relatively short time since the explosion in microarray-based genome-wide expression analysis, the field is undergoing another transformative advance through the application of high-throughput next-generation transcriptome sequencing technologies (RNA-Seq). This new platform provides a unique opportunity for researchers to connect gene expression with disease associated genetic and epigenetic variation by analyzing transcription at nucleotide resolution.

This revolutionary approach has many advantages over the existing microarray platforms. Read count data, for example, have a relatively high signal-to-noise ratio, which enables extraordinary dynamic range and the power to detect and quantify novel transcripts (Mortazavi, Williams, McCue, Schaeffer, & Wold, 2008; Ozsolak & Milos, 2011). By directly gathering base-resolution sequence information, RNA-Seq can also detect and quantify alternative isoforms induced by alternative splicing and alternative usage of promoters and polyadenylation (Trapnell et al., 2010; Wang et al., 2008). Alternative splicing is important in neural development, and aberrations in splicing events have been shown to be associated with neurological or neurodegenerative disease including schizophrenia and Alzheimer's disease (Black & Grabowski, 2003; Clinton, Haroutunian, Davis, & Meador-Woodruff, 2003; Dredge, Polydorides, & Darnell, 2001; Licatalosi & Darnell, 2006; Mazin et al., 2013). Many genes expressed in brains are differently spliced comparing other tissues and the splicing pattern in different brain regions can also vary (Wang et al., 2008), which indicates the functional importance of splicing in neural cells. Thus, examining alternative splicing will be essential for illuminating the full diversity of mechanisms underlying the pathogenesis of neurological disorders. Moreover, transcriptional analysis by deep sequencing facilitates the ascertainment of allele-specific gene expression by distinguishing heterozygous reads by single-nucleotide polymorphisms (SNPs) or other mutations transcribed into RNA (Pickrell et al., 2010; Zhang et al., 2009). Allele-expression imbalance occurs when *cis*-regulatory polymorphism or *cis*-acting quantitative trait loci (*cis*-eQTLs) affect *trans*-factor regulation. Alternatively, allele-expression imbalance can occur when there is differential chromatin structure between the maternal and paternal chromosomes, such as at imprinted loci. Nucleotide resolution expression analysis also enables the identification of RNA-editing events by comparing the transcriptional polymorphisms with the genomic sequence in the same individual (Park, Williams, Wold, & Mortazavi, 2012). RNA editing plays a critical role in brain development and abnormal RNA editing has been shown to be associated with neurological disorders and brain disease (Li & Church, 2013; Park et al., 2012). RNA-Seq has unprecedented power to discern these alterations and will be vital to revealing this functionally significant genomic modifications and their role in pathophysiology.

In this chapter, we summarize the recent progress in transcriptional analysis of neurological disorders facilitated by RNA-Seq and discuss the challenges and opportunities with a focus on schizophrenia. While the past a few years have witnessed the early exploration of transcriptome dynamics in schizophrenia and other neurological disorders using high-throughput sequencing technologies, we are only scratching the surface of their potential for shedding light on the mechanism of disease development. It is therefore tantalizing to speculate that the further implementation of these technologies will have tremendous prospects for the development of new diagnostic and treatment approaches.

2. RNA-Seq STUDIES ON NEUROLOGICAL DISORDERS

Several studies have applied RNA-Seq based advanced transcriptome analysis for the investigation of neurological disorders. We summarized some of these studies in Table 6.1 and discuss the experimental design and data analysis strategies employed.

2.1. Study design

The high heterogeneity of gene expression from human individuals requires a substantial cohort of neurological patients and matched controls for disease-related transcriptomic alteration profiling (Table 6.1), in order to account for biological variability and identify changes that are reproducible. Power analysis has also shown that increasing biological replicates rather than sequencing depth is more effective at identifying robust differential expression patterns in RNA-Seq studies (Hansen, Wu, Irizarry, & Leek, 2011). Unfortunately, the sample size is usually limited by the availability of postmortem tissue with appropriate controls and the relative cost of high-throughput sequencing. Most of the early studies employing high-throughput sequencing have been performed on limited number of samples (Xu et al., 2012). However, with the cost of sequencing going down, the assembly of large consortiums for neurological disease research may gather more substantial cohorts of samples to facilitate sequencing-based transcriptome studies with sufficient power.

Different anatomical regions in human brain, such as cerebral cortex and cerebellum, carry out different motoric, sensory, and cognitive functions, and all display region-specific gene expression patterns (Evans et al., 2003; Khaitovich et al., 2004). The cytoarchitectonics of cerebral cortex are further divided into Brodmann's areas (BAs), and these subregions have

Table 6.1 RNA-Seq studies on schizophrenia and other neurological disorders

Study	Analysis	Tissue used	Sample size	Platform
Wu et al. (2012)	Differential expression, promoter usage, and splicing	Postmortem superior temporal gyrus (STG or BA22)	9 SZ and 9 controls	Illumina GA II, 76 nt SR
Fillman et al. (2013)	Differentially expressed transcripts	Postmortem dorsolateral prefrontal cortex (DLPFC or BA46)	20 SZ and 20 controls	SOLiD v4, 50 nt SR
Mudge et al. (2008)	Differential expression and splicing	Cerebellar cortex	14 SZ and 6 controls	Illumina GA, 32–36 nt SR
Xu et al. (2012)	Differential gene expression	Blood lymphocytes	3 SZ and 3 controls, 1 SZ pool and 1 control pool	Illumina GA II, 2 × 43 nt PE for subjects, SR for pools
Hwang et al. (2013)	Differential expression and coexpression network	Postmortem hippocampus	14 SZ and 15 controls	Illumina GA, 2 × 50 nt PE
Hong, Chen, Jin, and Xiong (2013)	Coexpression network	Anterior cingulate cortex (ACC or BA24)	31 SZ, 25 BPD and 26 controls	Illumina HiSeq
Silberberg, Lundin, Navon, and Ohman (2012)	A-to-I RNA editing	Postmortem DLPFC or BA46	20 SZ, 20 BPD, and 20 controls	454 FLX on target genes, median 236 nt
Smith et al. (2013)	Allele-specific expression and RNA editing	Nine brain regions,[a] and DLPFC (BA46)	1 sample for each brain region, and 14 DLPFC samples	SOLiD 5500/4, PE

Continued

Table 6.1 RNA-Seq studies on schizophrenia and other neurological disorders—cont'd

Study	Analysis	Tissue used	Sample size	Platform
Lin et al. (2011)	Expression of coding gene, long noncoding RNA, pseudogenes, and splice variants	iPSCs and neural differentiation	1 iPSCs and 1 neural differentiation	Illumina HiSeq 2000, 2 × 100 nt PE
Lin et al. (2012)	Allele-specific expression	iPSCs and neural differentiation	1 iPSCs and 1 neural differentiation	Illumina HiSeq 2000, 2 × 100 nt PE

[a]The nine brain regions are BA10, BA22, BA24, insular cortex, amygdala, hippocampus, putamen, cerebellum, and raphe nuclei.
BA22, Brodmann area 22; BA46, Brodmann area 46; SZ, schizophrenia; BPD, bipolar disorder; SR, single reads; PE, paired end.

also been characterized by differences gene expression pattern, though not as dramatic as the anatomical regions (Khaitovich et al., 2004). Transcriptome profiling of postmortem brains with neurological diseases therefore usually focus on one of the brain regions or areas (Table 6.1). We have focused on the superior temporal gyrus (STG or BA22) and the dorsolateral prefrontal cortex (DLPFC or BA46) (Fillman et al., 2013; Wu et al., 2012). These studies are, however, still rather limited, as complex brain disorders are likely to affect many brain regions and evolve over considerable time periods. The analysis of these conditions therefore represents a multidimensional problem requiring very significant research investment.

For case–control studies based on human subjects, study design should strive to match in demographics and other extraneous parameters, to eliminate the effect of unwanted confounds. Demographic features such as gender (Vawter et al., 2004), age (Glass et al., 2013), and the cause of death, as well as postmortem interval (PMI) (Bauer, Gramlich, Polzin, & Patzelt, 2003) and RNA sample quality indicator (RQI) (Stan et al., 2006), have shown some correlation with gene expression or its measurement. In our RNA-Seq analyses of schizophrenia transcriptome (Fillman et al., 2013; Wu et al., 2012), cases were selected to minimize difference between cases and controls in age, gender, PMI, pH, and RNA integrity. Where these confounders correlate with gene expression, their effects should be regressed through analysis of covariates (ANCOVA) or similar approaches to account for their influence (Wu et al., 2012).

Considering the relatively low quality of RNAs from postmortem brain tissues and the sample size limited by matching demographics for case and control cohorts, performing RNA-Seq on peripheral tissues can serve as an alternative to achieve higher RNA quality and larger cohorts of samples. Cells from the blood tissue, such as lymphocytes (Xu et al., 2012) and peripheral blood mononuclear cells (PBMCs) (Beveridge & Cairns, 2012; Gardiner et al., 2012, 2013), are often used as a surrogate tissue for expression analysis. Wherever possible the transcriptome dynamics in these tissues can also be linked with brain imaging and assessments of neurocognitive phenotypes and may have utility as biomarkers of schizophrenia or associated subphenotypes.

2.2. Sequencing platforms and strategies

There are now a variety of platforms suitable for transcriptome analysis each with their advantages and disadvantages. With various research aims to profile neurological disease transcriptomes (Table 6.1), the selection of high-throughout sequencing platforms and strategies can be varied to suit the requirements of the study. With the capacity to identify alternative splicing, it is desirable to use longer reads and utilize a paired-end (PE) strategy. In the Illumina platforms, it is now possible to read up to 150 bp from each end of the cluster, although studies to date have not gone beyond 100 bp PE (Lin et al., 2011). Instruments such as 454 FLX and PacBio RS can generate longer reads and will facilitate unambiguous spliceoform identification. For example, PacBio referred to as the third-generation platform can generate reads with average length at 2–3 kb and can confidentially capture transcripts up to 1.5 kb (Au et al., 2013; Sharon, Tilgner, Grubert, & Snyder, 2013).

Measuring allele-specific gene expression is another capability of RNA-Seq, which facilitates the identification of expression eQTLs across heterozygous segments. This fantastic new insight into the genome is somewhat more demanding because it usually requires significantly more depth to resolve allele-specific transcripts in genes with low to modest expression levels. For example, the study performed by Lin et al. (2012) yielded more than 100 million pairs of reads per sample to study allele-specific gene expression more effectively than most studies around 20–40 million reads.

2.3. Data analysis

Typical RNA-Seq data analysis pipeline can be found in Fig. 6.1, and here we also emphasize some specific technical issues in case–control investigations with human subjects in studying neurological disorders.

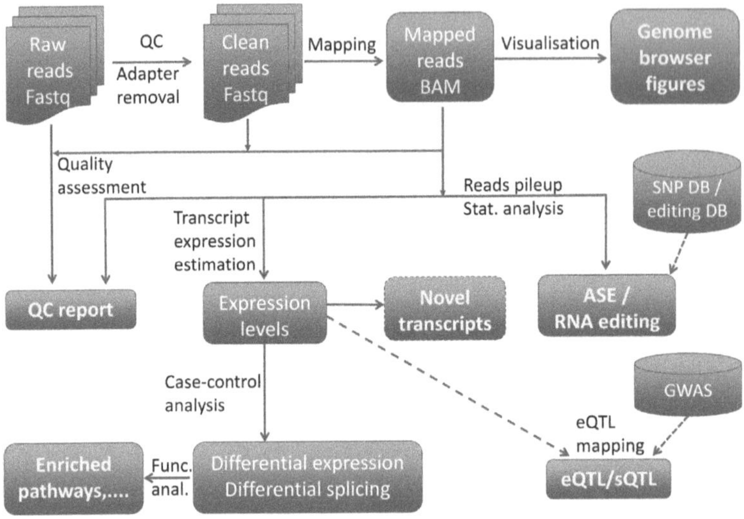

Figure 6.1 Typical RNA-seq data analysis pipeline. Fastq and BAM are file format to store reads and read mapping data, respectively; words highlighted in yellow (bold text in the print version) are outputs of the pipeline. QC, quality control; SNP, single nucleotide polymorphism; DB, database; ASE, allele-specific expression; GWAS, genome-wide association study; eQTL, expression quantitative trait loci; sQTL, splicing quantitative trait loci.

Mudge et al. applied two strategies to align RNA-Seq reads back to their original gene loci: align to the reference genome and to the reference transcriptome (RefSeq) (Mudge et al., 2008). Surprisingly, differentially regulated genes in the schizophrenia cohort detected by the two alignments had very limited overlaps—88 differentially expressed gene from genome alignment and 152 genes from transcriptome alignment, but only 25 genes in common, although most of the genes (96% of 215) showed consistent direction of expression change. This inconsistency could be explained by the incomplete annotation of RefSeq genes so that reads derived from unannotated exons and isoforms would be incorrectly aligned to other genes or be discarded without effective alignment (Mudge et al., 2008). Currently, there are considerably more software tools that perform better than those available at the time of the study; however, together with software for transcriptome reconstruction and expression-level estimation, large divergence still can be observed between different computational tools (Engstrom et al., 2013; Steijger et al., 2013). The high complexity of human transcriptomes imposes this technical limitation and cannot be easily solved with current sequencing technologies. To control possible false positives in RNA-Seq studies, validation with other technologies rather sequencing (e.g., qPCR) usually performed (Fillman et al., 2013; Hwang et al., 2013; Wu et al., 2012; Xu et al., 2012).

One other possible issue in RNA-Seq with postmortem tissues is RNA degradation in samples. When using poly-A selection protocol for sequencing library preparation, transcripts degraded from 5′-end can also be enriched, which results in extremely nonuniform read distribution along transcripts—reads will be biasedly mapped toward the 3′-untranslated regions (3′UTR). This characteristic would further cause large inaccuracy in isoform expression level estimation. Methods like NURD (Ma & Zhang, 2013) can remedy this situation by modeling this nonuniform read distribution with empirical methods.

To interpret differentially expressed/spliced genes, knowledge-based functional analysis links genes of similar function or located in the same pathway together, highlighting the functional importance of gene categories or pathways with enrichment of differentially regulated genes. Gene set enrichment analysis (GSEA) (Subramanian et al., 2005), a powerful functional analysis method, can gather subtle gene expression changes in the predefined gene sets to yield differentially regulated gene sets. Due to high heterogeneity of disease causality, gene expression alteration with respect to complex diseases including neurological diseases often has small effective size in the entire population, and thus GSEA is an important strategy for the functional analysis of transcriptome data in these disorders. Moreover, alternative splicing is increasingly believed to play functionally important roles in regulating gene expression in a tempospatial specific manner and is also involved in disease biogenesis and development. In order to functionally reconcile both these features of transcriptome alteration, we have developed method called SeqGSEA (Wang & Cairns, 2013) and implemented this as an R-package in Bioconductor (Wang & Cairns, 2014). We applied this analytical approach to RNA-Seq data in schizophrenia enabling both statistically significant and biologically meaningful gene sets associated to be derived where existing strategies focused on individual genes failed (Wang & Cairns, 2013).

While we found the SeqGSEA strategy to be the most effective on our data sets, functional analysis can also be determined without preexisting knowledge of functional annotation. Methodologies like coexpression network analysis generate graphic interpretation of gene expression changes with respect to diseases (Carter, Brechbuhler, Griffin, & Bond, 2004). Hwang et al. applied the coexpression network analysis to seek for biological interpretation from RNA-Seq data (Hwang et al., 2013). A recently published study proposed a new approach for construction of coexpression networks with RNA-Seq data, by accounting for exon-specific expression alteration caused by differential alternative splicing and allele-specific expression resulted from SNPs or RNA-editing, yielding more biologically relevant results (Hong et al., 2013).

3. QUANTIFYING TRANSCRIPTOME DYNAMICS IN NEUROLOGICAL DISORDERS

3.1. Gene/transcript expression

Microarray-based transcriptome profiling studies of schizophrenia carried out over the last decade have frequently reported differentially regulated genes related to synaptic plasticity (Arion, Unger, Lewis, Levitt, & Mirnics, 2007; Aston, Jiang, & Sokolov, 2004; Hakak et al., 2001; Haroutunian, Katsel, Dracheva, Stewart, & Davis, 2007; Mirnics, Middleton, Marquez, Lewis, & Levitt, 2000), glutamatergic and GABAergic neurotransmission (Arion et al., 2007; Hakak et al., 2001; Maycox et al., 2009), inflammatory/immune response (Saetre et al., 2007), and many others (Sequeira, Martin, & Vawter, 2012). More recently, RNA-Seq based studies have emerged (summarized in Table 6.2), most of which are consistent with previous results by microarrays, with strong support for the elevation of inflammatory/immune pathways (Fillman et al., 2013; Hwang et al., 2013; Xu et al., 2012). Genes and transcripts differentially regulated in schizophrenia according to RNA-Seq analysis are summarized below.

3.1.1 Synaptic plasticity and neurotransmission

To the best of our knowledge, the first RNA-Seq study in schizophrenia was carried out by Mudge et al. in 2008. They preformed Illumina RNA sequencing on postmortem cerebellar cortex from a cohort of schizophrenia patients including 14 individuals and 6 matched controls (Table 6.1), yielding 12.5–38.7 million 32–36 nt short reads for each sample, which resulted in 215 differentially expressed genes (Table 6.2). Functional analysis showed that the differentially expressed genes were overrepresented in function categories including Golgi apparatus and vesicular transport, and 23 genes were involved in a pathway responsible for presynaptic vesicular transport/Golgi function and postsynaptic neurotransmission, of which 20 genes were involved in Golgi-associated transport and presynaptic vesicular transport (13 upregulated and 7 downregulated). The other three genes were postsynaptic membrane genes (GABRA1 and CACNG2 upregulated and ZACN downregulated). While one of the gamma-aminobutyric acid (GABA)-mediated neuroinhibitory receptor, GABRA1, showed significant

Table 6.2 Numbers of genes reported by the reviewed RNA-seq studies with altered expression/splicing/ASE/editing in schizophrenia and other related disorders

Study	Disorder, tissue	Type of alteration	Number	Validated
Wu et al. (2012)	Schizophrenia, BA22	Differentially expressed genes	772 genes (374 up, 398 down)	3 genes (down)
Wu et al. (2012)	Schizophrenia, BA22	Differential promoter usage and splicing	1032 genes	–
Fillman et al. (2013)	Schizophrenia, BA46	Differentially expressed transcripts	798 transcripts (537 up and 261 down, 316 genes)	3 genes (up)
Mudge et al. (2008)	Schizophrenia, cerebellar cortex	Differentially expressed genes	215 genes (147 up, 68 down)[a]	–
Xu et al. (2012)	Schizophrenia, blood lymphocytes	Differentially expressed genes	218 genes (92 up, 126 down)	19 genes (9 up, 10 down; by pooled samples)
Hwang et al. (2013)	Schizophrenia, hippocampus	Differentially expressed genes	114 genes (123 up, 21 down)	6 genes (up)
Silberberg et al. (2012)	Schizophrenia and bipolar disorder, BA46	Genes with different RNA-editing rate	1 gene (SZ vs. controls) and 1 gene (BPD vs. controls) out of 7	–
Smith et al. (2013)	Brain regions	Genes with ubiquitous AEI	2 genes across 8 regions (NHP2L1, SLC1A3)	–
Lin et al. (2011)	iPSCs and neural differentiation	Differentially expressed genes	9008 (1443 lincRNA: 3055 up and 5953 down; lincRNA: 228 up and 220 down)	4 by RT-PCR, and 2 by q-PCR
Lin et al. (2012)	iPSCs and neural differentiation	Genes with allele-biased expression	48 schizophrenia genes in neurons, 17 in iPSCs	Most can be validated by Sanger sequencing

Continued

Table 6.2 Numbers of genes reported by the reviewed RNA-seq studies with altered expression/splicing/ASE/editing in schizophrenia and other related disorders—cont'd

Study	Disorder, tissue	Type of alteration	Number	Validated
Lin et al., 2012 (Lin et al., 2012)	iPSCs and neural differentiation	Genes with allele switching	42 genes	–

[a]Up-/downregulation is according to the alignment against transcriptome.
AEI, allelic expression imbalance; lincRNA, long intergenic noncoding RNA; iPSC, induced pluripotent stem cell.

increased expression (by ~50%) in schizophrenia cohort, majority of other GABA receptors were also increased (e.g., GABRB1, by ~170%) although not significantly.

More recently, we carried out RNA-Seq analysis of postmortem cortical gray matter from the superior temporal gyrus (STG or BA22) from nine schizophrenia patients and nine matched nonpsychiatric controls (Wu et al., 2012) (Table 6.1). This effort generated much longer reads (76 nt) than the previous study at similar sequencing depth (13.1–39.2 million reads per sample) and yielded 772 differentially expressed genes using the Illumina GAII. Gene ontology analysis showed that 40 and 42 core enrichment genes were related to synaptic vesicle trafficking and neurotransmission-related functions, respectively. Overlapped differentially expressed genes with previous microarray studies also showed relevance to GABA function, glutamate function, myelin and oligodendrocyte, etc. GABA is the principal inhibitory neurotransmitter in mammalian brains, and glutamate is the main excitatory neurotransmitter, also the precursor for GABA. It has shown that the two neurotransmitters are associated with synaptic plasticity in both presynaptic and postsynaptic process (Debanne, Daoudal, Sourdet, & Russier, 2003; Raimondo, Markram, & Akerman, 2012). Meanwhile, the main function of myelin and oligodendrocytes is to provide insulation to axons, which has also shown to be related to synaptic plasticity (Fields, 2005). Taking together, the two studies discussed above, although investigated in different brain regions (cerebellar cortex and cerebral cortex), have both reported the association between schizophrenia and synaptic plasticity/neurotransmission.

3.1.2 Inflammatory/immune pathways

In another study, we investigated postmortem dorsolateral prefrontal cortex (DLPFC or BA46) from 20 schizophrenia patients and 20 matched controls using RNA-Seq preformed on the Applied Biosystems SOLiD platform. This yielded 135 million 50 nt reads on average (Table 6.1) (Fillman et al., 2013). Functional analysis on 798 differentially regulated transcripts showed that inflammatory response was the key player among others, and genes in the inflammatory pathway, such as interleukin 6 (IL-6) and a serpin peptidase inhibitor (SERPINA3), were upregulated. It was found by different experiment technologies that microglial markers were significantly increased in schizophrenia. Microglial cells play the main immune defense role in the central nervous system, preventing most infections from reaching the vulnerable nervous tissue. Increased expression of microglial markers and increased density of microglia in schizophrenia largely supported a relationship between inflammation and the physiopathology of schizophrenia.

Meanwhile, the study by Hwang et al., which focused on the hippocampus, also emphasized the expression changes of immune genes in schizophrenia individuals. This study yielded paired-end reads (2×50 nt) by the Illumina GA sequencer from 14 individuals with schizophrenia and 15 matched controls (Table 6.1) (Hwang et al., 2013). Functional analysis and coexpression network analysis both showed the enrichment of inflammatory response genes. Similar as in DLPFC, these immune genes also showed increased expression in schizophrenia. However, contrary to their expectations, they did not find any immune-related proteins expressed in microglia in representative individuals with the disorder.

Prenatal infection showed positive correlations with schizophrenia developing (Brown & Derkits, 2010), and it may also result in an increase in peripheral body inflammatory makers for schizophrenia patients. This increase has been reported in many previous studies based on other technologies rather than RNA-Seq (Drexhage et al., 2010; Kurian et al., 2011; Miller, Buckley, Seabolt, Mellor, & Kirkpatrick, 2011) and was also confirmed by a recent RNA-Seq study by Xu et al. (2012). They executed paired-end (2×43 nt) Illumina sequencing in lymphocytes from blood of 3 schizophrenia patients and 3 controls, as well as 2 pools of 10 schizophrenia and 10 controls, respectively (Table 6.1). More than 200 differentially expressed transcripts were identified, and KEGG pathway analysis based on these candidate transcripts showed the enrichment in immune and inflammatory systems.

Summarizing the findings in the three papers above, it is clear that immune/inflammatory response is highly associated with the development of schizophrenia, and the exposure to virus or bacteria is implicated in the etiology of schizophrenia and many other psychiatric disorders (Torrey,

Miller, Rawlings, & Yolken, 1997). This consistent result also suggests that medical intervention targeted to immune/inflammatory pathways in high-risk individuals may be effective in preventing the disorder or improving the outcomes of patients with symptoms.

3.2. Alternative splicing

Alternative splicing is shown to be largely involved in high primates and plays important functions in regulating gene expression in a tissue-specific manner (Wang et al., 2008). Association between aberrant alternative splicing and human disease has also been reported (Caceres & Kornblihtt, 2002; Oldmeadow et al., 2014; Tazi, Bakkour, & Stamm, 2009). With RNA-Seq, alteration of splicing in schizophrenia can be more easily detected than using microarray, which serves another important layer in underlying mechanism in schizophrenia pathogenesis. Mudge et al. used a sophisticated read alignment approach to identify novel splicing isoforms associated schizophrenia and identified an intron-retained isoform in the PRODH gene locus, consistent with other independent reports (Mudge et al., 2008). We also analyzed schizophrenia-associated differential promoter usage and alternative splicing in the STG between cases and controls, and the results showed more 2000 and 1032 genes were associated alternative promoter usage and differential splicing in schizophrenia, respectively (Wu et al., 2012). This large numbers indicate the huge transcriptome dynamics in schizophrenia, which may be underappreciated in previous studies. Seven genes (GABRA5, HCRTR1, MBP, PRKG1, SYP, SYT1, and DCLK1) with differential promoter usage and two differentially spliced genes (PLP1 and DCLK1) were overlapped with core enrichment genes discussed in Section 3.1. Interestingly, biased allelic expression was also found in two of four expressed transcripts from the DCLK1 gene locus. The two genes with both altered overall expression levels and differential splicing (PLP1 and DCLK1) are of particular interest for future investigation. Both splice form of PLP1 encode important proteins to form the myelin protein complex, and the two forms switch during different stages of myelination (Jacobs, Bongarzone, Campagnoni, & Campagnoni, 2004; Jahn, Tenzer, & Werner, 2009; Sporkel, Uschkureit, Bussow, & Stoffel, 2002). DCLK1 is expressed in central nerves system and its splice variants play potential roles in neuronal migration and neural plasticity (Burgess & Reiner, 2000; Silverman et al., 1999).

3.3. Allele-specific expression

Another exciting dimension captured in RNA-Seq data is the transcript sequences themselves. This feature enables allele-specific discrimination of gene expression (Zhang et al., 2009). Biased allelic expression has shown to be relevant to human diseases in exemplified cases (Li et al., 2006) and can explain the contribution of SNPs and rare mutations to gene expression directly and enable the identification of eQTLs.

Smith et al. performed an RNA-Seq study focusing on allelic expression across nine human brain regions and found two genes (SLC1A3 and NHP2L1) with allelic expression imbalance through all regions and many other genes with regional specific biased allelic expression (Smith et al., 2013). In the analysis of DLPFC tissues from 14 individuals, they also observed 500 genes with biased allelic expression in more than samples. Intriguingly, GABA receptor subunit beta-1 (GABRB1) showed imbalanced allelic expression in two DLPFC samples out of five measured.

Lin et al. executed RNA-Seq-based allele-specific gene expression analysis on induced pluripotent stem cell (iPSC) and differentiating neurons. With stringent criteria to filter out low-quality data, the RNA-Seq data resulted in 314 genes on iPSCs and 801 genes on differentiating neurons that contain at least one high-quality SNP inducing expression imbalance. Of those, 181 were neuronal genes, which were overrepresented in schizophrenia and autism spectrum disorders candidates. Specifically, 48 schizophrenia genes showed allelic imbalance in neurons, while 17 in iPSCs. Sanger sequencing validated five schizophrenia and autism genes, some of which, however, could not be validated on a biological replicate. The sporadic differences in allelic expression imbalance in some tissues and genes indicate a caution in explaining functional SNPs or mutations in gene expression. In some cases, these differences may be attributed to epigenetic modifications/interactions particularly at imprinted loci.

3.4. RNA editing

Similarly, RNA-editing information can also be gleaned directly from reads derived through RNA-Seq (Peng et al., 2012). Smith et al. analyzed RNA editing in addition to allele-specific expression in their data set and discovered more sites undergoing RNA editing in DLPFC from 14 individuals (3249 sites) than comparing nine brain regions (2358 sites). Moreover, targeted sequencing in postmortem dorsolateral–prefrontal cortices from

20 schizophrenia, 20 bipolar disorder patients and 20 controls identified A-to-I RNA editing in 6 genes coding for neuroreceptor subunits using the 454 platform (Silberberg et al., 2012). Among 21 previously known and novel RNA editing sites on the 6 genes, only 1 site (in GRIA3) showed significantly lower editing ratio in schizophrenia comparing with controls and two sites (both in genes GRIK2, one marginally significant) in bipolar disorder comparing to controls. Interestingly, very large variance of editing level in schizophrenia has been observed on some editing sites, suggesting possible dysregulation in upstream where the capacity to maintain steady editing level was lost.

3.5. Integrative analysis

The analyses and results described above largely focused on only one dimension of transcriptome dynamics, ignoring, however, the interaction between different layers of regulation/function. Integrative analysis plays an irreplaceable role in connecting these components together. When we integrated differential expression and splicing from RNA-Seq data (Wang & Cairns, 2013, 2014), we found significant alterations in the total abundance and/or individual transcript composition in GABA receptor genes, although the overall expression levels of the GABA receptor genes were marginally altered. This finding is highly significant to our understanding of schizophrenia and demonstrates the power of integrative analysis for exploring complex disease mechanisms.

3.6. Noncoding RNA alterations in neurological disorders

In addition to protein coding genes, the dynamic transcriptome contains many other noncoding players that have historically not been well covered using microarray platforms. MicroRNAs (miRNAs) are a class of endogenous, functional, and small RNAs of length 21–24 nt, regulating up to hundreds of target genes after transcribed (Bartel, 2009; Shukla, Singh, & Barik, 2011). Abnormal miRNA expression causes various human diseases (Jiang et al., 2009), and thus disease-associated miRNAs can serve as biomarkers in disease diagnostics (Leidinger et al., 2013). RNA-Seq analysis offers some significant advantages over microarray for small RNA analysis and studies are beginning to emerge. High-throughput sequencing studies of small RNA are summarized in Table 6.3. Although there are only two such studies currently available, the two studies represent miRNA research in schizophrenia quite well—one was carried out on the postmortem brain tissue and

Table 6.3 Small RNA-Seq studies on schizophrenia and other neurological disorders

Study	Analysis	Tissue used	Sample size	Platform
Shi et al. (2012)	Differentially expressed miRNAs	Post-mortem prefrontal cortex (PFC or BA10) synaptosomes	Pooled samples from schizophrenia and controls	Illumina
Smalheiser et al. (2014)	Differentially expressed miRNAs/ miRNA biomarkers	Serum	115 schizophrenia and 40 controls	SOLiD

the other on blood focusing on biomarker discovery (Shi et al., 2012; Smalheiser et al., 2014).

Shi et al. performed two Illumina small RNA-Seq experiments: they first investigated miRNAs that were enriched in synaptosomes in normal human PFCs, and then they looked into miRNAs that had different abundance levels in schizophrenia synaptosomes comparing with normal ones. The results showed that a subset of miRNAs differentially enriched in synaptosomes comparing with the whole tissue, about 15% of miRNAs showing >1.5-fold enrichment and about 6% showing >2-fold. Interestingly, comparing with schizophrenia and control synaptosomes, they found that miRNAs highly enriched in normal synaptosomes were dramatically depleted in schizophrenia synaptosomes. For example, mir-219-5p showed the most highly enriched (fivefold) in synaptosomes relative to whole tissue, but in schizophrenia synaptosomes its abundance level decreased the most (70%). This phenomenon could partially explain the conflicting direction of miRNA changes in schizophrenia observed from the whole tissue and the synaptosomes and would also suggest the compartment-specific regulation of a subset of miRNAs, which may include miRNA transport, and differential processing or turnover (Goldie & Cairns, 2012).

In the other study by Smalheiser et al., serum samples from 115 subjects with schizophrenia and 40 healthy controls were analyzed to identify differentially expressed circulating miRNAs in schizophrenia (Table 6.3) (Smalheiser et al., 2014). From these data, they proposed serum miR-181b, miR-195, miR-219-2-3p, miR-1308, and let-7g could serve as biomarkers for schizophrenia diagnostics, with miR-181b achieving the highest prediction sensitivity (85%) and specificity (80%). It was noticed that some of the above circulating miRNAs exclusively expressed (miR-219-2-3p) or highly expressed (miR-181b and miR-195) in human brains, and therefore the source of such miRNAs in the serum potentially originated in the brain. These miRNAs were all found to be elevated in cortical gray matter in our studies of postmortem schizophrenia (Beveridge, Tooney, Gardiner, Carroll, & Cairns, 2010; Beveridge et al., 2008).

Long noncoding RNAs (lncRNAs) can also be profiled through transcriptome sequencing. Although this class of RNAs are of none or low protein-coding capacity, they are suggested to be functionally important in regulating other molecules. For example, a recent evidence showed that a lncRNA Gomafu was downregulated in cortical gray matter from postmortem superior temporal gyrus in schizophrenia, and this lncRNA can directly bind to splicing factors QKI and SRSF1, which suggested that the dysregulation of Gomafu may contribute to schizophrenia through alternative splicing in a few genes (Barry et al., 2014). Lin et al. (2011) performed RNA-Seq on iPSCs and differentiating neurons and discovered a number of lncRNAs differentially expressed during the transition besides many protein-coding genes, including a few schizophrenia candidate genes. Additional analysis showed that a few lncRNAs with expression increased in differentiating neurons were located near SNPs associated with schizophrenia and may be abnormally regulated in schizophrenia patients (Lin et al., 2011).

Circular RNAs (circRNAs) are formed by covalent linkage of both ends of linear RNA molecules as a consequence of back splicing events. Recently, properties and potential functions of circRNAs have been largely studied (Hansen et al., 2013; Memczak et al., 2013). As circRNAs are observed enriched in brain tissues, they would have particular associations with brain diseases and neurological disorders (Lukiw, 2013). Ghosal et al. provided a database with comprehensively annotated association between circRNAs and human disease using network-based computational predictions (Ghosal, Das, Sen, Basak, & Chakrabarti, 2013). Searching for schizophrenia-associated circRNAs on this database, five predictions are obtained, but all need further validation, possibly with RNA-Seq. Enrichment of circRNA in RNA-Seq analysis can be accomplished by an RNA exonuclease (RNase R) treatment prior to library construction (Jeck et al., 2013). This depletes linear RNA molecules and gives better coverage of these rare new regulatory RNAs.

4. DISCUSSION AND PERSPECTIVES

Recent progress in characterizing transcriptome dynamics in schizophrenia using RNA-Seq not only deepens our understanding on the expression changes of schizophrenia-associated genes but also extends our knowledge of schizophrenia-associated alternative splicing and, potentially, biased allelic gene expression and abnormal RNA editing. Prior to the wide application of high-throughput sequencing, transcriptome analysis in schizophrenia has already been profiled in numerous studies using microarray technologies. Much of the previously identified schizophrenia-associated changes have also been observed with RNA-Seq. For instance, we reported that half of the differentially expressed genes were previously observed in at least two microarray studies (Wu et al., 2012). However, a noticeable number of new genes with expression changes were only detectable by RNA-Seq, suggesting there is higher sensitivity in RNA-Seq because of its larger dynamic range and less bias, although some will be a result of heterogeneity. Enlarging the sample size in future studies would be desirable to provide a better coverage of variance in this complex population.

Although different brain regions have substantially different gene expression patterns, a few genes found significantly regulated toward the same direction. For example, neuropeptide B was upregulated in both the cerebellar cortex and the cerebral cortex (DLPFC), and interferon-induced transmembrane protein 2 (IFITM2) upregulated in both hippocampus and the DLPFC (Fillman et al., 2013; Hwang et al., 2013; Mudge et al., 2008). Moreover, cross-reference to microarray studies, 17 of 215 differentially expressed genes identified in cerebellar cortex have previously shown changes in cerebral cortex (DLPFC or superior temporal gurus) (Evans et al., 2003), including genes GABRA1, GOLGA1, and CACNG2. The same or similar schizophrenia-related expression changes of a few genes in different brain regions suggest that these genes might be regulated under a similar *trans* environment despite the regional difference, and the roles that they play might be similar across all these regions and need connecting all regions together to facilitate understanding.

To assess the linkage between the transcriptome dynamics and schizophrenia-associated SNPs or mutations, the most direct way is to check whether differentially regulated genes are overlapped with previous genome-wide association studies (GWASs). Allen et al. reported that several differentially expression genes in cerebellar cortices of schizophrenia subjects had also been previously shown association with schizophrenia in genetic association studies. The genes include CACNG2, GABRA1, GPSN2, HIRA, PSAP, RNABP5, and TCF4 (Allen et al., 2008). Xu et al. (2012) also showed a twofold enrichment than expectation of their detected differentially expressed genes overlapped with genetic association analysis results.

Such connections provide possible causal evidence of the functional SNPs/mutations, assuming that they have changed the local *cis*-regulatory elements. However, SNPs/mutations can also function through distal *cis*-elements or even *trans*-factors, which relationship could be constructed by analyzing eQTL.

The simultaneous availability of transcriptome dynamics profiling and genome-wide assay of genetic variation allows the mapping of eQTL, which would offer a more direct molecular basis for explaining the susceptibility SNPs/mutations identified through GWAS, rather than concealing biological mechanisms in the "black box" lying in between (Cookson, Liang, Abecasis, Moffatt, & Lathrop, 2009; Nica & Dermitzakis, 2013). Based on microarray technologies for gene expression profiling (Mirnics et al., 2000; Vawter et al., 2002), many eQTL studies on schizophrenia and other psychiatric disorders have already been performed (Kim, Cho, Lee, & Webster, 2012; Richards et al., 2012), not only validating the functional importance of particular SNPs/mutations in developing schizophrenia but also facilitating the discovery of molecular pathways relevant to the etiology of the disorder resulted from genetic polymorphism. Limited by microarrays, such analysis for QTLs was, however, restricted to only each gene's overall expression levels. Provided by RNA-Seq, multiple layers of transcriptome dynamics can also be involved in such analysis for a more comprehensive understanding of genetic effects on transcriptome alterations. The analysis of eQTL on isoform expression levels has already been proposed (Kwan et al., 2008) and, equivalently, exon eQTL to link exon-specific usage with genetic factors. In this sense, such strategies to bridge genetic variations and alternative splicing in transcriptome are call splicing QTL or sQTL (Oldmeadow et al., 2014). Although Oldmeadow et al. presented a paradigm for studying sQTL in schizophrenia, the sensitivity was plausibly limited by using exon-array data. Due to the advantages we discussed previously, RNA-Seq, rather than microarrays, provides more accurate measurements for all dimensions of transcriptome dynamics and thus can largely enhance the research on eQTL and sQTL, especially on large population studies (Montgomery et al., 2010; Pickrell et al., 2010).

Epigenetic and environmental factors may also affect gene expression dynamics (Jaenisch & Bird, 2003). Recently, many efforts on studying epigenetic effects on gene expression have been made (Dempster et al., 2011; Melas et al., 2012). Together with genetic and epigenetic effects, expression of regulatory genes and genes encoding regulatory proteins can largely determine the expression/splicing levels of their target genes and may further regulate RNA degradation and translocation, and protein synthesis and decay. To fully understand the transcriptome dynamics in schizophrenia and other neurological disorders, integrating different levels of data from its upstream and downstream would be critically important. Exploring the connection between different brain regions and peripheral tissues would also facilitate elucidating the underlying regulatory mechanisms and moreover the etiology of these disorders. Ultimately, new insight into the transcriptome and its underlying genetic and epigenetic influences will lead to new biomarkers and better diagnostic approaches, as well as new treatment strategies for improved clinical management of neurological and neuropsychiatric disorders.

REFERENCES

Allen, N. C., Bagade, S., McQueen, M. B., Ioannidis, J. P., Kavvoura, F. K., Khoury, M. J., et al. (2008). Systematic meta-analyses and field synopsis of genetic association studies in schizophrenia: The SzGene database. *Nature Genetics*, *40*(7), 827–834.

Arion, D., Unger, T., Lewis, D. A., Levitt, P., & Mirnics, K. (2007). Molecular evidence for increased expression of genes related to immune and chaperone function in the prefrontal cortex in schizophrenia. *Biological Psychiatry*, *62*(7), 711–721.

Aston, C., Jiang, L., & Sokolov, B. P. (2004). Microarray analysis of postmortem temporal cortex from patients with schizophrenia. *Journal of Neuroscience Research*, *77*(6), 858–866.

Au, K. F., Sebastiano, V., Afshar, P. T., Durruthy, J. D., Lee, L., Williams, B. A., et al. (2013). Characterization of the human ESC transcriptome by hybrid sequencing. *Proceedings of the National Academy of Sciences of the United States of America*, *110*(50), E4821–E4830.

Barry, G., Briggs, J. A., Vanichkina, D. P., Poth, E. M., Beveridge, N. J., Ratnu, V. S., et al. (2014). The long non-coding RNA Gomafu is acutely regulated in response to neuronal activation and involved in schizophrenia-associated alternative splicing. *Molecular Psychiatry*, *19*, 486–494.

Bartel, D. P. (2009). MicroRNAs: Target recognition and regulatory functions. *Cell*, *136*(2), 215–233.

Bauer, M., Gramlich, I., Polzin, S., & Patzelt, D. (2003). Quantification of mRNA degradation as possible indicator of postmortem interval—A pilot study. *Legal Medicine (Tokyo, Japan)*, *5*(4), 220–227.

Beveridge, N. J., & Cairns, M. J. (2012). MicroRNA dysregulation in schizophrenia. *Neurobiology of Disease*, *46*(2), 263–271.

Beveridge, N. J., Tooney, P. A., Carroll, A. P., Gardiner, E., Bowden, N., Scott, R. J., et al. (2008). Dysregulation of miRNA 181b in the temporal cortex in schizophrenia. *Human Molecular Genetics*, *17*(8), 1156–1168.

Beveridge, N. J., Tooney, P. A., Gardiner, E., Carroll, A. P., & Cairns, M. J. (2010). Schizophrenia is associated with an increase in cortical microRNA biogenesis. *Molecular Psychiatry*, *15*, 1176–1189.

Black, D. L., & Grabowski, P. J. (2003). Alternative pre-mRNA splicing and neuronal function. *Progress in Molecular and Subcellular Biology*, *31*, 187–216.

Brown, A. S., & Derkits, E. J. (2010). Prenatal infection and schizophrenia: A review of epidemiologic and translational studies. *The American Journal of Psychiatry*, *167*(3), 261–280.

Burgess, H. A., & Reiner, O. (2000). Doublecortin-like kinase is associated with microtubules in neuronal growth cones. *Molecular and Cellular Neurosciences*, *16*(5), 529–541.
Caceres, J. F., & Kornblihtt, A. R. (2002). Alternative splicing: Multiple control mechanisms and involvement in human disease. *Trends in Genetics*, *18*(4), 186–193.
Carter, S. L., Brechbuhler, C. M., Griffin, M., & Bond, A. T. (2004). Gene co-expression network topology provides a framework for molecular characterization of cellular state. *Bioinformatics*, *20*(14), 2242–2250.
Clinton, S. M., Haroutunian, V., Davis, K. L., & Meador-Woodruff, J. H. (2003). Altered transcript expression of NMDA receptor-associated postsynaptic proteins in the thalamus of subjects with schizophrenia. *The American Journal of Psychiatry*, *160*(6), 1100–1109.
Cookson, W., Liang, L., Abecasis, G., Moffatt, M., & Lathrop, M. (2009). Mapping complex disease traits with global gene expression. *Nature Reviews. Genetics*, *10*(3), 184–194.
Debanne, D., Daoudal, G., Sourdet, V., & Russier, M. (2003). Brain plasticity and ion channels. *Journal of Physiology, Paris*, *97*(4–6), 403–414.
Dempster, E. L., Pidsley, R., Schalkwyk, L. C., Owens, S., Georgiades, A., Kane, F., et al. (2011). Disease-associated epigenetic changes in monozygotic twins discordant for schizophrenia and bipolar disorder. *Human Molecular Genetics*, *20*(24), 4786–4796.
Dredge, B. K., Polydorides, A. D., & Darnell, R. B. (2001). The splice of life: Alternative splicing and neurological disease. *Nature Reviews. Neuroscience*, *2*(1), 43–50.
Drexhage, R. C., van der Heul-Nieuwenhuijsen, L., Padmos, R. C., van Beveren, N., Cohen, D., Versnel, M. A., et al. (2010). Inflammatory gene expression in monocytes of patients with schizophrenia: Overlap and difference with bipolar disorder. A study in naturalistically treated patients. *The International Journal of Neuropsychopharmacology*, *13*(10), 1369–1381.
Engstrom, P. G., Steijger, T., Sipos, B., Grant, G. R., Kahles, A., Alioto, T., et al. (2013). Systematic evaluation of spliced alignment programs for RNA-seq data. *Nature Methods*, *10*(12), 1185–1191.
Evans, S. J., Choudary, P. V., Vawter, M. P., Li, J., Meador-Woodruff, J. H., Lopez, J. F., et al. (2003). DNA microarray analysis of functionally discrete human brain regions reveals divergent transcriptional profiles. *Neurobiology of Disease*, *14*(2), 240–250.
Fields, R. D. (2005). Myelination: An overlooked mechanism of synaptic plasticity? *The Neuroscientist*, *11*(6), 528–531.
Fillman, S. G., Cloonan, N., Catts, V. S., Miller, L. C., Wong, J., McCrossin, T., et al. (2013). Increased inflammatory markers identified in the dorsolateral prefrontal cortex of individuals with schizophrenia. *Molecular Psychiatry*, *18*(2), 206–214.
Gardiner, E., Beveridge, N. J., Wu, J. Q., Carr, V., Scott, R. J., Tooney, P. A., et al. (2012). Imprinted DLK1-DIO3 region of 14q32 defines a schizophrenia-associated miRNA signature in peripheral blood mononuclear cells. *Molecular Psychiatry*, *17*(8), 827–840.
Gardiner, E. J., Cairns, M. J., Liu, B., Beveridge, N. J., Carr, V., Kelly, B., et al. (2013). Gene expression analysis reveals schizophrenia-associated dysregulation of immune pathways in peripheral blood mononuclear cells. *Journal of Psychiatric Research*, *47*(4), 425–437.
Ghosal, S., Das, S., Sen, R., Basak, P., & Chakrabarti, J. (2013). Circ2Traits: A comprehensive database for circular RNA potentially associated with disease and traits. *Frontiers in Genetics*, *4*, 283.
Glass, D., Vinuela, A., Davies, M. N., Ramasamy, A., Parts, L., Knowles, D., et al. (2013). Gene expression changes with age in skin, adipose tissue, blood and brain. *Genome Biology*, *14*(7), R75.
Goldie, B. J., & Cairns, M. J. (2012). Post-transcriptional trafficking and regulation of neuronal gene expression. *Molecular Neurobiology*, *45*(1), 99–108.
Hakak, Y., Walker, J. R., Li, C., Wong, W. H., Davis, K. L., Buxbaum, J. D., et al. (2001). Genome-wide expression analysis reveals dysregulation of myelination-related genes in

chronic schizophrenia. *Proceedings of the National Academy of Sciences of the United States of America*, *98*(8), 4746–4751.

Hansen, T. B., Jensen, T. I., Clausen, B. H., Bramsen, J. B., Finsen, B., Damgaard, C. K., et al. (2013). Natural RNA circles function as efficient microRNA sponges. *Nature*, *495*(7441), 384–388.

Hansen, K. D., Wu, Z., Irizarry, R. A., & Leek, J. T. (2011). Sequencing technology does not eliminate biological variability. *Nature Biotechnology*, *29*(7), 572–573.

Haroutunian, V., Katsel, P., Dracheva, S., Stewart, D. G., & Davis, K. L. (2007). Variations in oligodendrocyte-related gene expression across multiple cortical regions: Implications for the pathophysiology of schizophrenia. *The International Journal of Neuropsychopharmacology*, *10*(4), 565–573.

Hong, S., Chen, X., Jin, L., & Xiong, M. (2013). Canonical correlation analysis for RNA-seq co-expression networks. *Nucleic Acids Research*, *41*(8), e95.

Hwang, Y., Kim, J., Shin, J. Y., Kim, J. I., Seo, J. S., Webster, M. J., et al. (2013). Gene expression profiling by mRNA sequencing reveals increased expression of immune/inflammation-related genes in the hippocampus of individuals with schizophrenia. *Translational Psychiatry*, *3*, e321.

Jacobs, E. C., Bongarzone, E. R., Campagnoni, C. W., & Campagnoni, A. T. (2004). Embryonic expression of the soma-restricted products of the myelin proteolipid gene in motor neurons and muscle. *Neurochemical Research*, *29*(5), 997–1002.

Jaenisch, R., & Bird, A. (2003). Epigenetic regulation of gene expression: How the genome integrates intrinsic and environmental signals. *Nature Genetics*, *33*(Suppl.), 245–254.

Jahn, O., Tenzer, S., & Werner, H. B. (2009). Myelin proteomics: Molecular anatomy of an insulating sheath. *Molecular Neurobiology*, *40*(1), 55–72.

Jeck, W. R., Sorrentino, J. A., Wang, K., Slevin, M. K., Burd, C. E., Liu, J., et al. (2013). Circular RNAs are abundant, conserved, and associated with ALU repeats. *RNA*, *19*(2), 141–157.

Jiang, Q., Wang, Y., Hao, Y., Juan, L., Teng, M., Zhang, X., et al. (2009). miR2Disease: A manually curated database for microRNA deregulation in human disease. *Nucleic Acids Research*, *37*(Database issue), D98–D104.

Khaitovich, P., Muetzel, B., She, X., Lachmann, M., Hellmann, I., Dietzsch, J., et al. (2004). Regional patterns of gene expression in human and chimpanzee brains. *Genome Research*, *14*(8), 1462–1473.

Kim, S., Cho, H., Lee, D., & Webster, M. J. (2012). Association between SNPs and gene expression in multiple regions of the human brain. *Translational Psychiatry*, *2*, e113.

Kim, Y., Zerwas, S., Trace, S. E., & Sullivan, P. F. (2011). Schizophrenia genetics: Where next? *Schizophrenia Bulletin*, *37*(3), 456–463.

Kurian, S. M., Le-Niculescu, H., Patel, S. D., Bertram, D., Davis, J., Dike, C., et al. (2011). Identification of blood biomarkers for psychosis using convergent functional genomics. *Molecular Psychiatry*, *16*(1), 37–58.

Kwan, T., Benovoy, D., Dias, C., Gurd, S., Provencher, C., Beaulieu, P., et al. (2008). Genome-wide analysis of transcript isoform variation in humans. *Nature Genetics*, *40*(2), 225–231.

Leidinger, P., Backes, C., Deutscher, S., Schmitt, K., Mueller, S. C., Frese, K., et al. (2013). A blood based 12-miRNA signature of Alzheimer disease patients. *Genome Biology*, *14*(7), R78.

Li, J. B., & Church, G. M. (2013). Deciphering the functions and regulation of brain-enriched A-to-I RNA editing. *Nature Neuroscience*, *16*(11), 1518–1522.

Li, Y., Grupe, A., Rowland, C., Nowotny, P., Kauwe, J. S., Smemo, S., et al. (2006). DAPK1 variants are associated with Alzheimer's disease and allele-specific expression. *Human Molecular Genetics*, *15*(17), 2560–2568.

Licatalosi, D. D., & Darnell, R. B. (2006). Splicing regulation in neurologic disease. *Neuron*, *52*(1), 93–101.

Lin, M., Hrabovsky, A., Pedrosa, E., Wang, T., Zheng, D., & Lachman, H. M. (2012). Allele-biased expression in differentiating human neurons: Implications for neuropsychiatric disorders. *PLoS One*, 7(8), e44017.

Lin, M., Pedrosa, E., Shah, A., Hrabovsky, A., Maqbool, S., Zheng, D., et al. (2011). RNA-Seq of human neurons derived from iPS cells reveals candidate long non-coding RNAs involved in neurogenesis and neuropsychiatric disorders. *PLoS One*, *6*(9), e23356.

Lukiw, W. J. (2013). Circular RNA (circRNA) in Alzheimer's disease (AD). *Frontiers in Genetics*, *4*, 307.

Ma, X., & Zhang, X. (2013). NURD: An implementation of a new method to estimate isoform expression from non-uniform RNA-seq data. *BMC Bioinformatics*, *14*, 220.

Maycox, P. R., Kelly, F., Taylor, A., Bates, S., Reid, J., Logendra, R., et al. (2009). Analysis of gene expression in two large schizophrenia cohorts identifies multiple changes associated with nerve terminal function. *Molecular Psychiatry*, *14*(12), 1083–1094.

Mazin, P., Xiong, J., Liu, X., Yan, Z., Zhang, X., Li, M., et al. (2013). Widespread splicing changes in human brain development and aging. *Molecular Systems Biology*, *9*, 633.

McCarroll, S. A., & Hyman, S. E. (2013). Progress in the genetics of polygenic brain disorders: Significant new challenges for neurobiology. *Neuron*, *80*(3), 578–587.

Melas, P. A., Rogdaki, M., Osby, U., Schalling, M., Lavebratt, C., & Ekstrom, T. J. (2012). Epigenetic aberrations in leukocytes of patients with schizophrenia: Association of global DNA methylation with antipsychotic drug treatment and disease onset. *FASEB Journal*, *26*(6), 2712–2718.

Memczak, S., Jens, M., Elefsinioti, A., Torti, F., Krueger, J., Rybak, A., et al. (2013). Circular RNAs are a large class of animal RNAs with regulatory potency. *Nature*, *495*(7441), 333–338.

Miller, B. J., Buckley, P., Seabolt, W., Mellor, A., & Kirkpatrick, B. (2011). Meta-analysis of cytokine alterations in schizophrenia: Clinical status and antipsychotic effects. *Biological Psychiatry*, *70*(7), 663–671.

Mirnics, K., Middleton, F. A., Marquez, A., Lewis, D. A., & Levitt, P. (2000). Molecular characterization of schizophrenia viewed by microarray analysis of gene expression in prefrontal cortex. *Neuron*, *28*(1), 53–67.

Montgomery, S. B., Sammeth, M., Gutierrez-Arcelus, M., Lach, R. P., Ingle, C., Nisbett, J., et al. (2010). Transcriptome genetics using second generation sequencing in a Caucasian population. *Nature*, *464*(7289), 773–777.

Mortazavi, A., Williams, B. A., McCue, K., Schaeffer, L., & Wold, B. (2008). Mapping and quantifying mammalian transcriptomes by RNA-Seq. *Nature Methods*, *5*(7), 621–628.

Mowry, B. J., & Gratten, J. (2013). The emerging spectrum of allelic variation in schizophrenia: Current evidence and strategies for the identification and functional characterization of common and rare variants. *Molecular Psychiatry*, *18*(1), 38–52.

Mudge, J., Miller, N. A., Khrebtukova, I., Lindquist, I. E., May, G. D., Huntley, J. J., et al. (2008). Genomic convergence analysis of schizophrenia: mRNA sequencing reveals altered synaptic vesicular transport in post-mortem cerebellum. *PLoS ONE*, *3*(11), e3625.

Nica, A. C., & Dermitzakis, E. T. (2013). Expression quantitative trait loci: Present and future. *Philosophical Transactions of the Royal Society of London. Series B, Biological Sciences*, *368*(1620), 20120362.

Oldmeadow, C., Mossman, D., Evans, T. J., Holliday, E. G., Tooney, P. A., Cairns, M. J., et al. (2014). Combined analysis of exon splicing and genome wide polymorphism data predict schizophrenia risk loci. *Journal of Psychiatric Research*, *52*, 44–49.

Owen, M. J. (2012). Implications of genetic findings for understanding schizophrenia. *Schizophrenia Bulletin*, *38*(5), 904–907.

Ozsolak, F., & Milos, P. M. (2011). RNA sequencing: Advances, challenges and opportunities. *Nature Reviews. Genetics*, *12*(2), 87–98.

Park, E., Williams, B., Wold, B. J., & Mortazavi, A. (2012). RNA editing in the human ENCODE RNA-seq data. *Genome Research*, *22*(9), 1626–1633.

Peng, Z., Cheng, Y., Tan, B. C., Kang, L., Tian, Z., Zhu, Y., et al. (2012). Comprehensive analysis of RNA-Seq data reveals extensive RNA editing in a human transcriptome. *Nature Biotechnology*, *30*(3), 253–260.

Pickrell, J. K., Marioni, J. C., Pai, A. A., Degner, J. F., Engelhardt, B. E., Nkadori, E., et al. (2010). Understanding mechanisms underlying human gene expression variation with RNA sequencing. *Nature*, *464*(7289), 768–772.

Raimondo, J. V., Markram, H., & Akerman, C. J. (2012). Short-term ionic plasticity at GABAergic synapses. *Frontiers in Synaptic Neuroscience*, *4*, 5.

Richards, A. L., Jones, L., Moskvina, V., Kirov, G., Gejman, P. V., Levinson, D. F., et al. (2012). Schizophrenia susceptibility alleles are enriched for alleles that affect gene expression in adult human brain. *Molecular Psychiatry*, *17*(2), 193–201.

Saetre, P., Emilsson, L., Axelsson, E., Kreuger, J., Lindholm, E., & Jazin, E. (2007). Inflammation-related genes up-regulated in schizophrenia brains. *BMC Psychiatry*, 7, 46.

Sequeira, P. A., Martin, M. V., & Vawter, M. P. (2012). The first decade and beyond of transcriptional profiling in schizophrenia. *Neurobiology of Disease*, *45*(1), 23–36.

Sharon, D., Tilgner, H., Grubert, F., & Snyder, M. (2013). A single-molecule long-read survey of the human transcriptome. *Nature Biotechnology*, *31*(11), 1009–1014.

Shi, W., Du, J., Qi, Y., Liang, G., Wang, T., Li, S., et al. (2012). Aberrant expression of serum miRNAs in schizophrenia. *Journal of Psychiatric Research*, *46*(2), 198–204.

Shukla, G. C., Singh, J., & Barik, S. (2011). MicroRNAs: Processing, maturation, target recognition and regulatory functions. *Molecular and Cellular Pharmacology*, *3*(3), 83–92.

Silberberg, G., Lundin, D., Navon, R., & Ohman, M. (2012). Deregulation of the A-to-I RNA editing mechanism in psychiatric disorders. *Human Molecular Genetics*, *21*(2), 311–321.

Silverman, M. A., Benard, O., Jaaro, H., Rattner, A., Citri, Y., & Seger, R. (1999). CPG16, a novel protein serine/threonine kinase downstream of cAMP-dependent protein kinase. *The Journal of Biological Chemistry*, *274*(5), 2631–2636.

Smalheiser, N. R., Lugli, G., Zhang, H., Rizavi, H., Cook, E. H., & Dwivedi, Y. (2014). Expression of microRNAs and other small RNAs in prefrontal cortex in schizophrenia, bipolar disorder and depressed subjects. *PLoS ONE*, *9*(1), e86469.

Smith, R. M., Webb, A., Papp, A. C., Newman, L. C., Handelman, S. K., Suhy, A., et al. (2013). Whole transcriptome RNA-Seq allelic expression in human brain. *BMC Genomics*, *14*, 571.

Sporkel, O., Uschkureit, T., Bussow, H., & Stoffel, W. (2002). Oligodendrocytes expressing exclusively the DM20 isoform of the proteolipid protein gene: Myelination and development. *Glia*, *37*(1), 19–30.

Stan, A. D., Ghose, S., Gao, X. M., Roberts, R. C., Lewis-Amezcua, K., Hatanpaa, K. J., et al. (2006). Human postmortem tissue: What quality markers matter? *Brain Research*, *1123*(1), 1–11.

Steijger, T., Abril, J. F., Engstrom, P. G., Kokocinski, F., Akerman, M., Alioto, T., et al. (2013). Assessment of transcript reconstruction methods for RNA-seq. *Nature Methods*, *10*(12), 1177–1184.

Subramanian, A., Tamayo, P., Mootha, V. K., Mukherjee, S., Ebert, B. L., Gillette, M. A., et al. (2005). Gene set enrichment analysis: A knowledge-based approach for interpreting genome-wide expression profiles. *Proceedings of the National Academy of Sciences of the United States of America*, *102*(43), 15545–15550.

Tazi, J., Bakkour, N., & Stamm, S. (2009). Alternative splicing and disease. *Biochimica et Biophysica Acta*, *1792*(1), 14–26.

Tiwari, A. K., Zai, C. C., Muller, D. J., & Kennedy, J. L. (2010). Genetics in schizophrenia: Where are we and what next? *Dialogues in Clinical Neuroscience*, *12*(3), 289–303.

Torrey, E. F., Miller, J., Rawlings, R., & Yolken, R. H. (1997). Seasonality of births in schizophrenia and bipolar disorder: A review of the literature. *Schizophrenia Research*, *28*(1), 1–38.

Trapnell, C., Williams, B. A., Pertea, G., Mortazavi, A., Kwan, G., van Baren, M. J., et al. (2010). Transcript assembly and quantification by RNA-Seq reveals unannotated transcripts and isoform switching during cell differentiation. *Nature Biotechnology*, *28*(5), 511–515.

Vawter, M. P., Crook, J. M., Hyde, T. M., Kleinman, J. E., Weinberger, D. R., Becker, K. G., et al. (2002). Microarray analysis of gene expression in the prefrontal cortex in schizophrenia: A preliminary study. *Schizophrenia Research*, *58*(1), 11–20.

Vawter, M. P., Evans, S., Choudary, P., Tomita, H., Meador-Woodruff, J., Molnar, M., et al. (2004). Gender-specific gene expression in post-mortem human brain: Localization to sex chromosomes. *Neuropsychopharmacology*, *29*(2), 373–384.

Wang, X., & Cairns, M. J. (2013). Gene set enrichment analysis of RNA-Seq data: Integrating differential expression and splicing. *BMC Bioinformatics*, *14*(Suppl 5), S16.

Wang, X., & Cairns, M. J. (2014). SeqGSEA: A bioconductor package for gene set enrichment analysis of RNA-Seq data integrating differential expression and splicing. *Bioinformatics*, *30*, 1777–1779.

Wang, E. T., Sandberg, R., Luo, S., Khrebtukova, I., Zhang, L., Mayr, C., et al. (2008). Alternative isoform regulation in human tissue transcriptomes. *Nature*, *456*(7221), 470–476.

Wu, J. Q., Wang, X., Beveridge, N. J., Tooney, P. A., Scott, R. J., Carr, V. J., et al. (2012). Transcriptome sequencing revealed significant alteration of cortical promoter usage and splicing in schizophrenia. *PLoS One*, 7(4), e36351.

Xu, J., Sun, J., Chen, J., Wang, L., Li, A., Helm, M., et al. (2012). RNA-Seq analysis implicates dysregulation of the immune system in schizophrenia. *BMC Genomics*, *13*(Suppl. 8), S2.

Zhang, K., Li, J. B., Gao, Y., Egli, D., Xie, B., Deng, J., et al. (2009). Digital RNA allelotyping reveals tissue-specific and allele-specific gene expression in human. *Nature Methods*, *6*(8), 613–618.

CHAPTER SEVEN

The Central Role of Noncoding RNA in the Brain

Boris Guennewig[*,†,1], **Antony A. Cooper**[*,†,1]
*Garvan Institute of Medical Research, Darlinghurst, New South Wales, Australia
†St Vincent's Clinical School and BABS UNSW, Sydney, New South Wales, Australia
[1]Corresponding authors: e-mail address: b.guennewig@garvan.org.au; a.cooper@garvan.org.au

Contents

Abstract

Less than 3% of the human genome generates protein-coding transcripts; the majority, far from being strewn with evolutionary "junk," is dynamically transcribed into non(protein)-coding RNAs (ncRNAs). These ncRNAs provide another provide another, previously hidden, level of regulatory information that appears to be involved in hard- and

International Review of Neurobiology, Volume 116
ISSN 0074-7742
http://dx.doi.org/10.1016/B978-0-12-801105-8.00007-2

soft-wired epigentic processes. The extensive and intricate level of gene regulation provided by ncRNAs may be the major driver for the accelerated development of the human brain and its associated increase in complexity and cognition. Support for this is provided by the correlation between the evolutionary increase of complexity in the nonprotein-coding transcriptome paralleling cognitive evolution in primates, in contrast to the coincidently modest evolutionary changes of the protein-coding transcriptome. The essential role of these regulatory RNAs is reflected in almost every aspect in neuroscience, including chromatin modification, transcriptional regulation, alternative splicing, RNA editing and translation. Dissecting this plethora of regulatory networks and editing events, which are orchestrated through long and small noncoding RNAs, and their interaction with transcription factors, chromatin-modifying enzymes, and other protein effectors will provide essential insights into the transcriptional complexity and plasticity in the development and function of the human brain. Such complexity provides susceptibility to internal and external perturbations, which in rare cases might act as evolutionary catalysts, but in many cases could manifest as neuropsychiatric or neurodegenerative diseases. NcRNAs (especially lncRNAs) are therefore excellent candidates for both disease biomarkers and disease-ameliorating therapies.

1. INTRODUCTION

The established DNA → RNA → protein paradigm of biology is currently receiving a significant renovation to accommodate the recently discovered impact of regulatory RNA. This revolutionary shift has been facilitated by the genome-wide sequencing effort of multiple groups utilizing recent advances in sequencing technology. The discovery of introns in the 1970s as nontranslated (and thus noncoding) intragenic RNA was interpreted as "junk" RNA. This concept was questioned for the first time in the 1990s by the alternative proposal that intronic and other noncoding RNAs (ncRNAs) provide a highly parallel regulatory information network required for the development of multicellular organisms (Mattick, 1994). The full extent of ncRNA transcription is reflected in the fact that, while <3% of the human genome encodes protein-coding exons, >80% is transcribed into ncRNA (Djebali et al., 2012; Hangauer, Vaughn, & McManus, 2013).

A large proportion of ncRNAs show primate specificity and the rapid increase of ncRNA numbers, sequence, and expression complexity (Necsulea et al., 2014) parallels the rapid cognitive evolution of primates. This contrasts with only modest evolutionary changes in the protein-coding portion of the genome (Capra, Erwin, McKinsey, Rubenstein, & Pollard, 2013). It has therefore been postulated that the increased capacity in primate brains was facilitated by the rapid increase in the number of ncRNAs and their expanded degree of tissue-specific expression (Taft, Pheasant, & Mattick, 2007). Rapidly accumulating recent data (Kutter et al., 2012; Liu, Mattick, & Taft, 2013; Necsulea et al., 2014) indicate that ncRNAs (specifically lncRNAs) have expanded with increasing developmental complexity and might be the major driver of cognitive evolution.

The role of ncRNA is predominantly regulatory, playing complex and crucial roles in cell function, developmental regulation, evolutionary expansion, as well as disease initiation and progression. Current data indicate that ncRNAs achieve this by regulating gene expression at a number of levels including epigenetic modification, enhancer function, alternative RNA splicing, and translation (Mercer, Dinger, Trapnell, & Mattick, 2012; Mercer & Mattick, 2013a, 2013b; Ng, Lin, Soh, & Stanton, 2013). Expression profiling in different tissues and developmental systems has shown that the majority of ncRNA transcripts are actively regulated, spliced, and localized, especially in the brain, with tissue-specific and stimulus-dependent expression (Dinger et al., 2008; Dinger, Pang, et al., 2009; Mercer, Dinger, Sunkin, Mehler, & Mattick, 2008; Necsulea et al., 2014). In addition to encoding regulatory RNAs, noncoding regions in the genome can also serve other regulatory functions by providing binding sites for both ncRNA and regulatory proteins, such as transcription factors.

The human brain best exemplifies the maximal diversity and complexity of the human transcriptional landscape. To perform its functions, many of which appear either unique to or far more advanced in *Homo sapiens*. The human brain has developed as an intricately sophisticated network comprising ~90 billion neurons with trillions of synaptic connections. This remarkable organizational feat is achieved using a similar number of protein-coding genes as that of *C. elegans* (which has 302 neurons), through extensive enhancement in complexity of the human brain's transcriptome to provide a vast array of molecular and cellular specialization. ncRNAs, many of them lncRNAs specific to primate brains, contribute substantially to the brain's enhanced transcriptome providing complex regulation and modulation gene expression at almost every level. lncRNAs, themselves showing specific and highly regulated developmental and spatiotemporal expression, bind to chromatin-modifying proteins and recruit their catalytic activity to specific sites in the genome, to modulate chromatin states and impact regulation of gene expression. The abundance of lncRNAs (predicted to reach as many as 56,000 distinct transcripts; Xie et al., 2014) in combination with this regulatory potential suggests that lncRNAs may be part of a broad epigenetic regulatory network.

The brain's transcriptional diversity is also expanded by alternative splicing, which is profuse in the brain, of both mRNAs and lncRNAs. RNA editing, particularly of lncRNAs, additionally modulates the repertoire of targets and further expands the sphere of regulatory influence. These mechanisms likely underpin the dynamic changes in the transcriptome of individual neurons in response to developmental and external environmental cues, including memory formation and neuroplasticity.

The brain's enhanced sophistication mediated through ncRNA regulatory mechanisms also provides a vulnerability to dysfunction and disease. Dysregulation of ncRNA has already been identified as contributing to neurodegeneration and disorders of the brain.

The plethora of ncRNA functions and their potential role in the swift increase in size and complexity in primate/human brain will be highlighted in this review to showcase the central role of regulatory RNA in the brain.

2. THE LONG AND SHORT OF NONCODING RNAs

Defined by size, ncRNA molecules are classified as either short (<200 nt) or long (>200 nt). Long noncoding RNAs (lncRNAs) are a heterogeneous group (see below for classification) that can reach multiple kilobases in length and are predominantly located in the nucleus. Small noncoding RNAs (sncRNAs) include the well-studied transfer RNAs (tRNAs) and ribosomal RNAs (rRNAs) and the more recently discovered short small interference RNAs (siRNAs), piwi-interacting RNAs (piRNAs), small nucleolar RNAs (snoRNAs), y-RNAs, and the prominent microRNAs (miRNAs).

Bifunctional transcripts span these boundaries of coding/noncoding RNA or short/long ncRNAs in many ways that include ncRNA transcripts encoding previously undiscovered small proteins like the steroid receptor RNA activator and tumor protein (TP53) (Candeias, 2011; Cooper et al., 2011; Kageyama, Kondo, & Hashimoto, 2011; Kapranov et al., 2007). Bifunctionality also occurs among the ncRNA, where some lncRNAs can be processed into miRNAs (Dieci, Fiorino, Castelnuovo, Teichmann, & Pagano, 2007; Necsulea et al., 2014).

2.1. Short noncoding RNAs

Although <200 nt in their mature form, sncRNAs are often generated from longer primary transcripts, including introns, by endogenous RNA cleavage enzymes. Their final size, function, or localization pattern define their nomenclature. sncRNAs have been reported to play a major role in embryonic neuronal differentiation (Landgraf et al., 2007; Skreka et al., 2012) and a clear spatiotemporal differentiated expression in the human brain (Kang et al., 2011; Nowak & Michlewski, 2013; Ziats & Rennert, 2013). Inhibition of the sncRNA biogenesis results in a range of phenotypes, due to constrained differentiation in various tissues and mechanistic deficits as well as neurodegeneration (Cuellar et al., 2008; Davis et al., 2008; Haramati et al., 2010; Yoo et al., 2011).

Here, we will discuss the various functions of sncRNAs according to the role they are playing in gene expression (Fig. 7.1).

The 18 nt long class of transcription initiation RNAs (tiRNAs) map within 60–120 nt of transcription start sites (Taft, Glazov, Cloonan, et al., 2009). They are located in the nucleus and have been identified to bind to the sense and antisense DNA strand as well as up- and downstream to the transcription start site of their origin (Taft, Simons, et al., 2010). tiRNAs also show epigenetic modifications associated with transcription initiation indicated through enrichment with CCCTC-binding factor (Taft, Glazov, Cloonan, et al., 2009; Taft, Kaplan, Simons, & Mattick, 2009; Taft, Simons, et al., 2010). A body of work showed that these combined features affect the transcription level of the p21 locus and indicates that tiRNAs can be classified as another functional group of transcriptional regulators (Taft, Simons, et al., 2010). Splice site RNAs (spli-RNAs, 17–18 nt) show precise mapping to the splice donor site of internal exons in animals. Similar to tiRNAs, spli-RNAs are weakly expressed, but display developmental and tissue-specific expression (Taft, Simons, et al., 2010). Additionally, they present 3′ terminal guanine enrichment in their sequence, which might be explained due to the complementary 5′ consensus sequence of splice sites. Small nuclear RNAs (snRNA; 100–150 nt) or U-RNA are vital components of the spliceosome, where they contribute to the orchestration of transcript splicing (Guo et al., 2009).

Infrastructural ncRNAs are classified as ubiquitously expressed RNAs, which do not perform regulatory functions, e.g. transfer, ribosomal, mitochondrial, and snoRNAs. rRNA, despite being a hindrance due to their abundance in current sequencing approaches, is the core of the ribosome.

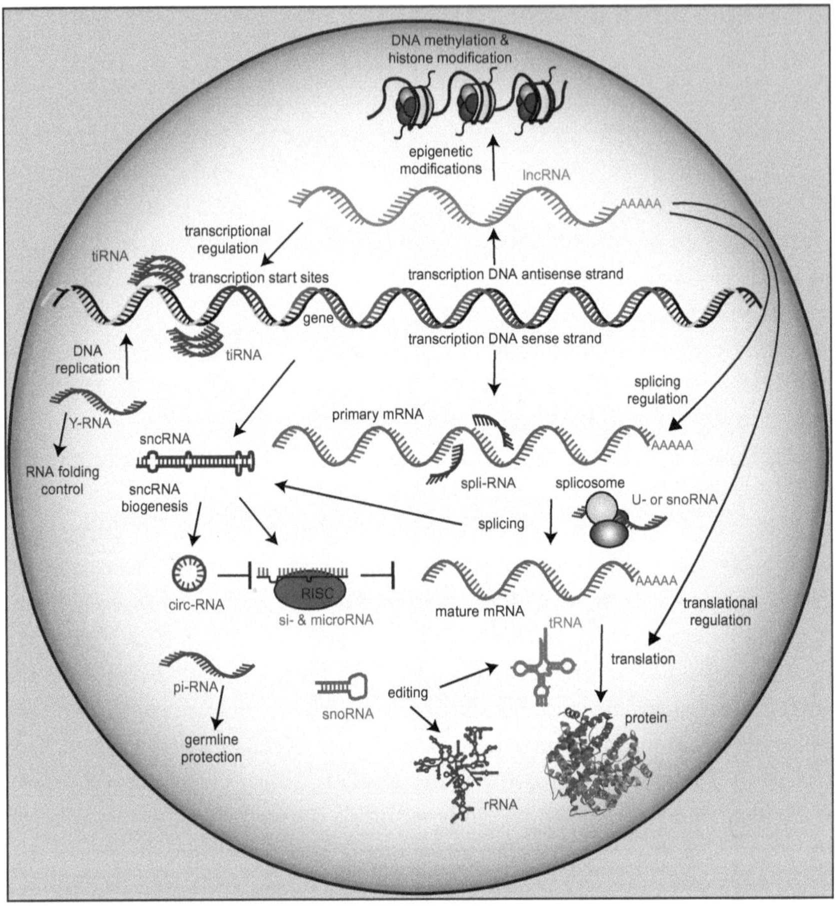

Figure 7.1 Schematic ncRNA roles and functions around a protein-coding *locus*. Depicted in this scheme is the central paradigm of biology, the transcription of a protein-coding gene, the splicing of the primary transcript, and its subsequent translation into proteins. Complementary depicted is the transcription from the antisense strand, which gives birth to a subfamily of lncRNAs and a canonical biogenesis of sncRNAs either from a noncoding gene or with introns as there origin. We show exemplary roles of lncRNA function that impacts upon diverse biological processes (Djebali et al., 2012; Hangauer et al., 2013; Mercer, Dinger, & Mattick, 2009; Zhang, Zhang, et al., 2012), including epigenetic modification (Gupta et al., 2010; Mercer, Dinger, Trapnell, & Mattick, 2012; Mercer & Mattick, 2013a; Ng et al., 2013), transcriptional regulation (Dinger et al., 2008; Dinger, Pang, et al., 2009; Feng et al., 2006; Mercer et al., 2008; Necsulea et al., 2014), splicing (Derrien et al., 2012; Ponjavic, Ponting, & Lunter, 2007; Smith, Gesell, Stadler, & Mattick, 2013; Yan et al., 2005), and translation (Dinger, Amaral, et al., 2009; ENCODE Project Consortium et al., 2012; Wang et al., 2005). We show the role of (i) tiRNA in transcriptional initiation upstream of transcription start sites (Taft, Glazov, Cloonan, et al., 2009), (ii) si- and miRNA in their function in posttranscriptional

rRNA introns can additionally be processed into miRNAs which unveil an additional layer of translational regulation (Son et al., 2013). Transfer RNAs or tRNA are adapter molecules with their well-described role in translation. These molecules undergo various editing events, with new modifications still being revealed to researchers even today (Kim et al., 2013). Mediators of these modifications are snoRNAs, which guide chemical modification of other RNA molecules including rRNA and tRNA (Bachellerie, Cavaillé, & Huttenhofer, 2002). snoRNAs can also act as precursors for miRNAs and sno-derived RNAs (sdRNAs) (Taft, Glazov, Lassmann, et al., 2009). These infrastructural ncRNAs are not restricted to the nuclear genome as the human mitochondrial genome encodes also ncRNAs (mito-RNAs), which to date comprise of 22 tRNAs and 2 rRNAs with a broad variance in their expression pattern (Mercer et al., 2011). Additionally, there are reports of various small mito-RNAs (Ro et al., 2013) and three long ncRNAs (Rackham et al., 2011) expressed in mitochondria.

Posttranscriptional sncRNAs often mediate their function via the RNA-induced silencing complex or RISC. miRNA and siRNA, which belong to a group of RNA interference molecules, are the best-studied regulatory ncRNAs to date. We describe the miRNA canonical biogenesis as one example for other sncRNAs here. It starts with the transcription of the primary miRNA (pri-miRNA) by either RNA polymerase II (Cai, Hagedorn, & Cullen, 2004; Lee et al., 2004) or III (Borchert, Lanier, & Davidson, 2006). They can originate from introns of annotated protein-coding genes (Kim & Kim, 2007) or from intergenic or antisense regions (Lagos-Quintana, Rauhut, Lendeckel, & Tuschl, 2001; Lau, Lim, Weinstein, & Bartel, 2001; Lee & Ambros, 2001; Mourelatos et al., 2002). Pri-miRNAs have a characteristic hairpin structure with unstructured flanking sequences. This results in an RNA "fold-back" structure (hairpin), which is recognized and processed by the RNase III DROSHA (Lee et al., 2003) and its cofactor DGCR8 (Denli, Tops, Plasterk, Ketting, &

regulation of protein expression and their recently discovered sponge inhibitors circRNA, (iii) snoRNA and their mediating function in the modification of infrastructural RNAs like rRNA and tRNA, and (iv) piRNA to protect the germline from transposable elements. In the process of splicing, we show the spli-RNAs, which show precise mapping to the splice donor site of internal exons in animals (Taft, Simons, et al., 2010) and snRNAs, like U2-snoRNA, which are vital components of the spliceosome (Guo, Karunatilaka, & Rueda, 2009; Padgett, Grabowski, Konarska, Seiler, & Sharp, 1986; Ruby, Jan, & Bartel, 2007).

Hannon, 2004; Gregory et al., 2004; Han et al., 2004; Landthaler, Yalcin, & Tuschl, 2004). The product of this endonucleolytic cleavage is a precursor miRNA (pre-miRNA) which is exported to the cytoplasm and is the major source of the double-stranded mature miRNAs (mat-miRs) resulting from RNase III Dicer1 cleavage (Grishok et al., 2001; Ketting et al., 2001; Knight & Bass, 2001; McLachlan, Pasquinelli, Bálint, Tuschl, & Zamore, 2001; Zhang, Kolb, Jaskiewicz, Westhof, & Filipowicz, 2004). The double-stranded RNA (dsRNA) species is unraveled and one of the strands is subsequently loaded into a complex containing one of the four human argonaute proteins (Wang et al., 2009) resulting in the RISC. The RISC mediates the repressive function of the miRNA through partial complementarity to targeted mRNA transcripts. Each of these so far 1872 identified pre-miRNA molecules (miRBase; Griffiths-Jones, Grocock, van Dongen, Bateman, & Enright, 2006) is capable of targeting thousands of mRNAs either with their 5p-, 3p-, or through dual activity (Guennewig et al., 2014). It is predicted that 60% of all expressed proteins are modulated through miRNAs. Although originally thought to be predominantly cytosolic, given both their biogenesis location and interaction with 3′ UTRs of coding mRNA, recent studies have shown that the majority of the mature miRNAs are located in the nucleus of neuronal stem cells (Jeffries, Fried, & Perkins, 2011). A current study unveiled the interaction of miRNA-9 with Malat-1 in the nucleus, which interconnects two ncRNA classes (miRNA and lncRNA) in novel manner (Leucci et al., 2013).

Y-RNAs are similar in size to miRNAs, but have a distinct biogenesis (Nicolas, Hall, Csorba, Turnbull, & Dalmay, 2012). They are associated with DNA replication and a part of the Ro60 complex (Hall, Turnbull, & Dalmay, 2013), which contributes to the quality control of RNA folding (Reinisch & Wolin, 2007).

Another group of ncRNAs are newly identified molecules called circular RNA (circRNA). These noncoding regulatory molecules have recently been identified to be spatiotemporal expressed and show stable circRNAs in a variety of tissue, species, and developmental stages (Memczak et al., 2013).

piRNAs are slightly longer (26–31 nt) than miRNA, with a Dicer-independent biogenesis mechanism that still has to be determined. Their conserved function is to protect germline cells from mobile elements like transposons. piRNAs are themselves generated from sense or antisense transposons and are therefore to a large extent sequence identical to their origin (Siomi, Sato, Pezic, & Aravin, 2011). Their biogenesis includes the interaction with the piwi-clad proteins whose slicer activity leads to the processing of the piRNA antisense strand, which then itself acts as a lead against the original sequence. Downregulation of proteins associated with piwi biogenesis leads to accumulation of retrotransposons and LINE elements in the germline with a potential effect of increased mutagenesis (Malone & Hannon, 2009).

These assorted regulatory functions of sncRNAs showcase the immense interlaced and adaptable complexity of these regulators, and their influence leads to multiple associations in a broad spectrum of disease (see section 11. ncRNAs and Disease).

2.2. Long noncoding RNAs

Distinct from sncRNAs, lncRNAs are RNA transcripts of >200 nt that function by binding RNA, DNA, or proteins to modulate transcription, alternative splicing, mRNA stability, mRNA translation, and epigenetic events, such as chromatin remodeling (Carrieri et al., 2012; Mercer et al., 2009; Qureshi & Mehler, 2013; Taft, Pang, Mercer, Dinger, & Mattick, 2010). Recent annotations of the human transcriptome estimate that the human genome encodes ~25,000 lncRNAs, a number larger than protein-coding genes, with recent indication of over 50,000 potential candidates (Carninci et al., 2005; Hangauer et al., 2013; Xie et al., 2014). lncRNA transcripts are in many cases produced like classic mRNAs via polymerase II activity, with such hallmarks as histone modification, transcription initiation, and elongation (5′ capping and polyadenylation) (Carninci et al., 2005; Derrien et al., 2012; Guttman & Rinn, 2012; Kapranov et al., 2007; Kutter et al., 2012; Liu et al., 2013; Necsulea et al., 2014; Ponjavic et al., 2007; Taft et al., 2007). Nonpolyadenylated lncRNAs have also been identified and point to the involvement of RNA polymerase III procession (Candeias, 2011; Cooper et al., 2011; Dieci et al., 2007; Kageyama et al., 2011; Kapranov et al., 2007; Necsulea et al., 2014) or that they are products of splicing (Johnston & Hobert, 2003; Yin et al., 2012). Most nuclear-encoded lncRNAs are restricted to the nucleus (Derrien et al., 2012; Kapranov et al., 2007) with exceptions showing functionality in the cytoplasm (Carrieri et al., 2012; Mercer et al., 2009; Qureshi & Mehler, 2013; Taft, Pang, et al., 2010).

The scarcity of protein-coding capacity, the lack of sequence conservation, the low fidelity/high promiscuity of RNA polymerase II (Ponjavic et al., 2007; Struhl, 2007), and the typically low expression levels relative to mRNA (Hangauer et al., 2013; He, Vogelstein, Velculescu, Papadopoulos, & Kinzler, 2008; Mattick & Makunin, 2006) have all contributed to the still active controversy questioning the relevance and functionality of lncRNA. Data supporting the likely functionality of lncRNAs include expression profiling in divergent tissues and developmental systems showing that the majority of lncRNA transcripts are expressed in a highly regulated manner (Dinger et al., 2008; Dinger, Pang, et al., 2009; Hangauer et al., 2013; Katayama et al., 2005; Mercer et al., 2008, 2012; Okada et al., 2008) and the evolutionary conservation of lncRNA promoters, splice sites, and secondary structure (Atkinson, Marguerat, & Bahler, 2012; Dinger, Amaral, Mercer, & Mattick, 2009; Ponjavic, Oliver, Lunter, & Ponting, 2009; Smith, Gesell, et al., 2013; Ulitsky, Shkumatava, Jan, Sive, & Bartel, 2011). Also regulatory molecules such as lncRNAs, especially when they act upstream (e.g., direct interaction with a specific genomic locus), need very few copies to generate an effect, which multiplies through a cascade (comparable to hormones or transcription factors). Indeed, although a field in its infancy, >150 lncRNAs already have an ascribed biological function and the number is growing rapidly (Amaral, Clark, Gascoigne, Dinger, & Mattick, 2011). A large fraction of lncRNAs (849 out of 1328 examined) are expressed in mouse brain tissues and are easily detectable by *in situ* hybridization in specific cells in, for example, the hippocampus, cortex, or cerebellum (Mercer et al., 2008).

3. TYPES AND FUNCTION OF lncRNAs

Like mRNA, lncRNAs themselves typically consist of multiple exons and undergo alternative splicing to produce a diverse range of transcripts (Derrien et al., 2012; Wang et al., 2005). Even though the highest proportion (42%) of lncRNAs consists of two exons, every combination imaginable is represented. Unlike other classes of ncRNAs, such as miRNAs and snoRNAs that have related functions, lncRNAs exhibit a broad range of functions that impact upon diverse biological processes (Djebali et al., 2012; Hangauer et al., 2013; Mercer et al., 2009; Zhang, Zhang, et al., 2012), including epigenetic modification (Gupta et al., 2010; Mercer et al., 2012; Mercer & Mattick, 2013a; Ng et al., 2013), transcriptional regulation (Dinger et al., 2008; Dinger, Pang, et al., 2009; Feng et al., 2006; Mercer et al., 2008; Necsulea et al., 2014), splicing (Derrien et al., 2012; Ponjavic et al., 2007; Smith, Gesell, et al., 2013; Yan et al., 2005), translation

(Dinger, Amaral, et al., 2009; ENCODE Project Consortium et al., 2012; Wang et al., 2005), and structure and organization of cellular components (Necsulea et al., 2014; Sunwoo et al., 2009).

3.1. Current lncRNA classification according to origin and function

A frequently employed classification system of lncRNAs is based on their genomic origin and/or their functional role. Since the functional characterization of these molecules is under extensive investigation with an increased publicationrate in recent years (255 (2011), 371 (2012), and 476 (2013) in a PubMed search for "long noncoding RNA"), we face an expanding nomenclature.

Intronic lncRNAs are transcribed from within an intron of a known protein-coding locus in either sense or antisense orientation. Due to their location, they are most often under the same transcriptional regulation as the surrounding gene, with very similar spatiotemporal expression pattern. Such intronic lncRNAs frequently function to stabilize or regulate alternative splicing of the flanking protein-coding transcript (Nakaya et al., 2007). In *cis*-acting examples such as ANRASSF1 and other intronic lncRNAs, the transcript recruits the polycomb repression complex to their promoter and facilitates an epigenetic modification, which leads to reduced expression patterns (Guil et al., 2012; Hangauer et al., 2013; Zhao et al., 2010). Intronic lncRNAs with *trans*-acting functions and differing expression pattern to their surrounding *loci* have been additionally reported (Hill et al., 2006; Mercer et al., 2008).

Antisense transcripts are lncRNAs solely transcribed antisense to a protein-coding region (including introns or not) with varying functions and tissue dependence (Carninci et al., 2005; He et al., 2008; Katayama et al., 2005; Okada et al., 2008). These lncRNAs are less frequently spliced and show lower abundance than their coding counterparts (He et al., 2008). Antisense transcripts most frequently function in a *cis*-regulatory manner, through blocking its counter strand transcript and therefore masking their features, be it protein binding, splicing, or other recognition sites similar to the way miRNA impacts its target sites (Buske, Mattick, & Bailey, 2011; Faghihi et al., 2010; Shearwin, Callen, & Egan, 2005; Werner & Sayer, 2009). The potential of building dsDNA:RNA triplexes that in turn inhibit transcription has also been reported, as well as the potential to recruit epigenetic factors or polysome to up- or downregulate its bound partner mRNA (Carrieri et al., 2012; Magistri, Faghihi, St Laurent, & Wahlestedt, 2012; Morris, 2011). An additional mechanism of action is the production of endosiRNA which, when in their mature state, either affects their sense strand or their origin (the antisense strand) (Tam et al., 2008; Watanabe et al., 2008; Werner & Sayer, 2009).

A subgroup of the antisense lncRNAs are called bidirectional lncRNAs, given they are not overlapping and are found head to head in a surrounding less than 1 kb to either side of the UTR ends of a protein-coding transcript. These molecules can share a common bidirectional promoter, but show discordant expression patterns, which speaks against sole open chromatin expression (Chakalova, Debrand, Mitchell, Osborne, & Fraser, 2005; Mercer et al., 2008; Struhl, 2007). A study found that these bidirectional promoters are well conserved between human and mouse indicating a crucial function during evolution (Trinklein et al., 2004). A famous example for a bidirectional transcript is the *FMR4* locus, which expresses FMR1 and shows downregulation in fragile X syndrome (Khalil, Faghihi, Modarresi, Brothers, & Wahlestedt, 2008; Nakaya et al., 2007). Another one is sox8OT, which is a bidirectional lncRNA with still unknown functions. The SOX8 locus additionally shows adjacent expression, which plays an important role in oligodendrocyte development (Mercer et al., 2010).

Antisense transcripts may also act by *trans*-regulation (regulation of a locus not immediately adjacent to the locus of their origin by antisense transcripts). A cross-species study predicted 25% of ESTs and their *trans* antisense transcripts are lncRNAs in humans (Engström et al., 2006), with functions in various mechanism through a sense–antisense binding mechanism. The nitric oxide synthase (NOS2A)-encoding mRNA is regulated by an antisense transcript transcribed from a *NOS* pseudogene termed *anti*-NOS2A. The anti-NOS2A RNA and NOS2A mRNA display reciprocal expression patterns, suggesting that lncRNA anti-NOS2A plays an important role in the regulation of the nitric oxide (NO) signaling pathway in the CNS (Korneev et al., 2008).

If an lncRNA is generated from a locus at least 1 kb distant from the next protein-coding sequence, it is called an intergenic long noncoding RNA (lincRNA). Initial nomenclature named these transcripts according to their nearest 3′ protein-coding protein gene. A subfamily of these lincRNAs are enhancer RNAs (eRNAs) named after their transcription origin, which is often the enhancer or promoter region of a gene. Their proposed mode of action is through transcription machinery recruiting or chromatin opening (Kim et al., 2010). Another function could be associated to transcripts acting in an activating fashion on their surrounding transcripts as described for activating lncRNAs (Ørom, Ørom, & Shiekhattar, 2011). *Trans*-acting lincRNAs include Neat1, which is a part of the assembly in paraspeckles (nuclear components associated with RNA editing) (Chen & Carmichael, 2009; Mercer et al., 2010) and a critical component of transcription through protein sequestration into paraspeckles (Hirose et al., 2014).

The final group is named sense-overlapping lncRNA. These molecules are considered variants of the coding transcript, since they show a partial overlap with coding regions. They can lack or contain predicted open reading frames, but typically show no protein expression. This is due to premature stop codons, which activate the nonsense-mediated degradation mechanism or have alternative reading frames, which inhibit the potential coding capacity of the lncRNA open reading frame. The FMR1 locus produces two sense-overlapping lncRNAs, which are differentially expressed in fragile X syndrome (Pastori et al., 2014).

4. RNA STRUCTURE

ncRNA functions are mediated through folded modular domains of RNA, which are the major mediators of their abilities to sense and guide their regulatory potential. The function of these modular RNA domains is primarily mediated through its secondary structure, which is based on hydrogen bonding of different moieties in the RNA nucleotides (mainly Watson–Crick base-pairing). The resulting bulges, loops, pseudoknots, and various helices encompass the basis for the tertiary structure. Tertiary structures gain stability through base-pair stacking (hydrophobic interactions) and other non-Watson–Crick base-pairing properties (Cruz & Westhof, 2009). Remarkable is that tertiary as well as the secondary structures are classifiable through a number of modules, which are influenced only to a certain extend by their neighboring sequences (Lescoute, Leontis, Massire, & Westhof, 2005). These modular entities, as well as the canonical antisense complementarity binding capacities, provide ncRNAs the ability to host a plethora of interactions, which is most commonly recognized among the miRNAs and their imperfect complementarity against an mRNA 3′ UTR. A good example of a structure-to-function relationship are ribozymes, which themselves have catalytic moieties to cleave either substrates or themselves to deliver their activity and are a crucial part in the theory of the RNA world (Wan, Kertesz, Spitale, Segal, & Chang, 2011). These interactions are extended by the countless RNA binding proteins, which modulate various aspects of coding as well as noncoding RNA functionality (Loughlin et al., 2012; Michlewski & Cáceres, 2010;

Ricci et al., 2014). Until recently, the structural properties of infrastructural RNAs like rRNA or tRNAs were in the focus of structural analysis studies, with recent reports to analysis the structure of short and long ncRNA like let-7 or SRA1 (Loughlin et al., 2012; Novikova, Hennelly, & Sanbonmatsu, 2012).

5. SPLICING

Almost all human transcripts can generate multiple distinct RNA transcript variants via alternative splicing of their exons and variable 5′ and 3′ untranslated regions (UTRs) (Derrien et al., 2012; Irimia & Blencowe, 2012; Kelemen et al., 2013; Sánchez-Pla, Reverter, Ruíz de Villa, & Comabella, 2012; Smith, Webb, et al., 2013; Taggart, DeSimone, Shih, Filloux, & Fairbrother, 2012). Alternative splicing, which is especially prevalent in the brain (Derrien et al., 2012; Dinger, Amaral, et al., 2009; ENCODE Project Consortium et al., 2012; Guo et al., 2009; Irimia & Blencowe, 2012; Kelemen et al., 2013; Mills & Janitz, 2012; Ponjavic et al., 2007; Sánchez-Pla et al., 2012; Smith, Gesell, et al., 2013; Taggart et al., 2012), shows regulation in both tissue- and/or cell-type-specific manner (Bernard et al., 2010; Grabowski, 2011; Kalsotra & Cooper, 2011; Necsulea et al., 2014; Tripathi et al., 2010). Alternative splicing might be the single biggest source of transcription diversity and enriches the transcript complexity. The selection of the branch point, which generates the splicing lariat, is considered to determine the possible splicing pattern (Barry et al., 2014; Beltran et al., 2008; Taggart et al., 2012). snRNAs, specifically U2-snoRNA, are important mediators in determining the splice isoform produced and are vital components of the spliceosome (Guo et al., 2009; Padgett et al., 1986; Ruby et al., 2007). Additional studies found that the lncRNA Malat1 can regulate alternative splicing through phosphorylation of the SR splicing factor (Berezikov et al., 2010; Bernard et al., 2010; Tripathi et al., 2010; Westholm, Ladewig, Okamura, Robine, & Lai, 2012; Westholm & Lai, 2011). Other lncRNA examples are Gomafu and ZEB2NAT, dysregulation of either leads to alternative splicing and might in the case of Gomafu cause an association with neurodegenerative disease (Barry et al., 2014; Beltran et al., 2008; Havens, Reich, Duelli, & Hastings, 2012).

Another important area where splicing plays a role is in ncRNA biogenesis. In the case of miRNAs, it is called the mirtron pathway, in which an miRNA is produced from a short miRNA intron (mirtron) where the intron constitutes the entire pre-miRNA. Their sequences are adjacent to intron–exon boundaries (Berezikov, Chung, Willis, Cuppen, & Lai, 2007; Padgett et al., 1986; Ruby et al., 2007). Mirtrons are less strongly expressed and are therefore thought to play more modest roles in biological processes. They are less conserved and have evolved faster than canonical miRNAs (Berezikov et al., 2007, 2010; Westholm et al., 2012; Westholm & Lai, 2011). The most prominent difference, however, is their biogenesis; rather than by DROSHA/DGCR8 processing, the pre-miRNA is produced by splicing. The mirtron is spliced out and forms a typical intron lariat, which is subsequently linearized by the debranching enzyme DBR1 (Havens et al., 2012). After these initial steps, the biogenesis of mirtrons and canonical miRNAs is identical; the pre-miRNA is further processed for incorporation into the RISC. The analysis of matched anatomical regions of human and rhesus macaque brain unveiled 16 mirtrons expressed in primate brains (Berezikov et al., 2007; Pan, Shai, Lee, Frey, & Blencowe, 2008). Even though primate brains have a lower number of short introns than invertebrates like flies and worms, they express more mirtrons, which provide additional hints that the diversity of ncRNAs might have contributed to primate evolution (Bass, 2002; Berezikov et al., 2007; Maas, Rich, & Nishikura, 2003).

Splicing, as the removal of lncRNA "introns" during transcript maturation, also plays a role in lncRNAs, specifically in the subgroup of intronic lncRNAs. The frequency of splicing events in noncoding transcripts is lower in comparison to protein transcripts (Dinger et al., 2008; Dinger, Pang, et al., 2009; Melcher et al., 1996; Mercer et al., 2008; Necsulea et al., 2014; Tilgner et al., 2012).

The first reports of alternative splicing in a tissue- and cell-specific manner (Pan et al., 2008; Paul & Bass, 1998) opens the discussion if the branch point selection is part of plasticity and inherits a regulatory code. Another possible role for alternative splicing is the context of brain development as well as human disease association; both points are fields of research, which are under strong investigation, but would exceed the scope of this chapter and are additionally discussed, in another chapter of this book.

6. NcRNA EDITING

The most abundant RNA editing in animals is the conversion from adenosine (A) into inosine (I) by adenosine deaminases (ADARs) (Bass, 2002; Luciano, Mirsky, Vendetti, & Maas, 2004; Maas et al., 2003). There are three ADARs reported in humans (ADAR1–3), with ADAR3 showing brain-specific expression (Melcher et al., 1996; Yang et al., 2006). Since the Watson–Crick base-pairing properties of inosine (I) are similar to those of guanosine (G), the editing process alters RNA/RNA interactions, affecting target recognition, alternative splicing, structure, and degradation, thereby altering gene expression. It is clear that a significant amount of RNA editing occurs in the human brain as all three ADARs are highly expressed there and the brain contains the highest level of inosine in the body (Blow et al., 2006; Paul & Bass, 1998).

Editing is reported in mRNA-coding regions that can change the codon and thereby alter the amino acid in the encoded protein. However, a far bigger proportion of editing occurs in ncRNAs. Editing can play a role in miRNAs, with the first reported editing of a pre-miRNA in miR-22 (Kawahara et al., 2007; Luciano et al., 2004) with following reports for miR-1-1, miR-142, miR-143, and miR-223 (Scadden, 2005; Yang et al., 2006) as well as miR-99a, miR-151, miR-197, miR-223, miR-376a, and miR-379 (Blow et al., 2006; Kawahara, Zinshteyn, Chendrimada, Shiekhattar, & Nishikura, 2007). A-to-I editing in the target recognizing region was shown to alter the suppressive activity of miR-376 in mice (Kawahara et al., 2008; Kawahara, Zinshteyn, Zinshteyn, et al., 2007). In addition to affecting miRNA targeting, ADAR-dependent editing can significantly impact the biogenesis of miRNAs. The editing of miR-142 led to inhibition of its cleavage by DROSHA and subsequent degradation (Peng, Cheng, Tan, Kang, & Tian, 2012; Scadden, 2005; Yang et al., 2006), while editing impaired Dicer1 cleavage of miR-151 (Kawahara, Zinshteyn, Chendrimada, Shiekhattar, & Nishikura, 2007; Prasanth et al., 2005). In contrast, editing by ADAR can also enhance miRNA biogenesis through acceleration of DROSHA cleavage (Audas, Jacob, & Lee, 2012; Kawahara et al., 2008).

Editing has been reported on several bases of lncRNAs JPX and Malat1 (Peng et al., 2012; Yang, Zhou, & Jin, 2013), although functional consequences have yet to be attributed. ncRNA transcript CTN-RNA shows nuclear retention after hyperediting in its 3′ region. This edited region is cleaved off under stress and the truncated transcript is exported to the cytoplasm and subsequently translated into a protein (Gustincich et al., 2006; Prasanth et al., 2005). This elegant way of regulation through nuclear retention seems to be more common than expected, with a report of a family of ncRNA immobilizing proteins in the nucleolus unveiling another ncRNA regulatory mechanism (Audas et al., 2012; Pollard et al., 2006).

Apart from the A-to-I editing, RNA can carry more than a hundred chemical modifications on their nucleotides (Cantara et al., 2011). Thus, RNA editing and/or modification is a further posttranscriptional regulatory layer in the ncRNA transcriptome as well as the epigenetic landscape, and likely contributes to plasticity, brain development, functional diversification, and along that line to neurodegenerative disease. The editing of 3′ UTRs, introns, and repeat elements is varied (reviewed in Skreka et al., 2012; Yang et al., 2013) and beyond the scope of this chapter.

7. EPIGENETIC MODIFICATIONS

At present, the most likely hypothesis of the role of lncRNAs in epigenetic modulations involves lncRNAs binding and guiding protein effector complexes to specific genomic locations where they execute their modulation on expression predominantly through epigenetic modifications (Mercer & Mattick, 2013a). This model finds its most prominent example in the imprinted lncRNA Xist, which acts through lncRNA-mediated chromatin regulation during X-chromosome inactivation (XCI), a process necessary in females to maintain equal X-chromosome expression (Leeb, Steffen, & Wutz, 2009). Xist achieves XCI by scanning and modifying the three-dimensional structure of the X-chromosome and detecting accessible sites to dock its silencing domain (Dimond & Fraser, 2013). This function is mediated through the A-repeats in Xist, which recruits Polycomb *Repressive* Complex 2 (PCR2) (Zhao, Sun, Erwin, Song, & Lee, 2008) and additionally relocates its functional domain over the remaining X-chromosome (Chaumeil, Le Baccon, Wutz, & Heard, 2006). Another well-studied lncRNA is Hotair, which showed association with chromatin remodeling proteins PCR2 (Rinn et al., 2007) and LSD1 (Tsai et al., 2010), which together contribute to specifying the histone modification on particular target genes. The recruitment of chromatin-modifying enzymes by lncRNAs such as Xist appears to be a frequent occurrence in lncRNA function. Alternative reported mechanisms include the inhibition of transcription due to a binding of the lncRNA to the promoter region in a reverse complementary fashion and subsequent methylation of histone H3K9 residues as demonstrated by lncRNA Airn (Santoro et al., 2013).

A similar functional pattern can be postulated for Meg3, an maternally imprinted lncRNA highly expressed in normal human brain, which shows a strong upregulation of genes involved in angiogenesis and microvessel formation upon Meg3 downregulation in mouse brains and an connection to

Huntington's disease (HD) and glioma (Aramayo, 2009; Gordon et al., 2010; Johnson, 2012; Zhang et al., 2010).

Similar functional patterns in which lncRNAs are used to guide the modulation of expression through histone modifications have been discovered during development (Guttman & Rinn, 2012; Khalil et al., 2009; Koziol & Rinn, 2010).

8. ncRNAs ARE INVOLVED IN NEURONAL DEVELOPMENT, MAINTENANCE, AND PLASTICITY

The vertebrate central nervous system (CNS), with its largest portion being the brain, is one of the most complex and diverse networks known. The CNS/brain contains a multitude of neuronal and glia cell types, and its development and differentiation require an enormous amount of coordination and both intrinsic and extrinsic stimuli to mature to its full form, function and potential. Fully developed human brains show prolific transcriptional activity of ncRNAs in their neurons (Gustincich et al., 2006; Kang et al., 2011; Ziats & Rennert, 2013). In humans, the brain went through a rapid evolution and specific accelerated genomic features are termed human accelerated regions. Tellingly, these regions are primarily (96%) located in nonprotein-coding regions of the genome (Cuellar et al., 2008; Davis et al., 2008; Haramati et al., 2010; Pollard et al., 2006).

sncRNAs have been reported to play a major role in embryonic neuronal differentiation with frequent discoveries of yet to be categorized noncanonical sncRNAs (Landgraf et al., 2007; Skreka et al., 2012) and a clear spatiotemporal differential expression in the human brain (Kang et al., 2011; Nowak & Michlewski, 2013; Ziats & Rennert, 2013). Inhibition of sncRNA biogenesis in mice, in most cases through Dicer knockout, leads to assorted phenotypes, due to constrained differentiation in various regions and mechanistic deficits, as well as neurodegeneration (Cuellar et al., 2008; Davis et al., 2008; Haramati et al., 2010; Yoo et al., 2011). A significant number of miRNAs are specifically expressed in the brain (Cabili et al., 2011; Hangauer et al., 2013; Landgraf et al., 2007; Mercer et al., 2008; Necsulea et al., 2014 and reviewed in Nowak & Michlewski, 2013) with hallmark miRNAs like miR-9* and miR-124, whose expression in fibroblasts leads to their conversion into neurons through chromatin remodeling (Yoo et al., 2011).

lncRNAs show strong enrichment of expression in the nervous system (Cabili et al., 2011; Eacker, Dawson, & Dawson, 2013; Hangauer et al.,

2013; Mercer et al., 2008; Necsulea et al., 2014) with high tissue-, developmental-, and subcellular compartment specificity (Derrien et al., 2012; Krol et al., 2010; Lipovich et al., 2014; Necsulea et al., 2014; Ponjavic et al., 2009), including brain-specific expression of ~40% of the most highly and differentially expressed lncRNAs (Derrien et al., 2012; Konopka et al., 2010).

Specific examples of lncRNA (*cis* or localized) action in the CNS include the transcription factor Emx2os, which is associated with cerebral cortex development and is activated by a short antisense transcript (Heins et al., 2001; Spigoni, Gedressi, & Mallamaci, 2010). Another prominent example is BDNF, which shows strong involvement in the nervous system with at least 12 isoforms and 4 antisense transcripts and a specific expression pattern in the brain. The inhibition of BDNF-AS leads to BDNF upregulation and therefore suggests potential pharmacological uses for lncRNAs (Bibel & Barde, 2000; Modarresi et al., 2012; Pruunsild, Kazantseva, Aid, Palm, & Timmusk, 2007). Additional brain-specific lncRNA examples are Nrgn and camk2n1 which are involved in corticogenesis (Ling et al., 2011).

A prominent CNS lincRNA is Neat2 (Malat1) (Mercer et al., 2010), which when knocked down in cultured hippocampal neurons decreased synaptic density (Bernard et al., 2010), but was dispensable in an *in vivo* mouse model with modest transcriptional *cis* effects on neighboring genes (Zhang, Arun, et al., 2012). A separate study performed differential expression analysis of mouse brain regions and identified 66 lincRNAs showing stronger transcriptional correlation to their neighboring protein-coding genes than nondifferentially expressed lincRNAs (Belgard et al., 2011).

An example of sense-overlapping lncRNA is EVF2, which regulates the transcription of Dlx5 and Dlx6 (DLX proteins are involved in mutiple functions in the forebrain), which is facilitated through the binding of Dlx2 (Bond et al., 2009; Feng et al., 2006). EVF2-deficient mice show reduced GABAergic receptors, which points to implications in autism, schizophrenia (SZ), epilepsy and tourette (Bond et al., 2009). SOX2OT has brain-specific expression and indicated association with adult neurogenesis and correlation with expression of *SOX2*, a gene required for stem-cell maintenance in the CNS (Amaral et al., 2009; Mercer et al., 2008).

There is determined interest to identify the function of lncRNAs in neuronal tissues and their roles in brain development, activity and stress response, memory formation, plasticity, cell identity, and how their dysfunction contribute to neurological diseases.

8.1. Stimuli depend on expression, specificity, and memory

A number of ncRNAs have recently been discovered to display expression dependent upon increased neuronal activity or stress-induced stimuli, suggesting a strong involvement in fine-tuning neuronal plasticity (reviewed in Banerjee, Neveu, & Kosik, 2009; Eacker et al., 2013). miRNAs show a higher metabolism (rapid decay and increased transcription) in neuronal cells than in nonneuronal cells, with an activity-dependent turnover pattern (Krol et al., 2010; Miura, Shenker, Andreu-Agullo, Westholm, & Lai, 2013). Articles of increased learning and memory function in mice upon inhibition of the canonical miRNA biogenesis through Dicer inactivation (Konopka et al., 2010; Nudelman et al., 2010), with linkage of the RISC pathway and an increase in protein synthesis have been reported previously (Banerjee et al., 2009; Rogelj, Hartmann, Yeo, Hunt, & Giese, 2003). This suggests that specific miRNAs are rapidly degraded upon neural activity, resulting in increased translation of their specific targets. These reports go hand in hand with recent evidence, which showed brain-specific lengthening of the 3′ UTRs in brains containing thousands of conserved miRNA target site, with enriched sites for neuronal-specific miRNAs (Lee et al., 2011; Miura et al., 2013). Not surprisingly, downregulation of miRNA upon stimulus is not the sole regulatory mechanism as miR-132 expression displays activity-dependent upregulation (Nudelman et al., 2010; Rajasethupathy et al., 2012). Similar modulation patterns (up- as well as downregulation) were reported for brain-specific snoRNAs in mice hippocampi after fear conditioning (Landry, Kandel, & Rajasethupathy, 2013; Rogelj et al., 2003). piRNAs are another class which has been shown to be of importance in the nervous systems (Gustincich et al., 2006; Kapranov et al., 2010; Lee et al., 2011). They are considered to act upon neural activation (Liu et al., 2013; Necsulea et al., 2014; Rajasethupathy et al., 2012; Taft et al., 2007) in their function as epigenetic modifiers and might therefore play a role in "memory storage" bridging generations (Landry et al., 2013; Pollard et al., 2006). lncRNA Gomafu shows activation-dependent transcription and involvement in alternative splicing in schizophrenia (Barry et al., 2014; Mercer et al., 2010).

The subsequent step from activation-dependent expression to modulation at the mRNA and protein level will most likely have its implications in memory formation and cognition. Long-term memory formation is considered as the stable strengthening of synaptic interactions and alterations in patterning between neurons in the brain. Given the increasing body of evidence for ncRNAs as modulators of activity-dependent gene expression, it very likely that this additional regulatory layer will be found to play a significant role in human brain plasticity.

9. EVOLUTIONARY ROLE OF ncRNAs AND PRIMATE SPECIFICITY

The contribution of ncRNAs to brain function is indicated by both their diversity and robust expression in the human brain (Chimpanzee Sequencing and Analysis Consortium, 2005; Gustincich et al., 2006; Kapranov et al., 2010; Nielsen et al., 2005). As highlighted below (Fig. 7.2), the correlation between the increasing number and diversity of ncRNAs and the developmental complexity of species is remarkable (Hu et al., 2011; Liu et al., 2013; Necsulea et al., 2014; Taft et al., 2007).

Particularly interesting is that the most accelerated evolutionary development could be mapped to regions in humans which are responsible for the expression of ncRNAs and connected to cortical development (Necsulea et al., 2014; Pollard et al., 2006). The lncRNAs HARF1 A and B, which are associated with forebrain development in early gestation, exemplify this idea (Pollard et al., 2006). Further, Chodroff et al. found similar expression patterns among lincRNAs in various vertebrates in close association to brain development (Chodroff et al., 2010).

The likely contribution of ncRNAs to human brain evolution includes both sncRNA and lncRNA. miRNA expression patterns between human and closely related chimpanzee differ by 11%, while between humans and macaques (old world monkeys) differ by 35%, with human-specific localization in neurons and an indication of changes after the split between human and Neanderthal lineage (Hu et al., 2011; Muslimov, Banker, Brosius, & Tiedge, 1998; Wang et al., 2002). The correlation between human and chimpanzee lncRNA brain expression is lower than between its coding counterparts between human and xenopus (Gemayel, Vinces, Legendre, & Verstrepen, 2010; Necsulea et al., 2014).

Additional potential explanations for an increased acceleration of evolution in certain neuronal regions among primates/humans could be (mosaic) copy number variations (McConnell et al., 2013) which might lead to increased organismal evolvability (Gemayel et al., 2010; Lin et al., 2011).

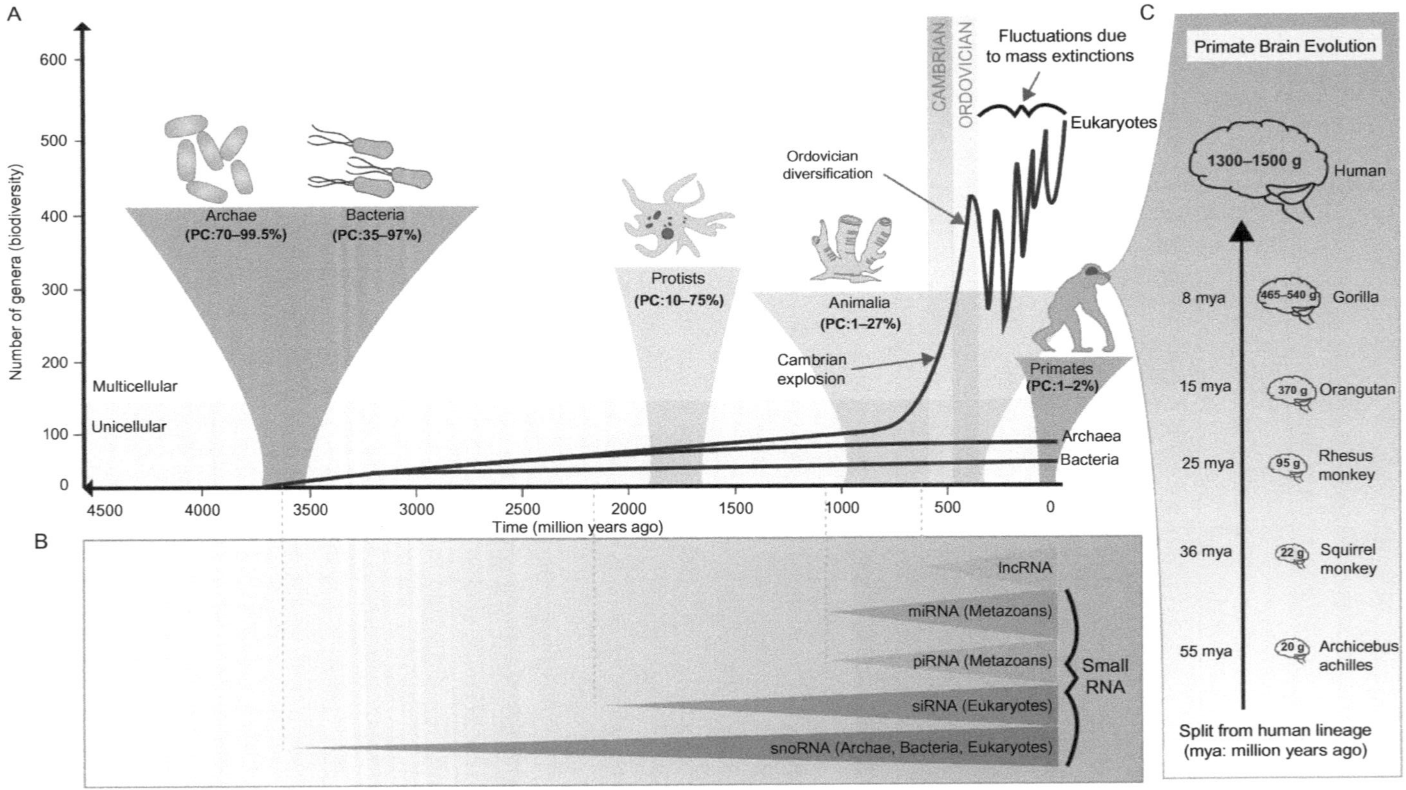
A
Number of genera (biodiversity)
600
500
400
300
200
100
0
Multicellular
Unicellular
Archae
(PC:70–99.5%)
Bacteria
(PC:35–97%)
Protists
(PC:10–75%)
Animalia
(PC:1–27%)
Primates
(PC:1–2%)
CAMBRIAN
ORDOVICIAN
Fluctuations due
to mass extinctions
Eukaryotes
Ordovician
diversification
Cambrian
explosion
Archaea
Bacteria
4500
4000
3500
3000
2500
2000
1500
1000
500
0
Time (million years ago)
B
lncRNA
miRNA (Metazoans)
piRNA (Metazoans)
siRNA (Eukaryotes)
snoRNA (Archae, Bacteria, Eukaryotes)
Small
RNA
C
Primate Brain Evolution
1300–1500 g
Human
8 mya
465–540 g
Gorilla
15 mya
370 g
Orangutan
25 mya
95 g
Rhesus
monkey
36 mya
22 g
Squirrel
monkey
55 mya
20 g
Archicebus
achilles
Split from human lineage
(mya: million years ago)

10. THE ROLE OF (RETRO)TRANSPOSONS AND PSEUDOGENES IN ncRNA EVOLUTION

Transposons, or transposable elements, are small DNA sequences present many times in the human genome. They can change their position within the genome, thereby creating or reversing mutations, and as a consequence of their insertion, change the genome size. Transposable elements exist as either retrotransposons that utilize RNA intermediates in the transposition process or as DNA transposons, which bypass the RNA intermediate step and are directly transferred as DNA into their new genomic location. Due to their insertion/excision character, transposable elements are considered mutagens whose often deleterious nature is proposed to be restricted in the germlines of eukaryotes by piRNAs. Thus, the abundance of piRNA in germline tissue may reflect the high levels of transposons and other repeat-derived RNA. The brain is the only nongermline-associated somatic tissue presently known to express piRNAs where they may act as a regulatory mechanism for L1 retrotransposons. The expression of L1 retrotransposons is regulated during neuronal differentiation and proposed to give rise to neuronal heterogeneity and somatic mosaicism in brain (Iyengar et al., 2014).

The contribution of transposable elements to evolution remains under lively discussion, although leaning toward a functional role. Transposable elements are highly abundant in human lncRNA, comprising ~30% of total lncRNA sequence, and are key to lineage-specific diversification of the vertebrate lncRNA portfolio (Hoekstra, 2013). Additionally, transposable elements in the lncRNAs have evolved under stronger constraints than intronic transposable element sequence or random DNA sequences, suggesting the

Figure 7.2 Depicting the correlation between the evolutionary increase of organismal complexity in the nonprotein-coding transcriptome and cognitive evolution. (A) Increase of biodiversity throughout biogenesis on earth and the protein-coding (PC) percentage in various domains. The domains of archae and bacteria show a high percentage of PC transcripts (Liu et al., 2013), whereas the arising of protest presents a significant lower PC level with an additional decrease in the domain of animalia. The Cambrian explosion led to a huge increase in biodiversity, which finally led to emergence of primates, whose genomes only contain around 2% PC regions. (B) The evolution and extension of noncoding families goes hand in hand with the diversification of genera. (C) The evolutionary increase of complexity in the non-PC transcriptome may be an underlying cause for the accelerated growth in primate brains and the result of higher-order cognition in humans. *Figure adapted from Barry (2014).*

transposable elements may functionally contribute to lncRNAs. Many of their features contain signals critical for lncRNA biogenesis including splicing, transcription, initiation, and polyadenylation, or alternatively transposable elements appear associated with to *cis* regulation of downstream lncRNA transcripts (Hoekstra, 2013).

Pseudogenes are gene remnants produced by the loss of regulatory sequences or inactivating mutations, often following retrotransposition events or gene duplication. Currently, >11,000 pseudogenes are annotated in humans of which a significant proportion of them are transcriptionally expressed in the brain and associated with open chromatin marks (Pei et al., 2012). Although viewed as junk, differentiation of human iPS cells into human neurons was accompanied by significant expression changes in over a thousand of these pseudogenes (Lin et al., 2011), suggesting a functional roles for these elements. One possible function of expressed pseudogenes is to act as decoys for regulatory entities (Hawkins & Morris, 2010; Poliseno et al., 2010; Tam et al., 2008; Watanabe et al., 2008). The sequence similarity/homology with a protein-coding gene provides pseudogene transcripts with the capacity to bind miRNAs targeted to the parent protein-coding transcript and thus act as an miRNA decoy or sponge to buffer or potentially regulate the activity of miRNAs on physiologically relevant targets (Johnsson et al., 2013; Wang et al., 2013).

11. ncRNAs AND DISEASE

lncRNAs can regulate gene expression through a wide range of mechanisms (e.g., chromatin modification, transcription, splicing, translation), and hence, it is not surprising that lncRNA dysfunction or mutations are associated with neurological and neurodegenerative disorders with many lncRNAs transcribed from *loci* associated with these diseases. The rapidly increasing identification and functional analysis of new lncRNAs will undoubtedly associate many more lncRNAs with diseases of the human brain.

miRNAs are well established as regulating human disease-associated genes, and polymorphisms in their 3′ UTR binding sites provide further possibilities for their dysregulation.

11.1. Alzheimer's disease

The proteolytic processing of the amyloid precursor protein (APP) to neurotoxic amyloid peptide Aβ42 in the brain by BACE1 (beta-secretase) is a significant contributor to Alzheimer's disease (AD). Elevated BACE1 expression and BACE1's intracellular processing access to APP, mediated by SORL1, can increase production of toxic Aβ42. BACE1 expression is regulated by both miRNAs and lncRNAs: miRNAs miR-107 and miR-29a/b-1 have binding sites within the BACE1 3′ UTR and are significantly decreased in sporadic AD patients, correlating with significantly increased BACE protein levels (Hébert et al., 2008; Wang et al., 2008; Zhang et al., 2013). lncRNA BACE1-AS is a transcript, antisense to BACE1 that regulates BACE1 mRNA and protein expression, and which is elevated in AD patients. BACE1-AS is proposed to act by masking a miRNA binding site, preventing miRNA-induced translational repression of BACE1 mRNA and resulting in increased BACE1 protein levels (Barry et al., 2014; Faghihi et al., 2008). Intriguingly, BACE1-AS is elevated upon exposure to Aβ42, producing a posttranscriptional feed-forward mechanism that results in further increased BACE1 and Aβ42.

By regulating membrane trafficking of APP, SORL1 modulates exposure of the substrate APP to BACE1 and thus Aβ42 production, with down-regulation of SORL1 leading to increased Aβ formation. lncRNA 51A, whose expression is upregulated in AD patients, is antisense to the first SORL1 intron and its expression shifts SORL1 mRNA splicing from the long protein variant SORL1-A to an alternatively spliced protein form. The resulting decreased production of SORL1-A is associated with significant overproduction of Aβ (Chubb, Bradshaw, Soares, Porteous, & Millar, 2008; Ciarlo et al., 2013). Thus, alterations in ncRNA expression can contribute to increased BACE1 and Aβ42 levels in sporadic AD (Kerin et al., 2012; Zhang et al., 2013).

11.2. Schizophrenia

Alternative splicing of mRNA is associated with the pathology of SZ with many of the genes associated with the disease displaying aberrant splicing patterns. Gomafu is an lncRNA highly regulated by neural activity, binds directly to splicing factors, and is significantly downregulated in the cortex of SZ patients. Modulation of Gomafu expression alters splicing patterns of at least two SZ-associated genes. Knockdown of Gomafu resulted in the upregulation of SZ pathology-related splice variants of DISC1 and ERBB4, matching the observation that overexpression of these same splice variants are associated with SZ. In contrast, Gomafu overexpression produced significant downregulation of the same disease-associated splice variants of both genes (Barry et al., 2014; Vaishnavi, Manikandan, Tiwary, & Munirajan, 2013). This suggests that the lncRNA Gomafu may contribute to the pathogenic splicing pattern of these key SZ genes.

Both DISC2 (disrupted in schizophrenia 2) and DISC1 are disrupted by a translocation in a large SZ cohort. DISC1 encodes a protein with roles in neurodevelopment whereas DISC2 encodes a putative ncRNA antisense to DISC1 and thus may contribute to the regulation of DISC1 (Chubb et al., 2008; Oiglane-Shlik et al., 2006).

11.3. Autism spectrum disorder

Accumulating evidence indicates that lncRNAs contribute to autism spectrum disorder (ASD) risk. Moesin regulates neuronal architecture and the 3.9 kb lncRNA MSNP1AS, transcribed in antisense to a moesin pseudogene, is 94% identical and antisense to moesin and can bind moesin mRNA (Kerin et al., 2012; Le Meur et al., 2005). Overexpression of MSNP1AS in cultured cells led to decreased moesin levels while MSNP1AS transcript levels were 12-fold higher in postmortem brain samples from ASD cases. High levels of the MSNP1AS transcript were associated with the presence of an ASD risk SNP and thus MSNP1AS is thus strongly positioned to be an lncRNA risk factor for ASDs.

The role of miRNAs in the pathophysiology of ASD has also emerged following the identification of copy number variants (CNVs) in the form of microdeletions and microduplications at multiple chromosomal loci implicated in autism. In addition to protein-coding genes being identified in the CNVs, miRNAs present in deleted and duplicated CNV loci may explain the difference in dosage of the crucial genes controlled by them (Bartolomei, 2013; Vaishnavi et al., 2013).

11.4. Parkinson's disease

α-Synuclein is an aggregation-prone neural protein that plays a central role in the pathogenesis of both sporadic and familial Parkinson's disease (PD). Elevated level of wild-type α-synuclein is disease causative and demonstrates the essential need for precise regulation of α-synuclein expression. miRNAs miR-7 and miR-153 have both been demonstrated to inhibit α-synuclein levels posttranscriptionally via binding sites in α-synuclein's long 3′ UTR. Modulation of miRNA levels causes reciprocal changes in α-synuclein mRNA/protein levels although it is unclear if PD patients display reduced levels of these miRNAs (Doxakis, 2010).

11.5. Angelman syndrome

Angelman syndrome (AS), a neurogenetic disorder characterized by severe intellectual and developmental disability, is typically caused by mutations or deletions of the maternally inherited UBE3A gene, which encodes an E3 ubiquitin ligase. UBE3A is biallelically expressed in most tissues, but is only maternally expressed in neurons due to epigenetic silencing of the paternal allele (Bartolomei, 2013; Oiglane-Shlik et al., 2006). Loss or inactivation of the maternal UBE3A gene leaves neurons devoid of functional UBE3A expression and results in AS. The antisense lncRNA transcript UBE3A-ATS acts in *cis* to regulate epigenetic silencing of the paternal allele (Chung, Rudnicki, Yu, & Margolis, 2011; Le Meur et al., 2005) by transcriptional interference between convergent UBE3A and UBE3A-ATS transcripts (Bartolomei, 2013; Rao, Benito, & Fischer, 2013). Depletion of UBE3A-ATS in the AS mouse model activated neural expression of UBE3A from the paternal chromosome, and reduced many disease-related symptoms in the mouse (Bartolomei, 2013; Leidinger et al., 2013), establishing the importance of this lncRNA in disease pathology.

11.6. Huntington's disease

HD is caused by a CAG trinucleotide repeat expansion within exon 1 of the HTT gene. The resulting elongated polyglutamine tract causes the mutant Huntington protein to aggregate and confer gain-of-function toxicity. The cellular consequences are many and complex and involve protein misfolding/impaired degradation as well as broad transcriptional dysregulation. A transcript antisense to HTT (Huntingtin antisense; HTTAS v1) overlaps the repeat expansion exon and its expression is reduced in human HD frontal cortex. Modulation of HTTAS v1 expression has reciprocal effects on HTT mRNA levels, strongly suggesting that HTTAS v1 regulates HTT expression (Chung et al., 2011).

11.7. ncRNA as biomarkers

The clinical diagnosis of neurodegenerative diseases such as AD and PD remains difficult and postmortem confirmation is typically required. Biomarkers for neurodegenerative and neurological disorders are therefore urgently needed to provide early and accurate diagnosis as well as for monitoring disease progression and therapeutic efficacy.

ncRNA networks are centrally involved in cellular regulatory mechanisms and as such, early signs of cellular dyshomeostasis may be reflected in changes in their expression pattern much earlier than presentation of clinical symptoms. The extensive range of ncRNA transcripts provides a spectrum of candidate transcripts from which to identify potential biomarkers. Furthermore, the opportunity exists to assess changes in a number of ncRNAs whose multiplexing combination could enhance diagnostic sensitivity and accuracy.

A further challenge in diagnosing brain disorders is the lack of access to the tissue directly affected. However, neurons and other cells of the CNS release miRNA and other small noncoding RNA extracellularly, often packaged in exosomes or microvesicles, that could act as indicators of pathological processes. Recent studies have demonstrated the existence of miRNAs in multiple body fluids including blood, CSF, and saliva are potential candidates for informative biomarker discovery (Rao et al., 2013). This approach is already showing success with a 12-miRNA signature obtained from blood showed >90% accuracy, specificity, and sensitivity in distinguishing AD patients from healthy controls (Leidinger et al., 2013).

12. PERSPECTIVES AND OUTLOOK

The human brain is based on a multitude of neuronal and glia cell types and its development and differentiation require an enormous amount of coordination and stimuli, both intrinsic and extrinsic, to blossom to its full form and function. Estimates are that >85 billion neurons and about 10-fold the number of synapses comprise this network. However, the increased intellectual ability in humans is unlikely to be that our brains simply represent a linearly scaled-up primate brain according to number and cell composition (Herculano-Houzel, 2009). Instead, it is likely that an increased regulatory capacity has provided our brains with the ability to form a far more complex network than that of other primates. The rapid expansion in ncRNAs, especially lncRNAs, during our recent primate evolution has likely provided this significantly enhanced regulatory ability and played a part in the evolution of human brain form and function (Liu et al., 2013; Taft et al., 2007). The ability of ncRNA to orchestrate the sophisticated regulation of gene expression, coupled with their own intricate spatiotemporal expression patterns in different cell types, and their activation in learning and memory as well as environmental response together provide a combinatorial matrix of the required complexity. Such intricate and sophisticated regulatory complexity in the human brain also likely provides the opportunity for intellectual enhancement, but its complexity also leaves it susceptible to dysregulation as represented by neurodegeneration or psychological disorders.

It is clear that ncRNA plays a central role in the functioning of the human brain, a role that will only be strengthened as more lncRNAs are discovered and their many functions unveiled. It is therefore clear that the full integration of ncRNAs into modern molecular neuroscience is critical for a complete understanding of the human brain and its disorders.

Since the divergence from the great apes and especially in recent decades, human brains are put under an enormous developmental pressure on the account that the numbers of extrinsic inputs in a single day in the life of human being living in an metropolitan area would easily accumulate to the amount of input of an entire life span two to three generations ago. In the light of activation-dependent expression to modulation and in combination with potential epigenetic modification and reports about the increased permeability of the Weismann barrier (Mattick, 2012), these findings open the door for speculation, on whether certain parts of our inheritance might be acquired rather than solely inherited and therefore following the rule Lamarck proposed in 1809.

Contrary to the speculative nature of the last paragraph, one aspect is becoming increasingly clear: the role of ncRNA in the transcriptome and especially in the brain is a central one and will gain more and more importance with the rapidly growing ncRNA-centric paradigm.

ACKNOWLEDGMENTS

We would like to thank Dr. Guy Barry for permission to adapt Fig. 7.2. We also thank Fabienne Bereiter and Kathryn Hill for their helpful suggestions and comments. This work was supported in part by grants from the NHMRC (APP1067350), Michael J. Fox Foundation for Rapid Response (MJFF22), and a Geoff and Dawn Dixon fellowship to B. G. and A. A. C. We gratefully acknowledge the Swiss National Science Foundation for the funding of B. G. (P2EZP3_152143). The authors declare no conflict of interest.

REFERENCES

Amaral, P. P., Clark, M. B., Gascoigne, D. K., Dinger, M. E., & Mattick, J. S. (2011). lncRNAdb: A reference database for long noncoding RNAs. *Nucleic Acids Research*, *39*(Database issue), D146–D151.

Amaral, P. P., Neyt, C., Wilkins, S. J., Askarian-Amiri, M. E., Sunkin, S. M., Perkins, A. C., et al. (2009). Complex architecture and regulated expression of the Sox2ot locus during vertebrate development. *RNA*, *15*(11), 2013–2027.

Aramayo, R. (Ed.), (2009). At least ten genes define the imprinted Dlk1-Dio3 cluster on mouse chromosome 12qF1. In *PLoS One*, *4*(2), e4352.

Atkinson, S. R., Marguerat, S., & Bahler, J. (2012). Exploring long non-coding RNAs through sequencing. *Seminars in Cell & Developmental Biology*, *23*(2), 200–205.
Audas, T. E., Jacob, M. D., & Lee, S. (2012). Immobilization of proteins in the nucleolus by ribosomal intergenic spacer noncoding RNA. *Molecular Cell*, *45*(2), 147–157.
Bachellerie, J. P., Cavaillé, J., & Huttenhofer, A. (2002). The expanding snoRNA world. *Biochimie*, *84*(8), 775–790.
Banerjee, S., Neveu, P., & Kosik, K. S. (2009). A coordinated local translational control point at the synapse involving relief from silencing and MOV10 degradation. *Neuron*, *64*(6), 871–884.
Barry, G. (2014). Integrating the roles of long and small non-coding RNA in brain function and disease. *Molecular Psychiatry*, *19*(4), 410–416.
Barry, G., Briggs, J. A., Vanichkina, D. P., Poth, E. M., Beveridge, N. J., Ratnu, V. S., et al. (2014). The long non-coding RNA Gomafu is acutely regulated in response to neuronal activation and involved in schizophrenia-associated alternative splicing. *Molecular Psychiatry*, *19*(4), 486–494.
Bartolomei, M. S. (Ed.), (2013). Truncation of Ube3a-ATS unsilences paternal Ube3a and ameliorates behavioral defects in the Angelman syndrome mouse model. In *PLoS Genetics*, *9*(12), e1004039.
Bass, B. L. (2002). RNA editing by adenosine deaminases that act on RNA. *Annual Review of Biochemistry*, *71*(1), 817–846.
Belgard, T. G., Marques, A. C., Oliver, P. L., Abaan, H. O., Sirey, T. M., Hoerder-Suabedissen, A., et al. (2011). A transcriptomic atlas of mouse neocortical layers. *Neuron*, *71*(4), 605–616.
Beltran, M., Puig, I., Peña, C., Miguel Garcia, J., Belen Alvarez, A., Peña, R., et al. (2008). A natural antisense transcript regulates Zeb2/Sip1 gene expression during Snail1-induced epithelial-mesenchymal transition. *Genes & Development*, *22*(6), 756–769.
Berezikov, E., Chung, W.-J., Willis, J., Cuppen, E., & Lai, E. C. (2007). Mammalian mirtron genes. *Molecular Cell*, *28*(2), 328–336.
Berezikov, E., Liu, N., Flynt, A. S., Hodges, E., Rooks, M., Hannon, G. J., et al. (2010). Evolutionary flux of canonical microRNAs and mirtrons in Drosophila. *Nature Genetics*, *42*(1), 6–9, author reply 9–10.
Bernard, D., Prasanth, K. V., Tripathi, V., Colasse, S., Nakamura, T., Xuan, Z., et al. (2010). A long nuclear-retained non-coding RNA regulates synaptogenesis by modulating gene expression. *The EMBO Journal*, *29*(18), 3082–3093.
Bibel, M., & Barde, Y. A. (2000). Neurotrophins: Key regulators of cell fate and cell shape in the vertebrate nervous system. *Genes & Development*, *14*(23), 2919–2937.
Blow, M. J., Grocock, R. J., van Dongen, S., Enright, A. J., Dicks, E., Futreal, P. A., et al. (2006). RNA editing of human microRNAs. *Genome Biology*, 7(4), R27.
Bond, A. M., Vangompel, M. J. W., Sametsky, E. A., Clark, M. F., Savage, J. C., Disterhoft, J. F., et al. (2009). Balanced gene regulation by an embryonic brain ncRNA is critical for adult hippocampal GABA circuitry. *Nature Neuroscience*, *12*(8), 1020–1027.
Borchert, G. M., Lanier, W., & Davidson, B. L. (2006). RNA polymerase III transcribes human microRNAs. *Nature Structural & Molecular Biology*, *13*(12), 1097–1101.
Buske, F. A., Mattick, J. S., & Bailey, T. L. (2011). Potential in vivo roles of nucleic acid triple-helices. *RNA Biology*, *8*(3), 427–439.
Cabili, M. N., Trapnell, C., Goff, L., Koziol, M., Tazon-Vega, B., Regev, A., et al. (2011). Integrative annotation of human large intergenic noncoding RNAs reveals global properties and specific subclasses. *Genes & Development*, *25*(18), 1915–1927.
Cai, X., Hagedorn, C. H., & Cullen, B. R. (2004). Human microRNAs are processed from capped, polyadenylated transcripts that can also function as mRNAs. *RNA*, *10*(12), 1957–1966.
Candeias, M. M. (2011). The can and can't dos of p53 RNA. *Biochimie*, *93*(11), 1962–1965.

Cantara, W. A., Crain, P. F., Rozenski, J., McCloskey, J. A., Harris, K. A., Zhang, X., et al. (2011). The RNA modification database, RNAMDB: 2011 update. *Nucleic Acids Research, 36*(Database issue), D195–201.

Capra, J. A., Erwin, G. D., McKinsey, G., Rubenstein, J. L. R., & Pollard, K. S. (2013). Many human accelerated regions are developmental enhancers. *Philosophical Transactions of the Royal Society of London. Series B, Biological Sciences, 368*(1632), 20130025.

Carninci, P., Kasukawa, T., Katayama, S., Gough, J., Frith, M. C., Maeda, N., et al. (2005). The transcriptional landscape of the mammalian genome. *Science, 309*(5740), 1559–1563.

Carrieri, C., Cimatti, L., Biagioli, M., Beugnet, A., Zucchelli, S., Fedele, S., et al. (2012). Long non-coding antisense RNA controls Uchl1 translation through an embedded SINEB2 repeat. *Nature, 491*(7424), 454–457.

Chakalova, L., Debrand, E., Mitchell, J. A., Osborne, C. S., & Fraser, P. (2005). Replication and transcription: Shaping the landscape of the genome. *Nature Reviews. Genetics, 6*(9), 669–677.

Chaumeil, J., Le Baccon, P., Wutz, A., & Heard, E. (2006). A novel role for Xist RNA in the formation of a repressive nuclear compartment into which genes are recruited when silenced. *Genes & Development, 20*(16), 2223–2237.

Chen, L.-L., & Carmichael, G. G. (2009). Altered nuclear retention of mRNAs containing inverted repeats in human embryonic stem cells: Functional role of a nuclear noncoding RNA. *Molecular Cell, 35*(4), 467–478.

Chimpanzee Sequencing and Analysis Consortium (2005). Initial sequence of the chimpanzee genome and comparison with the human genome. *Nature, 437*(7055), 69–87.

Chodroff, R. A., Goodstadt, L., Sirey, T. M., Oliver, P. L., Davies, K. E., Green, E. D., et al. (2010). Long noncoding RNA genes: Conservation of sequence and brain expression among diverse amniotes. *Genome Biology, 11*(7), R72.

Chubb, J. E., Bradshaw, N. J., Soares, D. C., Porteous, D. J., & Millar, J. K. (2008). The DISC locus in psychiatric illness. *Molecular Psychiatry, 13*(1), 36–64.

Chung, D. W., Rudnicki, D. D., Yu, L., & Margolis, R. L. (2011). A natural antisense transcript at the Huntington's disease repeat locus regulates HTT expression. *Human Molecular Genetics, 20*(17), 3467–3477.

Ciarlo, E., Massone, S., Penna, I., Nizzari, M., Gigoni, A., Dieci, G., et al. (2013). An intronic ncRNA-dependent regulation of SORL1 expression affecting Aβ formation is upregulated in post-mortem Alzheimer's disease brain samples. *Disease Models & Mechanisms, 6*(2), 424–433.

Cooper, C., Vincett, D., Yan, Y., Hamedani, M. K., Myal, Y., & Leygue, E. (2011). Steroid receptor RNA activator bi-faceted genetic system: Heads or tails? *Biochimie, 93*(11), 1973–1980.

Cruz, J. A., & Westhof, E. (2009). The dynamic landscapes of RNA architecture. *Cell, 136*(4), 604–609.

Cuellar, T. L., Davis, T. H., Nelson, P. T., Loeb, G. B., Harfe, B. D., Ullian, E., et al. (2008). Dicer loss in striatal neurons produces behavioral and neuroanatomical phenotypes in the absence of neurodegeneration. *Proceedings of the National Academy of Sciences of the United States of America, 105*(14), 5614–5619.

Davis, T. H., Cuellar, T. L., Koch, S. M., Barker, A. J., Harfe, B. D., McManus, M. T., et al. (2008). Conditional loss of Dicer disrupts cellular and tissue morphogenesis in the cortex and hippocampus. *The Journal of Neuroscience, 28*(17), 4322–4330.

Denli, A. M., Tops, B. B. J., Plasterk, R. H. A., Ketting, R. F., & Hannon, G. J. (2004). Processing of primary microRNAs by the microprocessor complex. *Nature Cell Biology, 432*(7014), 231–235.

Derrien, T., Johnson, R., Bussotti, G., Tanzer, A., Djebali, S., Tilgner, H., et al. (2012). The GENCODE v7 catalog of human long noncoding RNAs: Analysis of their gene structure, evolution, and expression. *Genome Research, 22*(9), 1775–1789.

Dieci, G., Fiorino, G., Castelnuovo, M., Teichmann, M., & Pagano, A. (2007). The expanding RNA polymerase III transcriptome. *Trends in Genetics*, *23*(12), 614–622.

Dimond, A., & Fraser, P. (2013). Molecular biology. Long noncoding RNAs Xist in three dimensions. *Science*, *341*(6147), 720–721.

Dinger, M. E., Amaral, P. P., Mercer, T. R., & Mattick, J. S. (2009). Pervasive transcription of the eukaryotic genome: Functional indices and conceptual implications. *Briefings in Functional Genomics & Proteomics*, *8*(6), 407–423.

Dinger, M. E., Amaral, P. P., Mercer, T. R., Pang, K. C., Bruce, S. J., Gardiner, B. B., et al. (2008). Long noncoding RNAs in mouse embryonic stem cell pluripotency and differentiation. *Genome Research*, *18*(9), 1433–1445.

Dinger, M. E., Pang, K. C., Mercer, T. R., Crowe, M. L., Grimmond, S. M., & Mattick, J. S. (2009). NRED: A database of long noncoding RNA expression. *Nucleic Acids Research*, *37*(Database issue), D122–D126.

Djebali, S., Davis, C. A., Merkel, A., Dobin, A., Lassmann, T., Mortazavi, A., et al. (2012). Landscape of transcription in human cells. *Nature*, *489*(7414), 101–108.

Doxakis, E. (2010). Post-transcriptional regulation of alpha-synuclein expression by mir-7 and mir-153. *Journal of Biological Chemistry*, *285*(17), 12726–12734.

Eacker, S. M., Dawson, T. M., & Dawson, V. L. (2013). The interplay of microRNA and neuronal activity in health and disease. *Frontiers in Cellular Neuroscience*, 7, 136.

ENCODE Project Consortium, Bernstein, B. E., Birney, E., Dunham, I., Green, E. D., Gunter, C., et al. (2012). An integrated encyclopedia of DNA elements in the human genome. *Nature*, *489*(7414), 57–74.

Engström, P. G., Suzuki, H., Ninomiya, N., Akalin, A., Sessa, L., Lavorgna, G., et al. (2006). Complex loci in human and mouse genomes. *PLoS Genetics*, *2*(4), e47.

Faghihi, M. A., Modarresi, F., Khalil, A. M., Wood, D. E., Sahagan, B. G., Morgan, T. E., et al. (2008). Expression of a noncoding RNA is elevated in Alzheimer's disease and drives rapid feed-forward regulation of beta-secretase. *Nature Medicine*, *14*(7), 723–730.

Faghihi, M. A., Zhang, M., Huang, J., Modarresi, F., van der Brug, M. P., Nalls, M. A., et al. (2010). Evidence for natural antisense transcript-mediated inhibition of microRNA function. *Genome Biology*, *11*(5), R56.

Feng, J., Bi, C., Clark, B. S., Mady, R., Shah, P., & Kohtz, J. D. (2006). The Evf-2 noncoding RNA is transcribed from the Dlx-5/6 ultraconserved region and functions as a Dlx-2 transcriptional coactivator. *Genes & Development*, *20*(11), 1470–1484.

Gemayel, R., Vinces, M. D., Legendre, M., & Verstrepen, K. J. (2010). Variable tandem repeats accelerate evolution of coding and regulatory sequences. *Annual Review of Genetics*, *44*(1), 445–477.

Gordon, F. E., Nutt, C. L., Cheunsuchon, P., Nakayama, Y., Provencher, K. A., Rice, K. A., et al. (2010). Increased expression of angiogenic genes in the brains of mouse meg3-null embryos. *Endocrinology*, *151*(6), 2443–2452.

Grabowski, P. (2011). Alternative splicing takes shape during neuronal development. *Current Opinion in Genetics & Development*, *21*(4), 388–394.

Gregory, R. I., Yan, K.-P., Amuthan, G., Chendrimada, T., Doratotaj, B., Cooch, N., et al. (2004). The microprocessor complex mediates the genesis of microRNAs. *Nature*, *432*(7014), 235–240.

Griffiths-Jones, S., Grocock, R. J., van Dongen, S., Bateman, A., & Enright, A. J. (2006). miRBase: microRNA sequences, targets and gene nomenclature. *Nucleic Acids Research*, *34*(Database issue), D140–D144.

Grishok, A., Pasquinelli, A. E., Conte, D., Li, N., Parrish, S., Ha, I., et al. (2001). Genes and mechanisms related to RNA interference regulate expression of the small temporal RNAs that control C-elegans developmental timing. *Cell*, *106*(1), 23–34.

Guennewig, B., Roos, M., Dogar, A. M., Gebert, L. F. R., Zagalak, J. A., Vongrad, V., et al. (2014). Synthetic pre-microRNAs reveal dual-strand activity of miR-34a on TNF-α. *RNA*, *20*(1), 61–75.

Guil, S., Soler, M., Portela, A., Carrère, J., Fonalleras, E., Gómez, A., et al. (2012). Intronic RNAs mediate EZH2 regulation of epigenetic targets. *Nature Structural & Molecular Biology*, *19*(7), 664–670.

Guo, Z., Karunatilaka, K. S., & Rueda, D. (2009). Single-molecule analysis of protein-free U2-U6 snRNAs. *Nature Structural & Molecular Biology*, *16*(11), 1154–1159.

Gupta, R. A., Shah, N., Wang, K. C., Kim, J., Horlings, H. M., Wong, D. J., et al. (2010). Long non-coding RNA HOTAIR reprograms chromatin state to promote cancer metastasis. *Nature*, *464*(7291), 1071–1076.

Gustincich, S., Sandelin, A., Plessy, C., Katayama, S., Simone, R., Lazarevic, D., et al. (2006). The complexity of the mammalian transcriptome. *The Journal of Physiology*, *575*(Pt. 2), 321–332.

Guttman, M., & Rinn, J. L. (2012). Modular regulatory principles of large non-coding RNAs. *Nature*, *482*(7385), 339–346.

Hall, A. E., Turnbull, C., & Dalmay, T. (2013). Y RNAs: Recent developments. *Biomolecular Concepts*, *4*(2), 103–110.

Han, J., Lee, Y., Yeom, K.-H., Kim, Y.-K., Jin, H., & Kim, V. N. (2004). The Drosha-DGCR8 complex in primary microRNA processing. *Genes & Development*, *18*(24), 3016–3027.

Hangauer, M. J., Vaughn, I. W., & McManus, M. T. (Eds.), (2013). Pervasive transcription of the human genome produces thousands of previously unidentified long intergenic noncoding RNAs. In *PLoS Genetics*, *9*(6), e1003569.

Haramati, S., Chapnik, E., Sztainberg, Y., Eilam, R., Zwang, R., Gershoni, N., et al. (2010). miRNA malfunction causes spinal motor neuron disease. *Proceedings of the National Academy of Sciences of the United States of America*, *107*(29), 13111–13116.

Havens, M. A., Reich, A. A., Duelli, D. M., & Hastings, M. L. (2012). Biogenesis of mammalian microRNAs by a non-canonical processing pathway. *Nucleic Acids Research*, *40*(10), 4626–4640.

Hawkins, P. G., & Morris, K. V. (2010). Transcriptional regulation of Oct4 by a long non-coding RNA antisense to Oct4-pseudogene 5. *Transcription*, *1*(3), 165–175.

He, Y., Vogelstein, B., Velculescu, V. E., Papadopoulos, N., & Kinzler, K. W. (2008). The antisense transcriptomes of human cells. *Science*, *322*(5909), 1855–1857.

Hébert, S. S., Horré, K., Nicolaï, L., Papadopoulou, A. S., Mandemakers, W., Silahtaroglu, A. N., et al. (2008). Loss of microRNA cluster miR-29a/b-1 in sporadic Alzheimer's disease correlates with increased BACE1/beta-secretase expression. *Proceedings of the National Academy of Sciences of the United States of America*, *105*(17), 6415–6420.

Heins, N., Cremisi, F., Malatesta, P., Gangemi, R. M., Corte, G., Price, J., et al. (2001). Emx2 promotes symmetric cell divisions and a multipotential fate in precursors from the cerebral cortex. *Molecular and Cellular Neuroscience*, *18*(5), 485–502.

Herculano-Houzel, S. (2009). The human brain in numbers: A linearly scaled-up primate brain. *Frontiers in Human Neuroscience*, *3*, 31.

Hill, A. E., Hong, J. S., Wen, H., Teng, L. H., McPherson, D. T., McPherson, S. A., et al. (2006). Micro-RNA-like effects of complete intronic sequences. *Frontiers in Bioscience*, *11*, 1998–2006.

Hirose, T., Virnicchi, G., Tanigawa, A., Naganuma, T., Li, R., Kimura, H., et al. (2014). NEAT1 long noncoding RNA regulates transcription via protein sequestration within subnuclear bodies. *Molecular Biology of the Cell*, *25*(1), 169–183.

Hoekstra, H. E. (Ed.), (2013). Transposable elements are major contributors to the origin, diversification, and regulation of vertebrate long noncoding RNAs. In *PLoS Genetics*, *9*(4), e1003470.

Hu, H. Y., Guo, S., Xi, J., Yan, Z., Fu, N., Zhang, X., et al. (2011). MicroRNA expression and regulation in human, chimpanzee, and macaque brains. *PLoS Genetics*, 7(10), e1002327.

Irimia, M., & Blencowe, B. J. (2012). Alternative splicing: Decoding an expansive regulatory layer. *Current Opinion in Cell Biology*, *24*(3), 323–332.

Iyengar, B. R., Choudhary, A., Sarangdhar, M. A., Venkatesh, K. V., Gadgil, C. J., & Pillai, B. (2014). Non-coding RNA interact to regulate neuronal development and function. *Frontiers in Cellular Neuroscience*, *8*, 47.

Jeffries, C. D., Fried, H. M., & Perkins, D. O. (2011). Nuclear and cytoplasmic localization of neural stem cell microRNAs. *RNA*, *17*(4), 675–686.

Johnson, R. (2012). Long non-coding RNAs in Huntington's disease neurodegeneration. *Neurobiology of Disease*, *46*(2), 245–254.

Johnsson, P., Ackley, A., Vidarsdottir, L., Lui, W.-O., Corcoran, M., Grandér, D., et al. (2013). A pseudogene long-noncoding-RNA network regulates PTEN transcription and translation in human cells. *Nature Structural & Molecular Biology*, *20*(4), 440–446.

Johnston, R. J., & Hobert, O. (2003). A microRNA controlling left/right neuronal asymmetry in Caenorhabditis elegans. *Nature Cell Biology*, *426*(6968), 845–849.

Kageyama, Y., Kondo, T., & Hashimoto, Y. (2011). Coding vs non-coding: Translatability of short ORFs found in putative non-coding transcripts. *Biochimie*, *93*(11), 1981–1986.

Kalsotra, A., & Cooper, T. A. (2011). Functional consequences of developmentally regulated alternative splicing. *Nature Reviews. Genetics*, *12*(10), 715–729.

Kang, H. J., Kawasawa, Y. I., Cheng, F., Zhu, Y., Xu, X., Li, M., et al. (2011). Spatio-temporal transcriptome of the human brain. *Nature*, *478*(7370), 483–489.

Kapranov, P., Cheng, J., Dike, S., Nix, D. A., Duttagupta, R., Willingham, A. T., et al. (2007). RNA maps reveal new RNA classes and a possible function for pervasive transcription. *Science*, *316*(5830), 1484–1488.

Kapranov, P., St Laurent, G., Raz, T., Ozsolak, F., Reynolds, C. P., Sorensen, P. H. B., et al. (2010). The majority of total nuclear-encoded non-ribosomal RNA in a human cell is "dark matter" un-annotated RNA. *BMC Biology*, *8*(1), 149.

Katayama, S., Tomaru, Y., Kasukawa, T., Waki, K., Nakanishi, M., Nakamura, M., et al. (2005). Antisense transcription in the mammalian transcriptome. *Science*, *309*(5740), 1564–1566.

Kawahara, Y., Megraw, M., Kreider, E., Iizasa, H., Valente, L., Hatzigeorgiou, A. G., et al. (2008). Frequency and fate of microRNA editing in human brain. *Nucleic Acids Research*, *36*(16), 5270–5280.

Kawahara, Y., Zinshteyn, B., Chendrimada, T. P., Shiekhattar, R., & Nishikura, K. (2007). RNA editing of the microRNA-151 precursor blocks cleavage by the Dicer-TRBP complex. *EMBO Reports*, *8*(8), 763–769.

Kawahara, Y., Zinshteyn, B., Zinshteyn, B., Sethupathy, P., Iizasa, H., Hatzigeorgiou, A. G., et al. (2007). Redirection of silencing targets by adenosine-to-inosine editing of miRNAs. *Science*, *315*(5815), 1137–1140.

Kelemen, O., Convertini, P., Zhang, Z., Wen, Y., Shen, M., Falaleeva, M., et al. (2013). Function of alternative splicing. *Gene*, *514*(1), 1–30.

Kerin, T., Ramanathan, A., Rivas, K., Grepo, N., Coetzee, G. A., & Campbell, D. B. (2012). A noncoding RNA antisense to moesin at 5p14.1 in autism. *Science Translational Medicine*, *4*((128), 128ra40.

Ketting, R. F., Fischer, S. E., Bernstein, E., Sijen, T., Hannon, G. J., & Plasterk, R. H. (2001). Dicer functions in RNA interference and in synthesis of small RNA involved in developmental timing in C. elegans. *Genes & Development*, *15*(20), 2654–2659.

Khalil, A. M., Faghihi, M. A., Modarresi, F., Brothers, S. P., & Wahlestedt, C. (2008). A novel RNA transcript with antiapoptotic function is silenced in fragile X syndrome. *PLoS One*, *3*(1), e1486.

Khalil, A. M., Guttman, M., Huarte, M., Garber, M., Raj, A., Rivea Morales, D., et al. (2009). Many human large intergenic noncoding RNAs associate with chromatin-modifying complexes and affect gene expression. *Proceedings of the National Academy of Sciences of the United States of America*, *106*(28), 11667–11672.

Kim, T. K., Hemberg, M., Gray, J. M., Costa, A. M., Bear, D. M., Wu, J., et al. (2010). Widespread transcription at neuronal activity-regulated enhancers. *Nature*, *465*(7295), 182–187.

Kim, Y.-K., & Kim, V. N. (2007). Processing of intronic microRNAs. *The EMBO Journal*, *26*(3), 775–783.

Kim, J., Xiao, H., Bonanno, J. B., Kalyanaraman, C., Brown, S., Tang, X., et al. (2013). Structure-guided discovery of the metabolite carboxy-SAM that modulates tRNA function. *Nature*, *498*(7452), 123–126.

Knight, S. W., & Bass, B. L. (2001). A role for the RNase III enzyme DCR-1 in RNA interference and germ line development in Caenorhabditis elegans. *Science*, *293*(5538), 2269–2271.

Konopka, W., Kiryk, A., Novak, M., Herwerth, M., Parkitna, J. R., Wawrzyniak, M., et al. (2010). MicroRNA loss enhances learning and memory in mice. *The Journal of Neuroscience*, *30*(44), 14835–14842.

Korneev, S. A., Korneeva, E. I., Lagarkova, M. A., Kiselev, S. L., Critchley, G., & O'Shea, M. (2008). Novel noncoding antisense RNA transcribed from human anti-NOS2A locus is differentially regulated during neuronal differentiation of embryonic stem cells. *RNA*, *14*(10), 2030–2037.

Koziol, M. J., & Rinn, J. L. (2010). RNA traffic control of chromatin complexes. *Current Opinion in Genetics & Development*, *20*(2), 142–148.

Krol, J., Busskamp, V., Markiewicz, I., Stadler, M. B., Ribi, S., Richter, J., et al. (2010). Characterizing light-regulated retinal microRNAs reveals rapid turnover as a common property of neuronal microRNAs. *Cell*, *141*(4), 618–631.

Kutter, C., Watt, S., Stefflova, K., Wilson, M. D., Goncalves, A., Ponting, C. P., et al. (2012). Rapid Turnover of Long Noncoding RNAs and the Evolution of Gene Expression. *PLoS Genetics*, *8*(7). http://dx.doi.org/10.1371/journal.pgen.1002841.

Lagos-Quintana, M., Rauhut, R., Lendeckel, W., & Tuschl, T. (2001). Identification of novel genes coding for small expressed RNAs. *Science*, *294*(5543), 853–858.

Landgraf, P., Rusu, M., Sheridan, R., Sewer, A., Iovino, N., Aravin, A., et al. (2007). A mammalian microRNA expression atlas based on small RNA library sequencing. *Cell*, *129*(7), 1401–1414.

Landry, C. D., Kandel, E. R., & Rajasethupathy, P. (2013). New mechanisms in memory storage: piRNAs and epigenetics. *Trends in Neurosciences*, *36*(9), 535–542.

Landthaler, M., Yalcin, A., & Tuschl, T. (2004). The human DiGeorge syndrome critical region gene 8 and its D. melanogaster homolog are required for miRNA biogenesis. *Current Biology*, *14*(23), 2162–2167.

Lau, N. C., Lim, L. P., Weinstein, E. G., & Bartel, D. P. (2001). An abundant class of tiny RNAs with probable regulatory roles in Caenorhabditis elegans. *Science*, *294*(5543), 858–862.

Lee, Y., Ahn, C., Han, J., Choi, H., Kim, J., Yim, J., et al. (2003). The nuclear RNase III Drosha initiates microRNA processing. *Nature Cell Biology*, *425*(6956), 415–419.

Lee, R. C., & Ambros, V. (2001). An extensive class of small RNAs in Caenorhabditis elegans. *Science*, *294*(5543), 862–864.

Lee, E. J., Banerjee, S., Zhou, H., Jammalamadaka, A., Arcila, M., Manjunath, B. S., et al. (2011). Identification of piRNAs in the central nervous system. *RNA*, *17*(6), 1090–1099.

Lee, Y., Kim, M., Han, J., Yeom, K.-H., Lee, S., Baek, S. H., et al. (2004). MicroRNA genes are transcribed by RNA polymerase II. *The EMBO Journal*, *23*(20), 4051–4060.

Leeb, M., Steffen, P. A., & Wutz, A. (2009). X chromosome inactivation sparked by non-coding RNAs. *RNA Biology*, *6*(2), 94–99.

Leidinger, P., Backes, C., Deutscher, S., Schmitt, K., Mueller, S. C., Frese, K., et al. (2013). A blood based 12-miRNA signature of Alzheimer disease patients. *Genome Biology*, *14*(7), R78.

Le Meur, E., Watrin, F., Landers, M., Sturny, R., Lalande, M., & Muscatelli, F. (2005). Dynamic developmental regulation of the large non-coding RNA associated with the mouse 7C imprinted chromosomal region. *Developmental Biology*, *286*(2), 587–600.

Lescoute, A., Leontis, N. B., Massire, C., & Westhof, E. (2005). Recurrent structural RNA motifs, isostericity matrices and sequence alignments. *Nucleic Acids Research*, *33*(8), 2395–2409.

Leucci, E., Patella, F., Waage, J., Holmstrøm, K., Lindow, M., Porse, B., et al. (2013). microRNA-9 targets the long non-coding RNA MALAT1 for degradation in the nucleus. *Scientific Reports*, *3*, 2535.

Lin, M., Pedrosa, E., Shah, A., Hrabovsky, A., Maqbool, S., Zheng, D., et al. (2011). RNA-Seq of human neurons derived from iPS cells reveals candidate long non-coding RNAs involved in neurogenesis and neuropsychiatric disorders. *PLoS One*, *6*(9), e23356.

Ling, K.-H., Hewitt, C. A., Beissbarth, T., Hyde, L., Cheah, P.-S., Smyth, G. K., et al. (2011). Spatiotemporal regulation of multiple overlapping sense and novel natural antisense transcripts at the Nrgn and Camk2n1 gene loci during mouse cerebral corticogenesis. *Cerebral Cortex*, *21*(3), 683–697.

Lipovich, L., Tarca, A. L., Cai, J., Jia, H., Chugani, H. T., Sterner, K. N., et al. (2014). Developmental changes in the transcriptome of human cerebral cortex tissue: Long non-coding RNA transcripts. *Cerebral Cortex*, *24*(6), 1451–1459.

Liu, G., Mattick, J. S., & Taft, R. J. (2013). A meta-analysis of the genomic and transcriptomic composition of complex life. *Cell Cycle*, *12*(13), 2061–2072.

Loughlin, F. E., Gebert, L. F. R., Towbin, H., Brunschweiger, A., Hall, J., & Allain, F. H.-T. (2012). Structural basis of pre-let-7 miRNA recognition by the zinc knuckles of pluripotency factor Lin28. *Nature Structural & Molecular Biology*, *19*(1), 84–89.

Luciano, D. J., Mirsky, H., Vendetti, N. J., & Maas, S. (2004). RNA editing of a miRNA precursor. *RNA*, *10*(8), 1174–1177.

Maas, S., Rich, A., & Nishikura, K. (2003). A-to-I RNA editing: Recent news and residual mysteries. *Journal of Biological Chemistry*, *278*(3), 1391–1394.

Magistri, M., Faghihi, M. A., St Laurent, G., & Wahlestedt, C. (2012). Regulation of chromatin structure by long noncoding RNAs: Focus on natural antisense transcripts. *Trends in Genetics*, *28*(8), 389–396.

Malone, C. D., & Hannon, G. J. (2009). Small RNAs as guardians of the genome. *Cell*, *136*(4), 656–668.

Mattick, J. S. (1994). Introns: Evolution and function. *Current Opinion in Genetics & Development*, *4*(6), 823–831.

Mattick, J. S. (2012). Rocking the foundations of molecular genetics. *Proceedings of the National Academy of Sciences of the United States of America*, *109*(41), 16400–16401.

Mattick, J. S., & Makunin, I. V. (2006). Non-coding RNA. *Human Molecular Genetics*, *15 Spec No 1*(90001), R17–R29.

McConnell, M. J., Lindberg, M. R., Brennand, K. J., Piper, J. C., Voet, T., Cowing-Zitron, C., et al. (2013). Mosaic copy number variation in human neurons. *Science*, *342*(6158), 632–637.

McLachlan, J., Pasquinelli, A. E., Bálint, E., Tuschl, T., & Zamore, P. D. (2001). A cellular function for the RNA-interference enzyme Dicer in the maturation of the let-7 small temporal RNA. *Science*, *293*(5531), 834–838.

Melcher, T., Maas, S., Herb, A., Sprengel, R., Higuchi, M., & Seeburg, P. H. (1996). RED2, a brain-specific member of the RNA-specific adenosine deaminase family. *The Journal of Biological Chemistry*, *271*(50), 31795–31798.

Memczak, S., Jens, M., Elefsinioti, A., Torti, F., Krueger, J., Rybak, A., et al. (2013). Circular RNAs are a large class of animal RNAs with regulatory potency. *Nature*, *495*(7441), 333–338.
Mercer, T. R., Dinger, M. E., & Mattick, J. S. (2009). Long non-coding RNAs: Insights into functions. *Nature Reviews. Genetics*, *10*(3), 155–159.
Mercer, T. R., Dinger, M. E., Sunkin, S. M., Mehler, M. F., & Mattick, J. S. (2008). Specific expression of long noncoding RNAs in the mouse brain. *Proceedings of the National Academy of Sciences of the United States of America*, *105*(2), 716–721.
Mercer, T. R., Dinger, M. E., Trapnell, C., & Mattick, J. S. (2012). Targeted RNA sequencing reveals the deep complexity of the human transcriptome. *Nature Biotechnology*, *30*(1), 99–104.
Mercer, T. R., & Mattick, J. S. (2013a). Structure and function of long noncoding RNAs in epigenetic regulation. *Nature Structural & Molecular Biology*, *20*(3), 300–307.
Mercer, T. R., & Mattick, J. S. (2013b). Understanding the regulatory and transcriptional complexity of the genome through structure. *Genome Research*, *23*(7), 1081–1088.
Mercer, T. R., Neph, S., Dinger, M. E., Crawford, J., Smith, M. A., Shearwood, A.-M. J., et al. (2011). The human mitochondrial transcriptome. *Cell*, *146*(4), 645–658.
Mercer, T. R., Qureshi, I. A., Gokhan, S., Dinger, M. E., Li, G., Mattick, J. S., et al. (2010). Long noncoding RNAs in neuronal-glial fate specification and oligodendrocyte lineage maturation. *BMC Neuroscience*, *11*, 14.
Michlewski, G., & Cáceres, J. F. (2010). Antagonistic role of hnRNP A1 and KSRP in the regulation of let-7a biogenesis. *Nature Structural & Molecular Biology*, *17*(8), 1011–1018.
Mills, J. D., & Janitz, M. (2012). Alternative splicing of mRNA in the molecular pathology of neurodegenerative diseases. *Neurobiology of Aging*, *33*(5), 1012.e11–1012.e24.
Miura, P., Shenker, S., Andreu-Agullo, C., Westholm, J. O., & Lai, E. C. (2013). Widespread and extensive lengthening of 3' UTRs in the mammalian brain. *Genome Research*, *23*(5), 812–825.
Modarresi, F., Faghihi, M. A., Lopez-Toledano, M. A., Fatemi, R. P., Magistri, M., Brothers, S. P., et al. (2012). Inhibition of natural antisense transcripts in vivo results in gene-specific transcriptional upregulation. *Nature Biotechnology*, *30*(5), 453–459.
Morris, K. V. (2011). The emerging role of RNA in the regulation of gene transcription in human cells. *Seminars in Cell & Developmental Biology*, *22*(4), 351–358.
Mourelatos, Z., Dostie, J., Paushkin, S., Sharma, A., Charroux, B., Abel, L., et al. (2002). miRNPs: A novel class of ribonucleoproteins containing numerous microRNAs. *Genes & Development*, *16*(6), 720–728.
Muslimov, I. A., Banker, G., Brosius, J., & Tiedge, H. (1998). Activity-dependent regulation of dendritic BC1 RNA in hippocampal neurons in culture. *The Journal of Cell Biology*, *141*(7), 1601–1611.
Nakaya, H. I., Amaral, P. P., Louro, R., Lopes, A., Fachel, A. A., Moreira, Y. B., et al. (2007). Genome mapping and expression analyses of human intronic noncoding RNAs reveal tissue-specific patterns and enrichment in genes related to regulation of transcription. *Genome Biology*, *8*(3), R43.
Necsulea, A., Soumillon, M., Warnefors, M., Liechti, A., Daish, T., Zeller, U., et al. (2014). The evolution of lncRNA repertoires and expression patterns in tetrapods. *Nature*, *505*(7485), 635–640.
Ng, S.-Y., Lin, L., Soh, B. S., & Stanton, L. W. (2013). Long noncoding RNAs in development and disease of the central nervous system. *Trends in Genetics*, *29*(8), 461–468.
Nicolas, F. E., Hall, A. E., Csorba, T., Turnbull, C., & Dalmay, T. (2012). Biogenesis of Y RNA-derived small RNAs is independent of the microRNA pathway. *FEBS Letters*, *586*(8), 1226–1230.
Nielsen, R., Bustamante, C., Clark, A. G., Glanowski, S., Sackton, T. B., Hubisz, M. J., et al. (2005). A scan for positively selected genes in the genomes of humans and chimpanzees. *PLoS Biology*, *3*(6), e170.

Novikova, I. V., Hennelly, S. P., & Sanbonmatsu, K. Y. (2012). Structural architecture of the human long non-coding RNA, steroid receptor RNA activator. *Nucleic Acids Research*, *40*(11), 5034–5051.

Nowak, J. S., & Michlewski, G. (2013). miRNAs in development and pathogenesis of the nervous system. *Biochemical Society Transactions*, *41*(4), 815–820.

Nudelman, A. S., DiRocco, D. P., Lambert, T. J., Garelick, M. G., Le, J., Nathanson, N. M., et al. (2010). Neuronal activity rapidly induces transcription of the CREB-regulated microRNA-132, in vivo. *Hippocampus*, *20*(4), 492–498.

Oiglane-Shlik, E., Talvik, T., Zordania, R., Poder, H., Kahre, T., Raukas, E., et al. (2006). Prevalence of Angelman syndrome and Prader-Willi syndrome in Estonian children: Sister syndromes not equally represented. *American Journal of Medical Genetics. Part A*, *140*(18), 1936–1943.

Okada, Y., Tashiro, C., Numata, K., Watanabe, K., Nakaoka, H., Yamamoto, N., et al. (2008). Comparative expression analysis uncovers novel features of endogenous antisense transcription. *Human Molecular Genetics*, *17*(11), 1631–1640.

Ørom, U. A., Ørom, U. A., & Shiekhattar, R. (2011). Long non-coding RNAs and enhancers. *Current Opinion in Genetics & Development*, *21*(2), 194–198.

Padgett, R. A., Grabowski, P. J., Konarska, M. M., Seiler, S., & Sharp, P. A. (1986). Splicing of messenger RNA precursors. *Annual Review of Biochemistry*, *55*(1), 1119–1150.

Pan, Q., Shai, O., Lee, L. J., Frey, B. J., & Blencowe, B. J. (2008). Deep surveying of alternative splicing complexity in the human transcriptome by high-throughput sequencing. *Nature Genetics*, *40*(12), 1413–1415.

Pastori, C., Peschansky, V. J., Barbouth, D., Mehta, A., Silva, J. P., & Wahlestedt, C. (2014). Comprehensive analysis of the transcriptional landscape of the human FMR1 gene reveals two new long noncoding RNAs differentially expressed in Fragile X syndrome and Fragile X-associated tremor/ataxia syndrome. *Human Genetics*, *133*(1), 59–67.

Paul, M. S., & Bass, B. L. (1998). Inosine exists in mRNA at tissue-specific levels and is most abundant in brain mRNA. *The EMBO Journal*, *17*(4), 1120–1127.

Pei, B., Sisu, C., Frankish, A., Howald, C., Habegger, L., Mu, X. J., et al. (2012). The GENCODE pseudogene resource. *Genome Biology*, *13*(9), R51.

Peng, Z., Cheng, Y., Tan, B., Kang, L., & Tian, Z. (2012). Comprehensive analysis of RNA-Seq data reveals extensive RNA editing in a human transcriptome. *Nature Biotechnology*, *30*(3), 253–260.

Poliseno, L., Salmena, L., Zhang, J., Carver, B., Haveman, W. J., & Pandolfi, P. P. (2010). A coding-independent function of gene and pseudogene mRNAs regulates tumour biology. *Nature*, *465*(7301), 1033–1038.

Pollard, K. S., Salama, S. R., Lambert, N., Lambot, M.-A., Coppens, S., Pedersen, J. S., et al. (2006). An RNA gene expressed during cortical development evolved rapidly in humans. *Nature*, *443*(7108), 167–172.

Ponjavic, J., Ponting, C. P., & Lunter, G. (2007). Functionality or transcriptional noise? Evidence for selection within long noncoding RNAs. *Genome Research*, *17*(5), 556–565.

Ponjavic, J., Oliver, P. L., Lunter, G., & Ponting, C. P. (2009). Genomic and transcriptional co-localization of protein-coding and long non-coding RNA pairs in the developing brain. *PLoS Genetics*, *5*(8), e1000617.

Prasanth, K. V., Prasanth, S. G., Xuan, Z., Hearn, S., Freier, S. M., Bennett, C. F., et al. (2005). Regulating gene expression through RNA nuclear retention. *Cell*, *123*(2), 249–263.

Pruunsild, P., Kazantseva, A., Aid, T., Palm, K., & Timmusk, T. (2007). Dissecting the human BDNF locus: Bidirectional transcription, complex splicing, and multiple promoters. *Genomics*, *90*(3), 397–406.

Qureshi, I. A., & Mehler, M. F. (2013). Long non-coding RNAs: Novel targets for nervous system disease diagnosis and therapy. *Neurotherapeutics*, *10*(4), 632–646.

Rackham, O., Shearwood, A.-M. J., Mercer, T. R., Davies, S. M. K., Mattick, J. S., & Filipovska, A. (2011). Long noncoding RNAs are generated from the mitochondrial genome and regulated by nuclear-encoded proteins. *RNA*, *17*(12), 2085–2093.

Rajasethupathy, P., Antonov, I., Sheridan, R., Frey, S., Sander, C., Tuschl, T., et al. (2012). A role for neuronal piRNAs in the epigenetic control of memory-related synaptic plasticity. *Cell*, *149*(3), 693–707.

Rao, P., Benito, E., & Fischer, A. (2013). MicroRNAs as biomarkers for CNS disease. *Frontiers in Molecular Neuroscience*, *6*, 39.

Reinisch, K. M., & Wolin, S. L. (2007). Emerging themes in non-coding RNA quality control. *Current Opinion in Structural Biology*, *17*(2), 209–214.

Ricci, E. P., Kucukural, A., Cenik, C., Mercier, B. C., Singh, G., Heyer, E. E., et al. (2014). Staufen1 senses overall transcript secondary structure to regulate translation. *Nature Structural & Molecular Biology*, *21*(1), 26–35.

Rinn, J. L., Kertesz, M., Wang, J. K., Squazzo, S. L., Xu, X., Brugmann, S. A., et al. (2007). Functional demarcation of active and silent chromatin domains in human HOX loci by noncoding RNAs. *Cell*, *129*(7), 1311–1323.

Ro, S., Ma, H.-Y., Park, C., Ortogero, N., Song, R., Hennig, G. W., et al. (2013). The mitochondrial genome encodes abundant small noncoding RNAs. *Cell Research*, *23*(6), 759–774.

Rogelj, B., Hartmann, C., Yeo, C. H., Hunt, S. P., & Giese, K. P. (2003). Contextual fear conditioning regulates the expression of brain-specific small nucleolar RNAs in hippocampus. *European Journal of Neuroscience*, *11*, 3089–3096.

Ruby, J. G., Jan, C. H., & Bartel, D. P. (2007). Intronic microRNA precursors that bypass Drosha processing. *Nature*, *448*(7149), 83–86.

Sánchez-Pla, A., Reverter, F., Ruíz de Villa, M. C., & Comabella, M. (2012). Transcriptomics: mRNA and alternative splicing. *Journal of Neuroimmunology*, *248*(1–2), 23–31.

Santoro, F., Mayer, D., Klement, R. M., Warczok, K. E., Stukalov, A., Barlow, D. P., et al. (2013). Imprinted Igf2r silencing depends on continuous Airn lncRNA expression and is not restricted to a developmental window. *Development*, *140*(6), 1184–1195.

Scadden, A. D. J. (2005). The RISC subunit Tudor-SN binds to hyper-edited double-stranded RNA and promotes its cleavage. *Nature Structural & Molecular Biology*, *12*(6), 489–496.

Shearwin, K. E., Callen, B. P., & Egan, J. B. (2005). Transcriptional interference—A crash course. *Trends in Genetics*, *21*(6), 339–345.

Siomi, M. C., Sato, K., Pezic, D., & Aravin, A. A. (2011). PIWI-interacting small RNAs: The vanguard of genome defence. *Nature Reviews. Molecular Cell Biology*, *12*(4), 246–258.

Skreka, K., Schafferer, S., Nat, I.-R., Zywicki, M., Salti, A., Apostolova, G., et al. (2012). Identification of differentially expressed non-coding RNAs in embryonic stem cell neural differentiation. *Nucleic Acids Research*, *40*(13), 6001–6015.

Smith, M. A., Gesell, T., Stadler, P. F., & Mattick, J. S. (2013). Widespread purifying selection on RNA structure in mammals. *Nucleic Acids Research*, *41*(17), 8220–8236.

Smith, R. M., Webb, A., Papp, A. C., Newman, L. C., Handelman, S. K., Suhy, A., et al. (2013). Whole transcriptome RNA-Seq allelic expression in human brain. *BMC Genomics*, *14*, 571.

Son, D. J., Kumar, S., Takabe, W., Kim, C. W., Ni, C.-W., Alberts-Grill, N., et al. (2013). The atypical mechanosensitive microRNA-712 derived from pre-ribosomal RNA induces endothelial inflammation and atherosclerosis. *Nature Communications*, *4*, 3000.

Spigoni, G., Gedressi, C., & Mallamaci, A. (2010). Regulation of Emx2 expression by antisense transcripts in murine cortico-cerebral precursors. *PLoS One*, *5*(1), e8658.

Struhl, K. (2007). Transcriptional noise and the fidelity of initiation by RNA polymerase II. *Nature Structural & Molecular Biology*, *14*(2), 103–105.

Sunwoo, H., Dinger, M. E., Wilusz, J. E., Amaral, P. P., Mattick, J. S., & Spector, D. L. (2009). MEN epsilon/beta nuclear-retained non-coding RNAs are up-regulated upon muscle differentiation and are essential components of paraspeckles. *Genome Research*, *19*(3), 347–359.

Taft, R. J., Glazov, E. A., Cloonan, N., Simons, C., Stephen, S., Faulkner, G. J., et al. (2009). Tiny RNAs associated with transcription start sites in animals. *Nature Genetics*, *41*(5), 572–578.

Taft, R. J., Glazov, E. A., Lassmann, T., Hayashizaki, Y., Carninci, P., & Mattick, J. S. (2009). Small RNAs derived from snoRNAs. *RNA*, *15*(7), 1233–1240.

Taft, R. J., Kaplan, C. D., Simons, C., & Mattick, J. S. (2009). Evolution, biogenesis and function of promoter-associated RNAs. *Cell Cycle*, *8*(15), 2332–2338.

Taft, R. J., Pang, K. C., Mercer, T. R., Dinger, M., & Mattick, J. S. (2010). Non-coding RNAs: Regulators of disease. *The Journal of Pathology*, *220*(2), 126–139.

Taft, R. J., Pheasant, M., & Mattick, J. S. (2007). The relationship between non-protein-coding DNA and eukaryotic complexity. *BioEssays*, *29*(3), 288–299.

Taft, R. J., Simons, C., Nahkuri, S., Oey, H., Korbie, D. J., Mercer, T. R., et al. (2010). Nuclear-localized tiny RNAs are associated with transcription initiation and splice sites in metazoans. *Nature Structural & Molecular Biology*, *17*(8), 1030–1034.

Taggart, A. J., DeSimone, A. M., Shih, J. S., Filloux, M. E., & Fairbrother, W. G. (2012). Large-scale mapping of branchpoints in human pre-mRNA transcripts in vivo. *Nature Structural & Molecular Biology*, *19*(7), 719–721.

Tam, O. H., Aravin, A. A., Stein, P., Girard, A., Murchison, E. P., Cheloufi, S., et al. (2008). Pseudogene-derived small interfering RNAs regulate gene expression in mouse oocytes. *Nature*, *453*(7194), 534–538.

Tilgner, H., Knowles, D. G., Johnson, R., Davis, C. A., Chakrabortty, S., Djebali, S., et al. (2012). Deep sequencing of subcellular RNA fractions shows splicing to be predominantly co-transcriptional in the human genome but inefficient for lncRNAs. *Genome Research*, *22*(9), 1616–1625.

Trinklein, N. D., Aldred, S. F., Hartman, S. J., Schroeder, D. I., Otillar, R. P., & Myers, R. M. (2004). An abundance of bidirectional promoters in the human genome. *Genome Research*, *14*(1), 62–66.

Tripathi, V., Ellis, J. D., Shen, Z., Song, D. Y., Pan, Q., Watt, A. T., et al. (2010). The nuclear-retained noncoding RNA MALAT1 regulates alternative splicing by modulating SR splicing factor phosphorylation. *Molecular Cell*, *39*(6), 925–938.

Tsai, M.-C., Manor, O., Wan, Y., Mosammaparast, N., Wang, J. K., Lan, F., et al. (2010). Long noncoding RNA as modular scaffold of histone modification complexes. *Science*, *329*(5992), 689–693.

Ulitsky, I., Shkumatava, A., Jan, C. H., Sive, H., & Bartel, D. P. (2011). Conserved function of lincRNAs in vertebrate embryonic development despite rapid sequence evolution. *Cell*, *147*(7), 1537–1550.

Vaishnavi, V., Manikandan, M., Tiwary, B. K., & Munirajan, A. K. (2013). Insights on the functional impact of microRNAs present in autism-associated copy number variants. *PLoS One*, *8*(2), e56781.

Wan, Y., Kertesz, M., Spitale, R. C., Segal, E., & Chang, H. Y. (2011). Understanding the transcriptome through RNA structure. *Nature Reviews. Genetics*, *12*(9), 641–655.

Wang, L., Guo, Z.-Y., Zhang, R., Xin, B., Chen, R., Zhao, J., et al. (2013). Pseudogene OCT4-pg4 functions as a natural micro RNA sponge to regulate OCT4 expression by competing for miR-145 in hepatocellular carcinoma. *Carcinogenesis*, *34*(8), 1773–1781.

Wang, H., Iacoangeli, A., Lin, D., Williams, K., Denman, R. B., Hellen, C. U. T., et al. (2005). Dendritic BC1 RNA in translational control mechanisms. *The Journal of Cell Biology, 171*(5), 811–821.

Wang, H. D., Iacoangeli, A., Popp, S., Muslimov, I. A., Imataka, H., Sonenberg, N., et al. (2002). Dendritic BC1 RNA: Functional role in regulation of translation initiation. *The Journal of Neuroscience, 22*(23), 10232–10241.

Wang, B., Li, S., Qi, H. H., Chowdhury, D., Shi, Y., & Novina, C. D. (2009). Distinct passenger strand and mRNA cleavage activities of human Argonaute proteins. *Nature Structural & Molecular Biology, 16*(12), 1259–1266.

Wang, W.-X., Rajeev, B. W., Stromberg, A. J., Ren, N., Tang, G., Huang, Q., et al. (2008). The expression of microRNA miR-107 decreases early in Alzheimer's disease and may accelerate disease progression through regulation of beta-site amyloid precursor protein-cleaving enzyme 1. *The Journal of Neuroscience, 28*(5), 1213–1223.

Watanabe, T., Totoki, Y., Toyoda, A., Kaneda, M., Kuramochi-Miyagawa, S., Obata, Y., et al. (2008). Endogenous siRNAs from naturally formed dsRNAs regulate transcripts in mouse oocytes. *Nature, 453*(7194), 539–543.

Werner, A., & Sayer, J. A. (2009). Naturally occurring antisense RNA: Function and mechanisms of action. *Current Opinion in Nephrology and Hypertension, 18*(4), 343–349.

Westholm, J. O., Ladewig, E., Okamura, K., Robine, N., & Lai, E. C. (2012). Common and distinct patterns of terminal modifications to mirtrons and canonical microRNAs. *RNA, 18*(2), 177–192.

Westholm, J. O., & Lai, E. C. (2011). Mirtrons: microRNA biogenesis via splicing. *Biochimie, 93*(11), 1897–1904.

Xie, C., Yuan, J., Li, H., Li, M., Zhao, G., Bu, D., et al. (2014). NONCODEv4: Exploring the world of long non-coding RNA genes. *Nucleic Acids Research, 42*(Database issue), D98–D103.

Yan, M.-D., Hong, C.-C., Lai, G.-M., Cheng, A.-L., Lin, Y.-W., & Chuang, S.-E. (2005). Identification and characterization of a novel gene Saf transcribed from the opposite strand of Fas. *Human Molecular Genetics, 14*(11), 1465–1474.

Yang, W., Chendrimada, T. P., Wang, Q., Higuchi, M., Seeburg, P. H., Shiekhattar, R., et al. (2006). Modulation of microRNA processing and expression through RNA editing by ADAR deaminases. *Nature Structural & Molecular Biology, 13*(1), 13–21.

Yang, Y., Zhou, X., & Jin, Y. (2013). ADAR-mediated RNA editing in non-coding RNA sequences. *Science China. Life Sciences, 56*(10), 944–952.

Yin, Q.-F., Yang, L., Zhang, Y., Zhang, Y., Xiang, J.-F., Wu, Y.-W., et al. (2012). Long noncoding RNAs with snoRNA ends. *Molecular Cell, 48*(2), 219–230.

Yoo, A. S., Sun, A. X., Li, L., Shcheglovitov, A., Portmann, T., Li, Y., et al. (2011). MicroRNA-mediated conversion of human fibroblasts to neurons. *Nature, 476*(7359), 228–231.

Zhang, B., Arun, G., Mao, Y. S., Lazar, Z., Hung, G., Bhattacharjee, G., et al. (2012). The lncRNA Malat1 is dispensable for mouse development but its transcription plays a cis-regulatory role in the adult. *Cell Reports, 2*(1), 111–123.

Zhang, B., Gaiteri, C., Bodea, L.-G., Wang, Z., McElwee, J., Podtelezhnikov, A. A., et al. (2013). Integrated systems approach identifies genetic nodes and networks in late-onset Alzheimer's disease. *Cell, 153*(3), 707–720.

Zhang, H., Kolb, F. A., Jaskiewicz, L., Westhof, E., & Filipowicz, W. (2004). Single processing center models for human Dicer and bacterial RNase III. *Cell, 118*(1), 57–68.

Zhang, X., Zhang, X., Rice, K., Wang, Y., Chen, W., Zhong, Y., et al. (2010). Maternally expressed gene 3 (MEG3) noncoding ribonucleic acid: Isoform structure, expression, and functions. *Endocrinology, 151*(3), 939–947.

Zhang, Z., Zhang, Z., Theurkauf, W. E., Theurkauf, W. E., Weng, Z., Weng, Z., et al. (2012). Strand-specific libraries for high throughput RNA sequencing (RNA-Seq) prepared without poly(A) selection. *Silence*, *3*(1), 9.

Zhao, J., Sun, B. K., Erwin, J. A., Song, J.-J., & Lee, J. T. (2008). Polycomb proteins targeted by a short repeat RNA to the mouse X chromosome. *Science*, *322*(5902), 750–756.

Zhao, J., Zhao, J., Ohsumi, T. K., Ohsumi, T. K., Kung, J. T., Kung, J. T., et al. (2010). Genome-wide identification of polycomb-associated RNAs by RIP-seq. *Molecular Cell*, *40*(6), 939–953.

Ziats, M. N., & Rennert, O. M. (2013). Identification of differentially expressed microRNAs across the developing human brain. *Molecular Psychiatry*, *19*, 848–852.

CHAPTER EIGHT

Genetics of Gene Expression in CNS

Ashutosh K. Pandey, Robert W. Williams[1]
Department of Genetics, Genomics and Informatics, Center for Integrative and Translational Genomics, University of Tennessee Health Science Center, Memphis, Tennessee, USA
[1]Corresponding author: e-mail address: rwilliams@uthsc.edu

Contents

Portions of this chapter are prepared by the UT-Battelle, LLC, Operator of Oak Ridge National Laboratory under Contract No. DE-AC05-00OR22725

International Review of Neurobiology, Volume 116
ISSN 0074-7742
http://dx.doi.org/10.1016/B978-0-12-801105-8.00008-4

Abstract

Transcriptome studies have revealed a surprisingly high level of variation among individuals in expression of key genes in the CNS under both normal and experimental conditions. Ten-fold variation is common, yet the specific causes and consequences of this variation are largely unknown. By combining classic gene mapping methods—family linkage studies and genomewide association—with high-throughput genomics, it is now possible to define quantitative trait loci (QTLs), single-gene variants, and even single SNPs and indels that control gene expression in different brain regions and cells. This review considers some of the major technical and conceptual challenges in analyzing variation in expression in the CNS with a focus on mRNAs, rather than noncoding RNAs or proteins. At one level of analysis, this work has been highly successful, and we finally have techniques that can be used to track down small numbers of loci that control expression in the CNS. But at a higher level of analysis, we still do not understand the genetic architecture of gene expression in brain, the consequences of expression QTLs on protein levels or on cell function, or the combined impact of expression differences on behavior and disease risk. These important gaps are likely to be bridged over the next several decades using (1) much larger sample sizes, (2) more powerful RNA sequencing and proteomic methods, and (3) novel statistical and computational models to predict genome-to-phenome relations

1. INTRODUCTION

For many years, gene mapping studies have focused on the identification of single-gene variants and molecular causes of diseases ranging from albinism and phenylketonuria to neurodegenerative diseases such as Huntington's and Alzheimer's disease (Charles, Moore, & Yates, 1992; Gusella et al., 1983; St George-Hyslop et al., 1987; Woo, Lidsky, Guttler, Thirumalachary, & Robson, 1984). The same linkage mapping methods that have been used to track down the CAG trinucleotide repeat expansion that causes Huntington's disease (MacDonald et al., 1993) can now be used to study the causes of variation in levels of microtraits, such as RNAs, metabolites, and proteins, in any tissue, organ, or cell. All that is required is a cohort of individuals and matched expression data for a specific brain region or cell type for each subject. A major goal of expression genetics research is to uncover primary and causal sequence variants that modulate expressions levels, but the long-term focus is on the complex hierarchical networks that link genetic variation, through mRNA and protein levels, to higher order phenotypes that influence disease risk and progression. If we understand the networks of causal linkages between differences in expression and differences in CNS function, then it may become possible to push just the right molecular buttons to prevent and cure many still intractable diseases of the brain.

Compared to a classic genetic analysis of a Mendelian trait such as Huntington's, there are two fundamental differences in mapping RNA or protein expression levels. First, the control of expression is usually genetically complex (polygenic), and large numbers of other genes and sequence variants (polymorphisms) can potentially influence expression of the target transcript or protein. For example, a group of cooperating transcription factors may control expression of a key transmitter receptor or an ion channel. These effects give rise to so-called *trans*-expression quantitative trait loci (*trans*-eQTLs) (Fig. 8.1A) that map far from the target gene itself—usually on different chromosomes. In contrast, expression of mRNAs may also be controlled by sequence variants that are in or very near to the parent gene itself (Fig. 8.1B and C). For example, a polymorphism in a promoter, enhancer, splice acceptor site, or the 3′ UTR of a gene may produce differences in transcriptional rates, mRNA stability, or ratios of alternative transcripts. When mapping the expression of mRNAs or proteins, this type of genetic "self-control" produces so-called *cis*-acting QTLs or *cis*-eQTLs (Schadt et al., 2003). *cis*-eQTLs are first-order local effects, whereas *trans*-eQTLs are second-order distant effects. In this review, we consider both the technical and conceptual utilities of *cis*- and *trans*-eQTLs. In short, *cis*-eQTLs can be used to evaluate the quality of expression data sets (more *cis*-eQTLs are always better), and validated *cis*- and *trans*-eQTLs can both be used as causal anchor points in genome-to-phenome studies (Ciobanu et al., 2010).

1.1. The history

Since the introduction of proteome and transcriptome methods in the mid-1990s, gene mapping methods have been applied to study progressively larger molecular data sets generated using segregating populations of F2 intercrosses, backcrosses, sets of recombinant inbred (RI) strains, genetic diversity panels, and families and cohorts of humans (Jansen & Nap, 2001; Li & Burmeister, 2005). Damerval and colleagues were the first to apply high-throughput methods to map what they called *protein quantity loci* in an F2 intercross of corn in 1994 (Damerval, Maurice, Josse, & de Vienne,

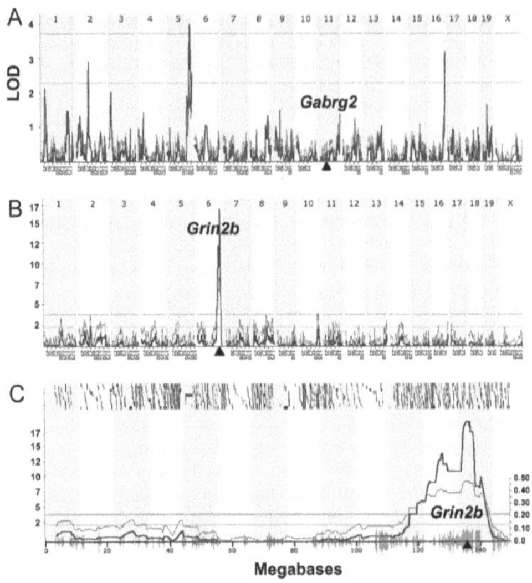

Figure 8.1 Linkage maps of *trans*- and *cis*-eQTLs in mouse hippocampus. (A) *Gabrg2* expression is controlled by a *trans*-eQTL on Chr 5 at 138 Mb (LOD score of 3.94 on *Y*-axis). The *Gabrg2* gene itself is located on Chr 11 at 41 Mb (triangle on the *X*-axis). (B) In contrast, *Grin2b* expression is controlled by a *cis*-eQTL with a peak LOD score of 17 located on Chr 6 at 135 Mb. This location corresponds precisely to the location of the *Grin2b* gene (triangle). (C) Magnified view of the *Grin2b cis*-eQTL that provides much more detail on the QTL map and its chromosomal context. The small shaded or colored blocks along the top represent genes on mouse Chr 6. Shading is used to encode the density of polymorphic SNPs within each gene. The horizontal lines provide genomewide significance thresholds for the QTL determined by permutation analysis (upper line at $p < 0.05$ and lower line at $p < 0.63$). The hash-lines overlapping the *X*-axis summarizes the density of SNPs along the chromosome. Regions of the genome that are identical by descent (i.e., not variable in the BXD family) have almost no *X*-axis hash. Finally, the so-called additive genetic effect (see Williams & Mulligan, 2012) is marked by the thinner line on the right-side *Y*-axis. All data here were generated in GeneNetwork (www.genentwork.org) using the BXD mouse *Hippocampus Consortium M430v2 (Jun06) PDNN* array data set (GeneNetwork.org, accession number GN112, $n = 67$, probe sets 1418177_at and 1457003_at).

1994). Their groundbreaking study is still a model of sophisticated genetic and genomic analysis. In 2002, microarray methods were exploited by Brem, Yvert, Clinton, and Kruglyak (2002) to study gene expression in budding yeast, and in 2003, Schadt et al. (2003) published a remarkable study on the genetic control of mRNA levels in three tissue types from three species: ear leaves of an F3 intercross of corn, livers of an F2 intercross between two strains of mice (C57BL/6J and DBA/2J), and blood cells from four large Mormon families. These landmark studies introduced much of the vocabulary and many of the types of analyses that are still used a decade later. In 2005, our group marked the first publication of a genetic analysis of expression in the CNS. We used a second-generation Affymetrix array—the U74v2—to estimate the expression of about 10,000 genes in whole brains of a set of 32 BXD-type RI strains of mice. A decade later, these eQTL methods have been applied to study over 20 brain regions in mice, rats, and humans using both arrays and RNA-seq, and most of these large eQTL data sets are accessible online for reanalysis and meta-analysis at the Gene-Network web site (www.genenetwork.org; see Williams & Mulligan, 2012 for a primer on using GeneNetwork).

1.2. How much variation is there in gene expression in brain?

Common and rare gene variants—SNPs, insertions, deletions, and inversions—are a major source of phenotypic diversity and of variation in gene expression in wild-type populations, model organisms, and human cohorts (Brem & Kruglyak, 2005; Brem et al., 2002; Damerval et al., 1994; Hubner et al., 2005; Massouras et al., 2012; Morley et al., 2004; Schadt et al., 2003; Storey et al., 2007; Turk et al., 2004). Variation in gene expression levels can be high. For example, in hippocampus of normal young strains of mice, the coefficient of variation (the standard deviation of strain means divided by the mean of all strains) averages about 7% (Fig. 8.2) but the range is often twofold or more. A significant fraction of this variation is under genetic control. Heritability of gene expression data is a function of the genetic diversity of the cases, the genetic complexity of the phenotype, the stability of the environment, and technical error and confounds. Heritability estimates are rough benchmarks that will be depressed by low gene expression (high noise), signal dilution due to cellular heterogeneity, poor technique, or specificity, and uncontrolled environmental factors. Conversely, heritability estimates will be inflated by poor experimental design (e.g., processing related individuals in single batches is a well-known statistical mistake) and by allele-specific hybridization or alignment artifacts.

Despite these problems, heritability estimates are used to gauge the likelihood of detecting one or more *cis*- or *trans*-eQTLs that modulate gene expression. It is not uncommon for variation in mRNA expression to have heritabilities in the range of 25–50% (Geisert et al., 2009) and to be under relatively strong genetic control in the CNS. This reflects modulation by a sizable number of sequence variants in upstream genes, including

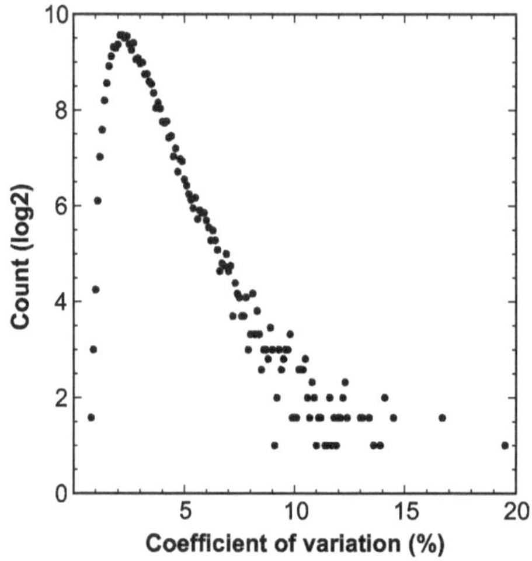

Figure 8.2 Expression variation in hippocampal mRNA expression. Log-transformed gene expression values were used to calculate the coefficient of variation (*X*-axis) across 99 genetically diverse strains. The *Y*-axis represents $\log_2$ of numbers of assays. A total of ~44,500 probe sets in the Hippocampus Consortium data (GN112) were used to generate this plot. Data points representing absolute counts of <2 were excluded.

transcription factors, RNA-binding proteins, transporters, and microRNAs (miRNAs) involved in degradation. Even house-keeping genes have not been spared, and genes such as *GAPDH*, *ACTB*, *SNRPD3*, *RAB7*, *PSMB2*, *GPI1*, *REEP5*, and *RAB7A* are surprisingly variable and heritable among individuals (Eisenberg & Levanon, 2013).

1.3. Brain gene expression studies—A summary

A growing number of eQTL studies have explored the genetics of expression both in whole brain and in specific brain regions of mice, rats, nonhuman primates, and humans over the last decade (Chesler et al., 2005; Heinzen et al., 2008; Hovatta et al., 2007; MacLaren & Sikela, 2005; Mulligan et al., 2012; Myers et al., 2007; Webster et al., 2009; Zou et al., 2012). Virtually, all of this work has been restricted to large and structurally heterogeneous nuclei and cortical regions. In this review, we focus on mouse, and even more specifically on the BXD family of mice for which there are remarkably deep and systematic expression data. Large open-access data are available online (Table 8.1) for neocortex (Gaglani, Lu, Williams, & Rosen, 2009), baso-lateral amygdala (Mozhui, Hamre, Holmes, Lu, & Williams, 2007), striatum (Rosen et al., 2009), nucleus accumbens (Wolen et al., 2012), hypothalamus (Mozhui, Lu, Armstrong, & Williams, 2012), hippocampus

Table 8.1 CNS eQTL data sets for BXD strains (see www.genenetwork.org for a complete list)

GN accession	CNS region	BXD *n*	Platform	mRNA assays	*cis*-eQTLs*	*trans*-eQTLs*
GN323	Amygdala (BLA)	54	Affy Mouse Gene 1.0 ST	34,760	1824	3058
GN394	Whole brain	28	SOLiD RNA-seq, (transcript level)	26,408	418	1075
GN123	Whole brain	39	Affy Mouse 430 2.0	45,101	2162	2771
GN56	Cerebellum	30	Affy Mouse 430 2.0	45,101	2559	6631
GN112	Hippocampus	67	Affy Mouse 430 2.0	45,101	4927	5520
GN281	Hypothalamus	46	Affy Mouse Gene 1.0 ST	34,760	1759	4230
GN375	Neocortex	43	Illumina Mouse WG-6 v2	45,281	2614	4120
GN156	Nucleus accumbens	34	Affy Mouse 430 2.0	45,101	3648	4624
GN135	Prefrontal cortex	27	Affy Mouse 430 2.0	45,101	2256	4392
GN399	Striatum	32	Affy Mouse 430 2.0	45,101	2115	3652
GN228	Ventral tegmental area	35	Affy Mouse 430 2.0	45,101	3156	3629
GN381	Midbrain	37	Agilent SurePrint G3 GE	55,681	6526	7258
GN302	Retina	73	Illumina Mouse WG-6 v2	45,281	3833	6941

**cis*-eQTLs defined as LOD >3 and within ±5 Mb of the parent gene. *trans*-eQTLs defined as LOD >3 and >10 Mb from the parent gene.

(Overall et al., 2009), midbrain (Ye et al., 2014), ventral tegmental area (Wolen et al., 2012), cerebellum (Robert W. Williams and Lu Lu, unpublished), and retina (Templeton et al., 2013). All of these data sets along with tools for gene mapping and eQTL analysis are accessible online (Williams & Mulligan, 2012). Given the high structural heterogeneity of the brain and the logistic difficulties of eQTL studies, there are still no data sets for many other key CNS regions including olfactory bulb, dorsal thalamus, globus pallidus, hindbrain, spinal cord, or dorsal root ganglia.

What is clear from these initial studies is that the genetic control of gene expression in different CNS regions is highly variable. Even when using the same cases and methods, only a small fraction of *cis*- and *trans*-eQTLs are well conserved across brain regions. In part this is due to differences in cellular demographics of brain regions, but it would not be surprising if even relatively homogeneous cell types in different regions (e.g., layer 5 projection neurons in different parts of neocortex) had variable eQTL patterns due to cell-extrinsic factors and axonal connectivity differences.

1.4. Missing pieces

There are still key missing pieces to the brain's gene expression puzzle. This should not be surprising given the difficulties and costs of eQTL studies of the CNS. Consider the short list that follows as a set of important and still open research areas.

1. *RNA-seq eQTL studies.* While RNA-seq technology has great promises (Hitzemann et al., 2013, 2014), the method has not been widely exploited yet for eQTL analysis (Hitzemann et al., 2013; Sun & Hu, 2013). The largest study in mouse that we know of for any CNS tissue is our own modest analysis of whole brains of ~30 genotypes of BXD strains (Li et al., 2010; Mulligan et al., 2012). This eQTL RNA-seq data set is accessible in GeneNetwork for analysis of 200,000 exons and 26,400 transcripts. The largest RNA-seq study of humans is the NIH Genotype-Tissue Expression (GTEx) program (GTEx Consortium, 2013). GTEx data sets for CNS regions are still small ($n < 30$ cases for most brain regions), but by 2016, there will be excellent data for more than 10 regions for several hundred humans.
2. *eQTL studies of alternative splicing.* There are no comprehensive eQTL studies of RNA splice variants in the CNS using array technology or RNA-seq. We know that a majority of genes expressed in brain have multiple isoforms and that eQTL analysis imperfectly combine isoforms into one or two mean "gene level" estimates of expression. The standard protocol used to convert mRNA to complementary DNA relies on a T7 polymerase that is specific to the poly-A tail of the 3′ UTR (Van Gelder et al., 1990). As a result, the great majority of array data only measure expression of the last few coding exons and the 3′ UTRs of mRNAs. New RNA amplification methods do not have this 3′ bias and the latest generation of arrays—so-called exon arrays and splice-junction arrays—can provide estimates of expression over a 500-fold range for exons and

splice-junction sites. It is ironic that just as arrays are reaching full maturity and sophistication, they are being pushed aside (prematurely in our view) by RNA-seq.

3. *Developmental eQTL studies.* There are virtually no developmental studies of genetic control of gene expression during development in any species or tissue type. We know of only two small studies in mouse. Glenn D. Rosen analyzed eQTLs in the neocortex of the BXD family at three stages—postnatal days 3, 14, and young adult (unpublished), and Daniel Goldowitz, Douglas J. Swanson, Thomas Ha (unpublished) are studying expression in cerebellum at eight stages—from embryonic day E12 to young adult. Given the dynamics of gene expression during development and the need to understand the coupling among expression, proliferation, differentiation, and cell death in brain, this is a potentially fascinating topic that warrants much more attention.
4. *Experimental eQTL studies.* There are only a handful of studies on changes in eQTLs in brain after experimental perturbations of any type. The reason is that these studies are doubly hard as they require genetically matched case and control cohorts and exceedingly careful experimental design to avoid statistical confounds. Two studies have evaluated the impact of ethanol treatment (in isolation and in combination) on expression in prefrontal cortex and whole brain (Vanderlinden et al., 2013; Wolen et al., 2012). What will be required to make these types of experimental eQTL studies more practical is a significant reduction in cost of transcriptome data sets and more sophisticated, accessible, and faster statistical workflows that incorporate linear mixed models.
5. *Heterogeneity of brain tissue.* No one has yet attempted an eQTL study of a single type of CNS cell. A genetic dissection of genetic variability of Purkinje cells, dopaminergic neurons, or a subtype of primarily motor neuron would be extremely interesting and could reveal the extent to which cellular heterogeneity obscures eQTL patterns. Use of averaged expression over many cell types may dilute expression variation and obscure genuine eQTLs (more on this topic below). Given the rapid progress in single-cell genomic methods (Islam et al., 2014; Jaitin et al., 2014), these critical studies will certainly be accomplished in the next decade, but getting down to this level will probably come at the cost of increased technical noise. Large sample sizes may be a necessity.
6. *miRNA eQTL studies.* Finally, no one has yet evaluated the extent, causes, and consequences of miRNA expression variability at the population

level. A large number of miRNAs are expressed in brain and serve as important regulators of gene expression (Bak et al., 2008; Shao et al., 2010). There is a growing body of evidence demonstrating important roles of miRNAs in brain development (Shao et al., 2010; Somel et al., 2011; Ziats & Rennert, 2013). A few eQTL studies have used RNA-seq to sequence small RNA molecules from lymphoblastoid cell lines and adipose tissues (Lappalainen et al., 2013; Parts et al., 2012). Parsons and colleagues used RT-PCR to quantify hippocampal expression of five mature miRNAs in the BXDs ($n=24$) (Parsons et al., 2012). The largest study in mouse that we know of is our own RNA-seq analysis of hippocampal miRNA expression differences in 45 genotypes of BXD strains (AK Pandey, K Hamre, L Lu, unpublished). A systematic eQTL study ideally would involve joint analysis of miRNA and mRNA expression from matched biological samples. Such work would be extremely helpful in revealing the shared genetic control of miRNA and mRNA transcripts.

1.5. Genetic architecture of expression traits

One of the main surprises of the genetics of gene expression is that it has nearly the same level of complexity as higher order behavioral traits. *cis*-eQTLs represent one welcome exception to this complexity—they are relatively common, have strong effect sizes, and are easy to validate and interpret (Peirce et al., 2006), albeit with some difficulties related to hybridization artifacts (Ciobanu et al., 2010). While the specific SNP or indel that causes a *cis*-eQTL may not be known, there is a very strong prior probability that polymorphisms in or near the parent gene are responsible for almost all *cis*-eQTLs, and this can be provedn using allele-specific assays in reciprocal F1 hybrids (more on this below). In contrast, the mapping, analysis, and validation of *trans*-eQTLs are far more interesting and complicated. A single mRNA can be associated with several *trans*-eQTLs. These associations define the core elements of molecular networks. But a side effect of the polygenic nature of *trans*-eQTLs is that individual effects and the matched logarithm of odds (LOD) scores of *trans*-eQTLs are typically much smaller than those of *cis*-eQTL (Peirce et al., 2006). Figure 8.3 plots the abundance of *cis*- and *trans*-eQTLs at different LOD scores. *trans*-eQTLs with smaller effect sizes and lower LOD scores are numerous (left side of Fig. 8.3), but the ratio of *cis*- to *trans*-eQTLs increases steeply with the LOD score.

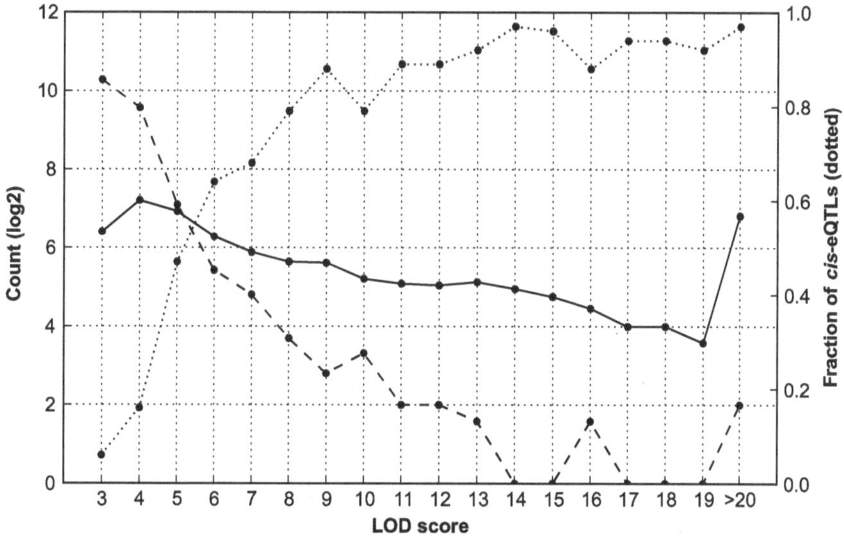

Figure 8.3 Comparison of LOD scores of *cis*- and *trans*-eQTLs. Numbers of *cis*-eQTLs (solid) and *trans*-eQTLs (dashed) are plotted on the left *Y*-axis as a function of LOD score (*X*-axis). The fraction of *cis*-eQTLs (dotted) are plotted on the right *Y*-axis. *cis*-eQTLs are defined as those eQTLs within 5 Mb of the parent gene. *trans*-eQTLs are usually on different chromosomes or more than 10 Mb of the parent gene. A total of 44,500 probe sets in the Hippocampus Consortium data were analyzed.

The naive hope that *trans*-eQTLs would often turn out to be polymorphic transcription factors or RNA metabolism genes has not been borne out by a decade of research. In retrospect, this is perhaps not surprising, since expression of transcription factors, RNA-binding proteins, and miRNAs will themselves be under intense genetic control leading to a regression of causality and an increase in complexity. It is also possible that the genetic complexity of *trans*-eQTL effects is an artifact caused by the high cellular heterogeneity of brain regions. The problem may be analogous to trying to follow one conversation in a noisy restaurant with a single microphone placed high above the crowd. If cellular complexity explains the problem, then it should be much more effective to dissect and make sense of patterns of eQTLs in relatively homogenous parts of the CNS such as the cerebellum (~90% of all cells are granule cells) or the dorsal striatum (~60% of all cells are medium spiny neurons) than in heterogeneous tissues such as whole brain, neocortex, or hippocampus. We evaluated the impact of cellular heterogeneity and possible signal dilution on the detection of *cis*-eQTLs using the cerebellum as a test case (Fig. 8.4). The cerebellum makes up 12% of the mouse brain—52 mg versus 430 mg—an eightfold dilution. *cis*-eQTLs with

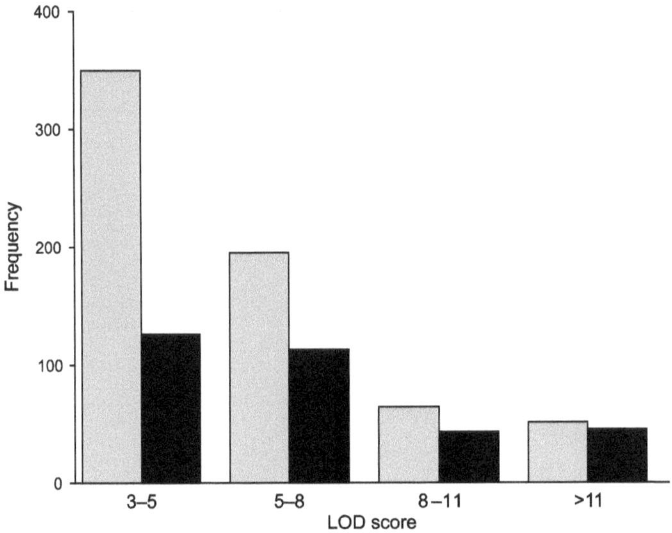

Figure 8.4 Impact of cellular heterogeneity on the detection of *cis*-eQTLs. Numbers of *cis*-eQTLs identified in the cerebellum (gray bars) are plotted as a function of LOD score. A subset of these cerebellar *cis*-eQTLs were also identified in matched whole-brain data (black bars). A total of ~44,500 probe sets in the GE-NIAAA cerebellum (GN72, $n=28$) and whole-brain (GN123, $n=30$) set were used for this analysis.

strong effects in the cerebellum can be easily reidentified using comparable sample sizes and the same array type in whole-brain data (compare gray cerebellum bars vs. black whole-brain bars). However, more than half of small and modest effect eQTLs in cerebellum do not have a large enough signal to be detected in whole brain (left-most bars in Fig. 8.4).

Cellular heterogeneity can also reduce expression correlations between associated genes and transcripts. For example, *Fev* and *Slca64* are two genes with expression in serotonin neurons in midbrain with tightly correlated expression ($r=0.88$). However, transcripts of these two genes have almost no correlation in whole-brain data ($r=0.15$). To the best of our knowledge, there have been no systematic attempts to relate the complexity of *cis-* or *trans*-eQTLs to levels of cellular heterogeneity except in the hematopoietic stem cell lineage (Gerrits et al., 2009), but current data for the BXD strains certainly make this a tractable problem.

This dilution effect does not imply that every cell type has to be isolated for eQTL analysis, but it does mean that the signal-to-noise ratio of mRNA measurements needs to be optimized for mapping. Resampling to reduce noise may often be more effective than the finest laser microdissection.

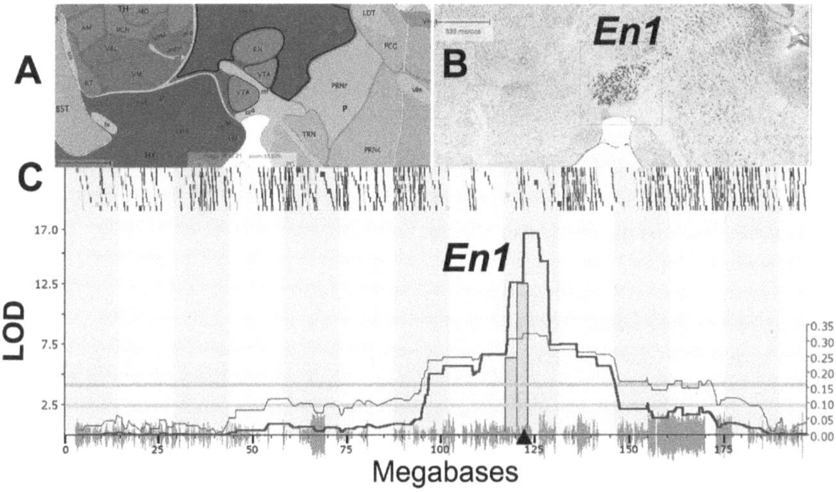

Figure 8.5 Highly selective but diluted expression of *En1* (*Engrailed 1*) in midbrain. (A) Sagittal section of the brainstem and ventral tegmental area (VTA) from the Allen Brain Atlas (www.brain-map.org). (B) Matched *in situ* hybridization image of *En1* expression in serotonin neurons from the Allen Brain Atlas. (C) *En1* expression in the midbrain is controlled by a *cis*-eQTL on Chr 1 with a LOD of 16.7 (*Y*-axis). This location corresponds to the location of *En1* itself (triangle on the *X*-axis). Dilution is clearly not a factor in this instance.

For example in a recent study of the whole midbrain, we measured mRNA levels using four arrays for each genotype. By averaging across these biological replicates, we reduced measurement noise and mapped strong *cis*-eQTLs that originate from the small population of serotonin neurons (Fig. 8.5). Engrailed 1 (*En1*) is a gene with highly selective expression in a few thousand serotonin neurons, and despite ~500-fold dilution in midbrain, *En1* maps as a very strong *cis*-eQTL (LOD ~17).

1.6. RNA-seq to the rescue?

Recent progress in high-throughput sequencing has substantially improved the assessment of individual variation at multiple levels including whole genome, transcriptome (RNA-seq), and even the metagenome. RNA-seq uses high-throughput sequencing to profile gene expression and potentially provides more accurate estimates of transcript abundance over a wider dynamic range than arrays. RNA-seq should eventually facilitate detection of eQTLs with small effects and should also provide insight into the control of alternative splicing and polyadenylation, making it useful to study tissues such as brain (Blencowe, 2006; Dredge, Polydorides, & Darnell, 2001; Grabowski & Black, 2001; Johnson et al., 2009; Ule et al., 2005; Xu, Modrek, & Lee, 2002). Another remarkable feature of RNA-seq is the ability to assess genome-wide allele-specific expression (ASE) by exploiting isogenic F1 hybrids (Bell, Kane, Rieseberg, & Adams, 2013; Korir & Seoighe, 2014; McManus et al., 2010; Rozowsky et al., 2011; Skelly, Johansson, Madeoy, Wakefield, & Akey, 2011).

1.7. RNA-seq data generation

Until recently, generating high-quality RNA-seq data for a large number of samples was not feasible due to technical complexity and cost. However, over the past 6 years, RNA-seq is emerging as a viable alternative to exon arrays for eQTL studies. An RNA-seq sample "library" can now be prepared and sequenced at a depth of 20–30 million reads for the same cost of an exon array (about $400). Ten million RNA-seq reads have been shown to provide roughly similar dynamic range as arrays (Montgomery et al., 2010). To prevent wastage of sequencing resources due to highly abundant ribosomal RNA (rRNA), RNA libraries are either selectively enriched for mRNAs using poly(A)+ selection methods or depleted of rRNA. Selective enrichment of mRNAs with poly(A) tails is done using poly-T oligo-attached magnetic beads. Alternatively, rRNA can be depleted through a hybridization approach. A comparison of these methods showed higher yield of exon reads from poly(A) enrichment (60% of total reads) compared to rRNA subtraction (~30% of total reads) (Cui et al., 2010). The rRNA-depleted libraries generate higher numbers of intron and intergenic reads (~25% and ~45% of total reads) compared to poly(A) methods (~15% and 23%). Although expression estimates from both methods were highly correlated, poly(A) method seems to be a more suitable choice for eQTL studies. Achieving higher numbers of reads in exons is critical to the detection of small expression differences and provides higher power to detect ASE difference using F1 hybrids. The downside of poly(A) enrichment is that it ignores a small number of mRNAs lacking poly(A) tail and most noncoding RNAs.

2. GENETIC RESOURCES FOR eQTL ANALYSIS IN MICE

Mapping eQTLs involves linkage analysis between variation in expression and genetic polymorphisms (markers) that segregate in a family, cohort, or population of individuals. F2 intercrosses, sets of RI strains, and heterogeneous stock (HS) have been used to map eQTLs (Churchill, Gatti, Munger, & Svenson, 2012; Hitzemann, Belknap, & McWeeney, 2008; Svenson et al., 2012; Taylor et al., 1999; Threadgill, Miller, Churchill, & de Villena, 2011; Valdar et al., 2006). *cis*-eQTLs are relatively easily identifiable compared to *trans*-eQTLs due to their strong effects and prior knowledge about their location. Mapping and narrowing down *trans*-eQTLs to single genes remains challenging because *trans*-eQTLs have small effects and require a large sample size for detection, but also because mapping precision of the currently available mouse crosses is poor (particularly in F2s), generally in the range of tens to hundreds of genes. In this section, we consider advantages and disadvantages of different crosses currently available for eQTL mapping. Data sets generated from different crosses can now be easily combined using a number of statistical methods, and there are good reasons to combine the best of each of the mapping resources described below in eQTL mapping.

2.1. Intercross progeny

Test-cross progeny—either F2 intercrosses or backcrosses—is the traditional mapping population described in the experiments of Gregor Mendel. Generating an F2 intercross is a simple two-generation affair. Two distinct strains (often inbred strains) are bred to produce the first filial (F1) generation. F1s are mated to generate a cohort of F2s—usually several hundred individuals. Alternatively, F1s can be backcrossed to either parent to generate a backcross. Meiotic recombination in the F1s produces genetically diverse F2 individuals that segregate for gene variants and heritable phenotypes. Lusis, Schadt, and colleagues have successfully exploited large mouse F2 populations to study eQTLs in the brain and other tissues (Lum et al., 2006; Yang et al., 2006), and many of their data sets are available in Gene-Network. Each member of an intercross needs to be genotyped at 100–200 markers. Large sample sizes ($n > 100$) are often required to map eQTLs in F2s because recombination density per animal is low. For the same reason, F2 crosses often lack adequate positional precision. This makes narrowing down a *trans*-eQTL to a single gene almost impossible.

2.2. RI strains

RI strains have been used widely for mapping of both Mendelian and quantitative traits and, for reasons described below, are advantageous for eQTL studies. RI strains are families of fully inbred strains that are produced by intercrossing two parental strains, followed by repeated sibling matings for at least 20 generations. Each RI strain represents a unique and fixed chromosomal mosaic of the parental genomes. Once an RI strain is fully inbred and genotyped, it can be used as an immortal and genetically defined resource. RI strains are ideal for developmental and experimental eQTL studies because the same genotype can be studied at many time points and under many conditions. They are also ideal for studies of gene-by-environmental interactions because cases and controls can be matched. Finally, in the context of noisy eQTL experiments, one can resample a given strain and brain regions (multiple independent biological replicates) to reduce technical and unintended experimental variability (as in Fig. 8.5). But the most important advantage of RI strains is that legacy phenotypes and eQTL studies can be combined to assemble massive phenomes. The current champion in terms of phenome depth is the BXD family of RI strains—a group of individuals for whom there are now close to 100 independent eQTL studies.

The historical disadvantage of RI strains was their limited numbers and modest power and precision of associated QTL studies. Throughout most of the 1990s, there were fewer than 30 strains per family. Now however, three mouse RI families—BXD, LXS, and the Collaborative Cross (CC), each have well over 60 strains (Williams et al., 2004; Williams, Gu, Qi, & Lu, 2001). The main disadvantage of RI strains is not QTL power or precision but steadily rising costs of acquisition and maintenance of large numbers of strains. It is now possible to achieve average eQTL precision of ±1–4 Mb across most of the genome with a set of only 60–80 RI strains (Fig. 8.6), even without any replication within strain. Thus, RI strains together with whole-genome sequence data of parental strains can be used to achieve single-gene resolution.

2.3. The BXD family

The BXDs—an RI set made by crossing C57BL/6J (B) with DBA/2J (D)—is the largest and oldest RI set. They have been used to study complex traits since the mid-1970s and the genetics of gene expression since the early 2000s.

In addition to the remarkably deep phenome data sets available for the BXDs, a further advantage is that both parents have been fully sequenced (Keane et al., 2011; Waterston et al., 2002). A complete compendium of

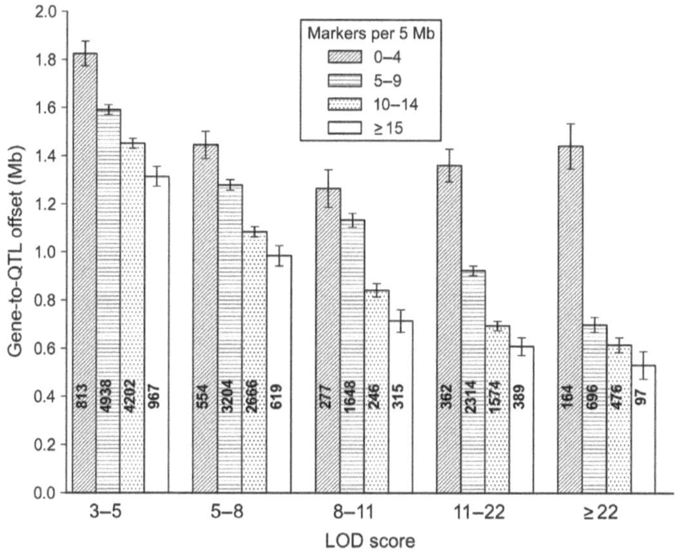

Figure 8.6 QTL mapping precision of the BXD family. The precision of QTL mapping was estimated empirically by measuring the distance between the marker closest to the *cis*-QTL peak and the location of the parent gene (specifically, the position of the proximal-most nucleotide in each probe set associated with a *cis*-eQTL). A total of ~27,500 *cis*-eQTLs were used for this analysis (GN206, $n = 67$ BXD strains). Average precision in megabases (*Y*-axis) is plotted as a function of LOD score (*X*-axis). Each bar includes a count of *cis*-eQTLs and the SEM. Four levels of shadings were used to evaluate the effects of marker density (per 5-Mb bin) on precision. Precision varies from ±1.2 to 1.8 Mb for *cis*-eQTLs with modest LOD scores to ±0.5 Mb for *cis*-eQTL with high LOD scores in regions with high marker density. Precision would be improved by a factor of two by including all strains.

B versus *D* sequence variants is available online and can be used to track down causal SNPs, indels, and CNVs. It is possible to use reverse genetic methods with the BXDs and to look up those phenotypes that map to the location of a particular sequence variant (Carneiro et al., 2009). The current BXD panel contains around 80 lines that are almost fully inbred and available from the Jackson Laboratory, and another set of 40–50 that are being inbred by Robert W. Williams and Lu Lu at UTHSC.

2.4. Heterogeneous stock

HS mice (Hitzemann et al., 2008) and rats (Baud et al., 2013) were created by repeated random mating of stock animals. Unlike RI strains that descend from two parental strains, some HS crosses have incorporated as many as eight founder strains. This adds a high level of genetic diversity to HS progeny. The high recombinant density of HS increases the resolution of QTL and eQTL mapping. Huang and colleagues used an HS cross to map eQTLs in several tissues with an average precision of 2.45 and 3.75 Mb for *cis*- and *trans*-eQTLs, respectively (Huang et al., 2009). However, eQTL mapping of HS progeny is not straightforward (Hitzemann et al., 2014). The family structure causes genotype correlations that can produce spurious eQTLs. As a result, sophisticated statistical approaches such as mixed model associations have been designed for QTL analysis in heterogeneous populations (Brem & Kruglyak, 2005; Huang et al., 2009; Saarikangas et al., 2008; Valdar, Flint, & Mott, 2003). High level of genetic diversity may also cause spurious *cis*-eQTLs due to hybridization artifacts (microarray) or allelic bias in aligning RNA-seq reads. The eQTL study mentioned earlier (Huang et al., 2009) found a significant enrichment of SNPs in probes corresponding to large-effect *cis*-eQTLs. HS also requires high-density genotyping and large sample sizes to map moderate effect eQTLs.

2.5. The Collaborative Cross

The CC is multiparental RI set derived from eight genetically diverse strains (Churchill et al., 2004) that combine features of an HS with those of a conventional RI family. The parents of the CC include five common inbred strains (A/J, C57BL/6J, 129S1/SvImJ, NOD/ShiLtJ, and NZO/HlLtJ) and three wild-derived strains (CAST/EiJ, PWK/PhJ, and WSB/EiJ), and consequently, it has much higher genetic complexity than a normal RI panel. Like an HS panel, the CC can have high-potential eQTL precision at a given sample size (Aylor et al., 2011; Philip et al., 2011). Aylor and colleagues used the CC to map *cis*-eQTLs at a resolution of <1 Mb (Aylor et al., 2011). A 1-Mb interval in the CC will contain nearly 10 × as many sequence variants as the BXDs. In some cases, this will be a major advantage, but in other cases, it will make it hard to find the causative SNPs. As of 2014, 70 CC are ready for distribution. This is a resource that is now ready for prime time.

3. GENETIC MAPPING METHODS

Several statistical approaches have been developed for genomewide linkage analysis of traditional phenotypes. The same approaches can be used to map eQTLs. These approaches range from single marker tests (*t*-test, ANOVA, and simple regression analysis) to multiple locus mapping methods. The only major difference is that eQTL studies involve tens of thousands of expression traits and require fast algorithms. Since an eQTL study tests for thousands of markers and many thousands of molecular traits, associations must be statistically adjusted to account for multiple testing at two levels of analysis (Chesler et al., 2005).

3.1. Single marker test

As the name suggests, the single marker test considers an individual marker or SNP without regard to information about adjacent markers. Single marker tests can be as simple as *t* test between two sets of expression values, where each set represents expression values for a distinct genotype. Analysis of variance (ANOVA) can also be used. The advantage of ANOVA is that it can incorporate covariates such as sex and nongenetic variables such as environmental effects and technical error.

3.2. Interval mapping

Interval mapping is widely used for QTL mapping of F2 and RI crosses and human linkage analysis. Interval mapping interrogates region between the two adjacent marker loci to precisely determine the location of the QTL. It imputes the genotypes at intervals (for, e.g., every 1 cM) between each pair of adjacent markers and tests for the presence of QTL. The results are expressed as LOD scores. This score represents the ratio of likelihoods of a statistical model that includes a genetic effect at a particular locus versus a model that does not include that effect (the null expectation). A downside of interval mapping is that it is can be computationally intensive. Haley and Knott (1992) devised accurate and computationally tractable regression-based methods to compute interval maps. Their method is now widely used in eQTL analysis because of its speed. It is now possible to map an entire transcriptome in short time, and single mRNAs and proteins can be mapped with up to 10,000 permutation tests to compute genomewide significance in less than a minute.

3.3. Composite interval mapping

The mapping methods discussed so far assume that a QTL acts independently. However, a QTL can be linked to or interact epistatically with other QTLs. Composite interval mapping (Jansen & Stam, 1994; Zeng, 1994) combines interval mapping with multiple-marker regression analysis, which controls for the effect of a known QTL. In short, composite interval mapping uses a subset of significantly associated markers as covariates and controls for variation produced at these controlled markers. Thus, composite interval mapping helps to detect weaker but biologically relevant QTLs. This is particularly important for mapping small-effect eQTLs, as strong *cis*-eQTLs often mask secondary eQTLs.

3.4. Evaluation of mapping precision

Mapping precision for a genetic population can be empirically determined by measuring the offset distance between *cis*-eQTL peaks and locations of the parent genes. *cis*-eQTLs are essentially used as positional "gold standards," and any errors in this assumption will tend to be conservative and degrade apparent precision. In other words, *cis*-eQTL offsets will be conservatively biased estimates. We evaluated the QTL mapping precision of the BXD family using a hippocampus exon array data set generated using 67 BXD strains. We divided the genome into 5-Mb bins, and counted the number of markers within each bin. Bins with similar marker density—less than 5 markers/5 Mb, 5–9, 10–15, and greater than 15 markers—were pooled. *cis*-eQTLs were operationally defined as QTLs with LOD score above 3 and an offset distance of less than ±5 Mb. Strong *cis*-eQTLs with LOD scores greater than 22 (genomewide p value $<10^{-6}$) have a mean gene-to-QTL peak distance of 1.44, 0.70, 0.61, and 0.53 Mb for genomic regions with different marker densities (Fig. 8.6). Similar results were seen when the offset distance of less than ±10 Mb was used to define *cis*-eQTL.

4. RNA-seq eQTL STUDIES

Only a handful of eQTL studies have yet exploited RNA-seq (Montgomery et al., 2010; Pickrell et al., 2010). Most have unfortunately involved immortalized human lymphoblastoid cells—a cell type that is riddled with chromosomal abnormalities that make analysis problematic. As far as we know, the only accessible RNA-seq eQTL study of the brain is our own modest study of the BXD strains, in which we generated an average of 30 million 50 nt reads for each of 28 BXD strains and both parents (Li et al., 2010; Mulligan et al., 2012) but without any biological replication. Each fragment library was generated using a pool of RNAs from three or more cases using an rRNA depletion method and a protocol that preserved strand polarity. We mapped 350 transcripts with LOD scores above 5, of which 225 were *cis*-eQTLs with a median LOD of 6.4 and an average gene-to-marker offset of ±3 Mb, whereas 125 were *trans*-eQTLs with a median

LOD score of 5.3. The precision of these QTLs is impressive, given the small sample size and the lack of biological replication. The precision of *trans*-eQTLs can be expected to be closely matched to those of *cis*-eQTLs—perhaps ±3.5 Mb—a small enough interval to begin candidate gene analysis. Among the most interesting *trans*-eQTLs relevant to CNS function are *Atf4, Atp2b1, Atp13a2, Atrx, Cacnb4, Foxa1, Foxc1, Gap43, Kcnk10, Lifr, Ntsr2, Per3, Pdyn, Pou5f1,* and *Ptprz1*. Those with strong *cis*-eQTLs are *Mrps5* (Houtkooper et al., 2013), *Alad, Ckb, Glo1* (Williams et al., 2009), *Ntn4, Prdx2, and Sae1.*

5. PROS AND CONS OF ARRAYS AND RNA-seq FOR eQTL STUDIES

5.1. Advantages of arrays

For fully sequenced model organisms, commercial arrays now have essentially comprehensive coverage of protein-coding RNAs. The latest arrays from Affymetrix also include a nearly complete set of probes for miRNAs, exons, and even splice junctions. This intense focus on the core subset of RNAs can be an advantage in many situations—particularly in eQTL studies in which investigators are interested on the impact of mRNA variation on phenotypes. The modest dynamic range of arrays relative to RNA-seq is rarely an issue in eQTL studies, as the key variable is variation across individuals rather than across transcripts. In fact, many eQTL studies discard mean expression values and retain only the offset from the mean (the *z*-score). A second advantage of arrays is that every transcript has its own "real estate" on the array. Even those genes with low expression have an opportunity to produce a hybridization signal. In contrast, RNA-seq count data have a highly skewed distribution, and a small number of genes account for a large fraction of the reads, and many transcripts have no or very low read counts (<5). Consequently, the power to detect differential expression among shorter and more modestly expressed genes is poor due to high Poisson noise (Bullard, Purdom, Hansen, & Dudoit, 2010). A final pragmatic advantage of arrays is that the analysis workflow is far less computationally intensive and can be performed on a desktop computer. A strong case can still be made in favor of arrays for large eQTL studies.

5.2. Advantages of RNA-seq

RNA-seq offers advantages relative to arrays and can provide more accurate estimates of isoform abundance over a wider dynamic range. Dynamic range is only limited by the RNA complexity of samples (library complexity) and the depth of sequencing. In a small study, Fu and colleagues compared RNA-seq and array data with protein levels in cerebellar cortex and found a slightly better relation between RNA-seq and protein (Fu et al., 2009). The higher dynamic range of RNA-seq could potentially facilitate detection of eQTLs associated with transcripts that have either low or very high expression. Saturation of signal at the high end or at the low end of the expression spectrum could obscure genetic expression differences. However, to the best of our knowledge, this has never been tested. In Fig. 8.7, we compare *cis*-eQTL effect sizes from the RNA-seq data discussed earlier with a matched Affymetrix M430 array data set. We extracted all *cis*-eQTLs in both data sets—around 2000 and 3000, respectively—and compared their LOD scores. For those transcripts with *cis*-eQTLs in both data sets, there was no advantage to the RNA-seq in terms of effect size or LOD scores of the eQTLs. In this particular case, four arrays per strain outperformed 30 million reads per strain.

The second factor is more compelling—RNA-seq enables expression quantification of novel transcripts and transcripts not represented on arrays. RNA-seq analysis tools such as Cufflinks (Roberts, Pimentel, Trapnell, & Pachter, 2011) can utilize reads to annotate novel transcripts by performing reference-based *de novo* assembly of transcripts. This is particularly important in analyzing a complex tissue such as brain known to have high frequency of alternative splicing events (Pan, Shai, Lee, Frey, & Blencowe, 2008). However, most RNA-seq analysis ignores reads mapped to unannotated regions in the reference genome, somewhat reducing the significance of this advantage.

Third, RNA-seq is a hybridization-free approach and does not suffer from confounds such as cross hybridization and artifacts due to variants in probe sequences. Probes with variants—SNPs and small indels—influence hybridization kinetics and cause incorrect detection of the expression level of genes. Ciobanu and colleagues found 25% of apparent *cis*-modulation detected in the hippocampus was caused by probe variants rather than genuine mRNA quantitative differences (Ciobanu et al., 2010). RNA-seq suffers from a similar problem of allele bias inherent when aligning reads to a single reference genome. However, alignment methods allow for mismatches and are less sensitive to sequence differences (see below).

Finally, RNA-seq makes it possible to study complex transcriptional events including alternative splicing and polyadenylation. Hybridization-based approaches use probes targeting small regions of the transcript, mostly single

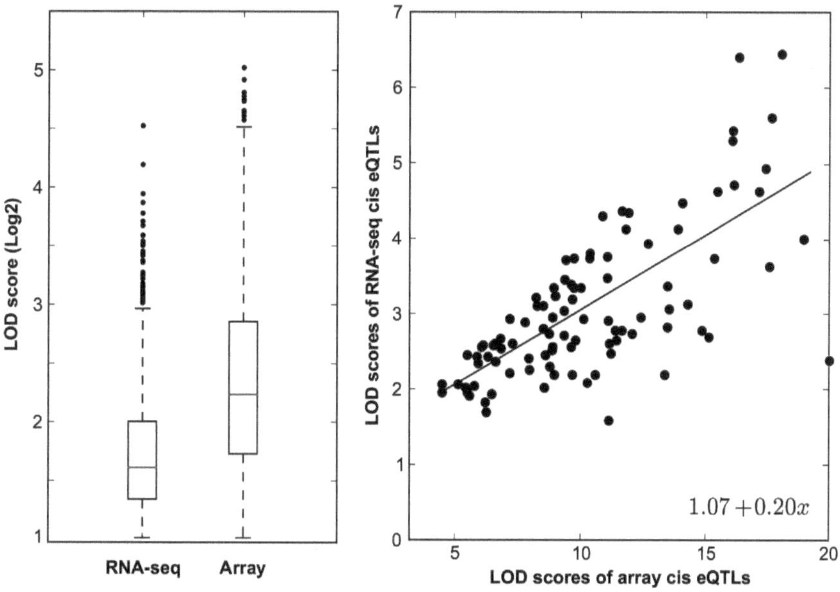

Figure 8.7 Comparison of *cis*-eQTLs identified by arrays and RNA-seq. (A) The box plots show LOD score distributions (*Y*-axis) for RNA-seq ($n=1779$) versus arrays ($n=2839$). *cis*-eQTLs are defined as those eQTLs having an LOD score of >2.0 and within 5 Mb of the parent gene. A total of ~44,500 assays in a whole-brain array data set (GN123, $n=30$) were compared to ~200,000 exons in a whole-brain RNA-seq data set (GN394, $n=28$). (B) Scatter plot of LOD scores for 105 *cis*-eQTLs shared by array and RNA-seq data sets.

exons. As a result, they have not been used to study splicing extensively. High-density exon arrays consisting of both exon and splice junction probes are now available; although they still suffer from systematic errors that lead to over-estimation of alternative splicing (Gaidatzis, Jacobeit, Oakeley, & Stadler, 2009; Laderas et al., 2011). Splicing array also require *a priori* information of isoforms for probe design. In contrast, RNA-seq reads mapping to splice junctions provide direct evidence of splicing. These reads can be quantified to roughly estimate the relative abundance of alternative isoforms. Additionally, they can be combined with read distribution across different exons to precisely quantify alternatively spliced isoforms of a gene (Griffith et al., 2010; Jiang & Wong, 2009; Trapnell et al., 2010). Of course, the near term technical goal is to sequence and count entire mRNA molecules—so-called single-molecule sequencing. Once this goal has been reached, it will be possible to use quantitative genetic methods to study splice isoform usage in brain regions and even single cells.

6. RNA-seq READ ALIGNMENT AND NORMALIZATION

Accurate estimation of transcript abundance is critical for the success of eQTL studies. Unbiased alignment of short sequences and correct normalization of RNA-seq counts are important for accurate estimates. There are now accurate RNA-seq aligners that can align reads in a splice-aware manner, but normalization methods for RNA-seq counts are still evolving. Details of RNA-seq data analysis are covered in Chapter 2. Here, we will only discuss data analysis issues that are important for conducting a more error-free eQTL study.

6.1. Allelic bias in read mapping

Though RNA-seq is a hybridization-free approach, it still requires read mapping—essentially digital hybridization—against the reference genome. Unlike hybridization, RNA-seq alignment is not particularly sensitive to the presence of variants because the algorithms allow for specified numbers of mismatches and gaps. Nevertheless, a significant allelic bias can still occur when aligning reads with large numbers of genetic variants relative to a single reference genome (Degner et al., 2009). This is because reads with reference alleles will match the reference genome precisely, whereas reads that contain nonreference alleles will not. For example, reads originating from *B* haplotypes in BXD strains will show better alignment rate to the reference genome (B genome) than the *D* haplotypes. This bias will appear as a strong *cis*-eQTL, similar to ones produced by hybridization artifacts in arrays. This bias can be more of a problem in complicated crosses with multiple and highly divergent haplotypes such as HS and CC.

Degner and colleagues (2009) proposed using a masked reference genome in which each polymorphic position (SNP) in the reference genome is masked with a third allele that is neither the reference allele nor the allele from the nonreference haplotype. Their approach cannot reduce allelic bias completely; moreover, it increases the number of unmapped reads because SNP masking this way adds an obligatory mismatch in the alignment. As a result, expression values of genes with large number of variants are underestimated. The advantage of their method is that it only requires read alignment against a single genome. A more widely used method (Blencowe, 2006; Rozowsky et al., 2011) to reduce allelic bias is to create strain-specific genomes for alignment by substituting reference alleles with known variants. Reads are aligned to both the reference genome and substituted strain-specific genomes. Aligned reads are then combined in a nonredundant manner to estimate transcript abundance. If sequence variants between a nonreference genome and the reference genome are not known, RNA-seq can also be used to detect variants. In this case, RNA-seq reads are as usual aligned against the reference genome and sequence variants are generated for the nonreference case. The locations of these variants can be used to identify genomic regions corresponding to different haplotypes. RNA-seq reads can then be realigned in a haplotype-sensitive manner to reduce bias.

6.2. Correct normalization of RNA-seq counts

Choice of normalization procedure can significantly affect the outcomes of gene expression studies. Similar to arrays, normalized estimates of expression are necessary to evaluate whether differences among samples used for eQTL mapping are genuine. Systematic biases in RNA-seq experiments can result from differences in the sequencing depth of the sample libraries and differences in the length of genes. Deeply sequenced libraries generate more reads per gene than less deeply sequenced libraries. Similarly, longer genes will have more aligned reads compared to shorter genes. Mortazavi and colleagues introduced a normalization method called the *reads per kilobase per million* mapped reads (RPKM) that rescales counts to correct for differences in library size and gene length (Mortazavi, Williams, McCue, Schaeffer, & Wold, 2008). RPKM values enable between- and within-library sample comparisons. Sandberg and colleagues found that RPKM values that take only exonic reads into account correlate better with qRT-PCR data (Ramskold, Wang, Burge, & Sandberg, 2009). Recent work has criticized RPKM normalization when detecting for differentially expressed transcripts (Bullard et al., 2010; Dillies et al., 2013; Rapaport et al., 2013). Dudoit and colleagues found a bias that favors longer transcripts with small differences over shorter transcripts with large differences (Bullard et al., 2010).

The total number of RNA-seq counts generated for a gene depends not only on its expression level, length, and depth of sequencing but also on the library composition and complexity that is used for sequencing. Differences in library composition between samples can contribute to high levels of variability and spuriously high variance in expression. To address this problem, normalization methods such as *Trimmed Mean of M-values* (implemented in edgeR Bioconductor package) (Robinson, McCarthy, & Smyth, 2010) and DESeq (Anders & Huber, 2010) have been proposed. Both model RNA-seq counts using a negative binomial distribution and have outperformed several other methods in studies focused on differential expression (Bullard et al., 2010; Dillies et al., 2013; Rapaport et al., 2013). No comparison of normalization methods with respect to eQTL mapping has yet been performed. Traditional eQTL mapping methods based on linear regression are highly sensitive to outliers and work best when the expression data are roughly normally distributed. Unfortunately, raw RNA-seq expression counts are strongly right skewed, with a few transcripts having extraordinarily high expression. Thus, RNA-seq counts must be transformed in order to apply linear regression or equivalent approaches for eQTL mapping. A $\log_2$ transform is often suitable. Pickerell and colleagues used a normal quantile transformation of RNA-seq counts in their eQTL study (Pickrell et al., 2010). Another study showed that modeling RNA-seq counts using a discrete distribution, such as a negative binomial or a beta binomial distribution, results in higher statistical power in eQTL mapping (Sun, 2012).

7. eQTL MAPPING OF ALTERNATIVE SPLICING AND POLYADENYLATION

Alternative splicing and alternative polyadenylation increase the complexity of the transcriptome and diversity protein isoforms. Xu and colleagues showed that brain is highly enriched for alternative splice forms (Xu et al., 2002). Similarly, highly expressed genes in mammalian brain are known to have unusually long 3′ UTRs (Miura, Shenker, Andreu-Agullo, Westholm, & Lai, 2013). Alternative splicing plays an important role in neuronal differentiation, synaptic transmission, and plasticity. Splicing differences and mutations have been linked with several disorders (Faustino & Cooper, 2003; Nissim-Rafinia & Kerem, 2005). Alternative polyadenylation also plays a role in the stability and localization of the mRNA through interactions with RNA-binding proteins, ribosomes, miRNA, etc. Several association studies have utilized exon arrays and RNA-seq to study variability and heritability of splicing in human lymphoblastoid cell lines (Montgomery et al., 2010; Nembaware et al., 2008; Pickrell et al., 2010). These studies have confirmed genetic control of splicing variation in humans. However, eQTLs regulating alternative splicing in brain have not been studied extensively. To the best of our knowledge, the only eQTL study that has systematically investigated splicing eQTLs genomewide is by Heinzen and colleagues. They studied expression in human cortex and peripheral blood using the Affymetrix Human Exon 1.0 ST array

(Heinzen et al., 2008) and identified 23 and 84 associations at the transcript and the exon level, respectively. This is likely to be a massive underestimate of the actual number of splice isoforms that are under genetic control.

8. RNA-seq FOR ALLELE-SPECIFIC EXPRESSION

A remarkable feature of RNA-seq is its ability to assay genomewide ASE using isogenic F1 hybrids (Bell et al., 2013; Korir & Seoighe, 2014; McManus et al., 2010; Rozowsky et al., 2011; Skelly et al., 2011) made by crossing inbred parents. RNA-seq can reliably distinguish mRNA representing the alternative alleles and can be used to detect unequal production of alleles. An advantage of using F1 animal for ASE analysis is that the two alleles in these animals share all environmental and trans-acting influences. As a result, any genetic expression differences in heterozygotes must be attributed to the local allele-specific endogenous effect.

8.1. Key factors in design of genomewide ASE

A key factor to consider for measuring *cis*-eQTLs on a genomic scale is the presence of appropriate coding variants—usually SNPs—to assay allelic imbalance. Another factor is the sequencing depth needed to detect differences with good statistical power. Fontanillas and colleagues showed that the read depth required to detect an allelic imbalance depends on the size of difference in the allelic expression (Fontanillas et al., 2010). They determined that 50 reads per SNP is enough to provide 60% statistical power for larger than twofold differences in expression. Small allelic expression differences of less than 1.25 fold will require more than 500 reads to reach the same power.

ASE can be used to identify imprinted genes by comparing ratios of expression in reciprocal F1 crosses. The reciprocal F1 females are genetically identical but the polarity of parents differ (e.g., B mother to D father, or D mother to B father). An initial RNA-seq study of this type reported an implausibly high number of imprinted genes in the CNS (Griffith et al., 2010). Correct modeling of biological and technical variation brought this estimate down to less than 100 genes (DeVeale, van der Kooy, & Babak, 2012). Several other factors can contribute to error in estimating ASE. Allelic bias in read mapping to a single reference genome has already been discussed. Other mapping artifacts can also produce false positives include using nonunique reads (reads that can be mapped to multiple locations) and reads that map to low-complexity genomic regions. PCR amplification bias during library preparation can also cause false-positive allelic imbalances.

8.2. Advantages and disadvantages of ASE

An advantage of using ASE-based approach to identify *cis*-eQTLs on a genomic scale is that it requires relatively few samples. Additionally, it does not depend on arbitrary window cut-off as used in eQTL mapping. A disadvantage of ASE analysis over eQTL mapping is its complete inability to locate *trans*-eQTLs. Babak and colleagues compared array-based eQTL mapping with RNA-seq ASE to detect *cis*-eQTLs (Babak et al., 2010). They found an extensive agreement between *cis*-eQTL results. For genes showing discrepancies between methods, RNA-seq more frequently matched subsequent validation using conventional qRT-PCR protocols.

9. CONCLUSIONS

The last decade has seen a rapid growth in the number of eQTL studies of the CNS and large efforts to accumulate massive gene expression data sets across multiple brain regions and cell types. There are two very general findings. First, *cis*-eQTLs have large effects, are often replicable across different data sets and even regions, and are comparatively easy to validate and interpret. However, these first-order *cis*-effects usually do not expose critical gene–gene interactions that define molecular networks. They can, however, be used as seeds to define downstream effects on protein levels and higher order behavioral traits (Ciobanu et al., 2010). Second, *trans*-eQTLs usually have smaller effects, are harder to validate, and often do not replicate well across different data sets. But they are also most interesting because they can define gene–gene interactions (e.g., Ciobanu et al., 2010, figure 4). Although *trans*-eQTLs are common in expression data sets, they are hard to pin down to single causative genes. But this problem is being resolved. For example, eQTL mapping resolution in the CC will soon provide 1-Mb resolution (5–10 genes in mouse). The BXD family can already routinely achieve a resolution of 2–5 Mb (around 10–50 genes, Fig. 8.6) with high power. Once they are more fully developed, these RI families will each contain 150 strains and they should routinely achieve single-gene resolution suitable for high power *trans*-eQTL studies of any part of the CNS.

RNA-seq offers the high dynamic range and resolution essential for capturing small expression differences. Additionally, this method can be used for isoform-specific eQTL mapping. While RNA-seq offers great promises, it has not yet been widely exploited for eQTL mapping studies. This is primarily because of the cost and complexity of library preparation and the high bioinformatics overhead required to process and analyze data. Rapid technical advances have dramatically reduced both types of costs, and the interpretation of RNA-seq is now becoming much more tractable.

Gene expression studies of the brain are particularly challenging due to the extreme cellular heterogeneity. There are probably well over 7000 statistically distinct cell types with unique mRNA and protein expression profiles in brain. This estimate is based on the well-known cellular complexity of retina—a CNS tissue in which there are ~70 cell types in mouse and human (Marc, Jones, Lauritzen, Watt, & Anderson, 2012)—and a conservative multiplier of 100 for the effective number of equally complex CNS regions. The high level of still undefined cellular and molecular heterogeneity in the brain is a major issue that still confounds neuroscience. For eQTL studies, the use of averaged expression over diverse cell types dilutes but does not eliminate the important genetic signals. The ultimate genetic studies of gene expression will require extremely efficient workflows to quantify mRNAs, proteins, and metabolites for hundreds of cells belonging to thousands of unique CNS cell types. This may now seem daunting, but rapid progress in single-cell genomics methods (Islam et al., 2014; Jaitin et al., 2014) will make this just as practical in a decade as whole-genome sequencing is today.

ACKNOWLEDGMENTS

We thank Drs. Megan K. Mulligan, Snigdha Roy, and Evan G. Williams for their critical reading of drafts of this review. We also thank a group of highly supportive colleagues for generating many massive eQTL data sets over the past decade that are now easily accessible to the neurogenetics research community. Our thanks to Drs. Lu Lu, Glenn D. Rosen, Michael F. Miles, Eldon E. Geisert, Randy D. Blakely, Daniel Goldowitz, Gerd Kempermann, Rupert W. Overall, Boris Tabakoff, Divyen Patel, Kristin M. Hamre, and Robert J. Rooney. Much of the work reviewed here has been generously supported by NIAAA's Integrative Neuroscience Initiative on Alcoholism (U01 AA016662, U01 AA013499, U24 AA013513, U01 AA014425) and by the UT Center for Integrative and Translational Genomics. We also acknowledge support from UT-Battelle, operator of ORNL, under Contract No. DE-AC05-00OR22725, with the U.S. Department of Energy.

REFERENCES

Anders, S., & Huber, W. (2010). Differential expression analysis for sequence count data. *Genome Biology*, *11*(10), R106. http://dx.doi.org/10.1186/gb-2010-11-10-r106.

Aylor, D. L., Valdar, W., Foulds-Mathes, W., Buus, R. J., Verdugo, R. A., Baric, R. S., et al. (2011). Genetic analysis of complex traits in the emerging Collaborative Cross. *Genome Research*, *21*(8), 1213–1222. http://dx.doi.org/10.1101/gr.111310.110.

Babak, T., Garrett-Engele, P., Armour, C. D., Raymond, C. K., Keller, M. P., Chen, R., et al. (2010). Genetic validation of whole-transcriptome sequencing for mapping expression affected by cis-regulatory variation. *BMC Genomics*, *11*, 473. http://dx.doi.org/10.1186/1471-2164-11-473.

Bak, M., Silahtaroglu, A., Moller, M., Christensen, M., Rath, M. F., Skryabin, B., et al. (2008). MicroRNA expression in the adult mouse central nervous system. *RNA*, *14*(3), 432–444. http://dx.doi.org/10.1261/rna.783108.

Baud, A., Hermsen, R., Guryev, V., Stridh, P., Graham, D., McBride, M. W., et al. (2013). Combined sequence-based and genetic mapping analysis of complex traits in outbred rats. *Nature Genetics*, *45*(7), 767–775. http://dx.doi.org/10.1038/ng.2644.

Bell, G. D., Kane, N. C., Rieseberg, L. H., & Adams, K. L. (2013). RNA-seq analysis of allele-specific expression, hybrid effects, and regulatory divergence in hybrids compared with their parents from natural populations. *Genome Biology and Evolution*, *5*(7), 1309–1323. http://dx.doi.org/10.1093/gbe/evt072.

Blencowe, B. J. (2006). Alternative splicing: New insights from global analyses. *Cell*, *126*(1), 37–47. http://dx.doi.org/10.1016/j.cell.2006.06.023.

Brem, R. B., & Kruglyak, L. (2005). The landscape of genetic complexity across 5700 gene expression traits in yeast. *Proceedings of the National Academy of Sciences of the United States of America*, *102*(5), 1572–1577. http://dx.doi.org/10.1073/pnas.0408709102.

Brem, R. B., Yvert, G., Clinton, R., & Kruglyak, L. (2002). Genetic dissection of transcriptional regulation in budding yeast. *Science*, *296*(5568), 752–755. http://dx.doi.org/10.1126/science.1069516.

Bullard, J. H., Purdom, E., Hansen, K. D., & Dudoit, S. (2010). Evaluation of statistical methods for normalization and differential expression in mRNA-Seq experiments. *BMC Bioinformatics*, *11*, 94. http://dx.doi.org/10.1186/1471-2105-11-94.

Carneiro, A. M., Airey, D. C., Thompson, B., Zhu, C. B., Lu, L., Chesler, E. J., et al. (2009). Functional coding variation in recombinant inbred mouse lines reveals multiple serotonin transporter-associated phenotypes. *Proceedings of the National Academy of Sciences of the United States of America*, *106*(6), 2047–2052. http://dx.doi.org/10.1073/pnas.0809449106.

Charles, S. J., Moore, A. T., & Yates, J. R. (1992). Genetic mapping of X linked ocular albinism: Linkage analysis in British families. *Journal of Medical Genetics*, *29*(8), 552–554.

Chesler, E. J., Lu, L., Shou, S., Qu, Y., Gu, J., Wang, J., et al. (2005). Complex trait analysis of gene expression uncovers polygenic and pleiotropic networks that modulate nervous system function. *Nature Genetics*, *37*(3), 233–242. http://dx.doi.org/10.1038/ng1518.

Churchill, G. A., Airey, D. C., Allayee, H., Angel, J. M., Attie, A. D., Beatty, J., et al. (2004). The Collaborative Cross, a community resource for the genetic analysis of complex traits. *Nature Genetics*, *36*(11), 1133–1137. http://dx.doi.org/10.1038/ng1104-1133.

Churchill, G. A., Gatti, D. M., Munger, S. C., & Svenson, K. L. (2012). The Diversity Outbred mouse population. *Mammalian Genome*, *23*(9–10), 713–718. http://dx.doi.org/10.1007/s00335-012-9414-2.

Ciobanu, D. C., Lu, L., Mozhui, K., Wang, X., Jagalur, M., Morris, J. A., et al. (2010). Detection, validation, and downstream analysis of allelic variation in gene expression. *Genetics*, *184*(1), 119–128. http://dx.doi.org/10.1534/genetics.109.107474.

Cui, P., Lin, Q., Ding, F., Xin, C., Gong, W., Zhang, L., et al. (2010). A comparison between ribo-minus RNA-sequencing and polyA-selected RNA-sequencing. *Genomics*, *96*(5), 259–265. http://dx.doi.org/10.1016/j.ygeno.2010.07.010.

Damerval, C., Maurice, A., Josse, J. M., & de Vienne, D. (1994). Quantitative trait loci underlying gene product variation: A novel perspective for analyzing regulation of genome expression. *Genetics*, *137*(1), 289–301.

Degner, J. F., Marioni, J. C., Pai, A. A., Pickrell, J. K., Nkadori, E., Gilad, Y., et al. (2009). Effect of read-mapping biases on detecting allele-specific expression from RNA-sequencing data. *Bioinformatics*, *25*(24), 3207–3212. http://dx.doi.org/10.1093/bioinformatics/btp579.

DeVeale, B., van der Kooy, D., & Babak, T. (2012). Critical evaluation of imprinted gene expression by RNA-Seq: A new perspective. *PLoS Genetics*, *8*(3), e1002600. http://dx.doi.org/10.1371/journal.pgen.1002600.

Dillies, M. A., Rau, A., Aubert, J., Hennequet-Antier, C., Jeanmougin, M., Servant, N., et al. (2013). A comprehensive evaluation of normalization methods for illumina high-throughput RNA sequencing data analysis. *Briefings in Bioinformatics*, *14*(6), 671–683. http://dx.doi.org/10.1093/bib/bbs046.

Dredge, B. K., Polydorides, A. D., & Darnell, R. B. (2001). The splice of life: Alternative splicing and neurological disease. *Nature Reviews. Neuroscience*, *2*(1), 43–50. http://dx.doi.org/10.1038/35049061.

Eisenberg, E., & Levanon, E. Y. (2013). Human housekeeping genes, revisited. *Trends in Genetics*, *29*(10), 569–574. http://dx.doi.org/10.1016/j.tig.2013.05.010.

Faustino, N. A., & Cooper, T. A. (2003). Pre-mRNA splicing and human disease. *Genes & Development*, *17*(4), 419–437. http://dx.doi.org/10.1101/gad.1048803.

Fontanillas, P., Landry, C. R., Wittkopp, P. J., Russ, C., Gruber, J. D., Nusbaum, C., et al. (2010). Key considerations for measuring allelic expression on a genomic scale using high-throughput sequencing. *Molecular Ecology*, *19*(Suppl. 1), 212–227. http://dx.doi.org/10.1111/j.1365-294X.2010.04472.x.

Fu, X., Fu, N., Guo, S., Yan, Z., Xu, Y., Hu, H., et al. (2009). Estimating accuracy of RNA-Seq and microarrays with proteomics. *BMC Genomics*, *10*, 161. http://dx.doi.org/10.1186/1471-2164-10-161.

Gaglani, S. M., Lu, L., Williams, R. W., & Rosen, G. D. (2009). The genetic control of neocortex volume and covariation with neocortical gene expression in mice. *BMC Neuroscience*, *10*, 44. http://dx.doi.org/10.1186/1471-2202-10-44.

Gaidatzis, D., Jacobeit, K., Oakeley, E. J., & Stadler, M. B. (2009). Overestimation of alternative splicing caused by variable probe characteristics in exon arrays. *Nucleic Acids Research*, *37*(16), e107. http://dx.doi.org/10.1093/nar/gkp508.

Geisert, E. E., Lu, L., Freeman-Anderson, N. E., Templeton, J. P., Nassr, M., Wang, X., et al. (2009). Gene expression in the mouse eye: An online resource for genetics using 103 strains of mice. *Molecular Vision*, *15*, 1730–1763.

Gerrits, A., Li, Y., Tesson, B. M., Bystrykh, L. V., Weersing, E., Ausema, A., et al. (2009). Expression quantitative trait loci are highly sensitive to cellular differentiation state. *PLoS Genetics*, *5*(10), e1000692. http://dx.doi.org/10.1371/journal.pgen.1000692.

Grabowski, P. J., & Black, D. L. (2001). Alternative RNA splicing in the nervous system. *Progress in Neurobiology*, *65*(3), 289–308.

Griffith, M., Griffith, O. L., Mwenifumbo, J., Goya, R., Morrissy, A. S., Morin, R. D., et al. (2010). Alternative expression analysis by RNA sequencing. *Nature Methods*, 7(10), 843–847. http://dx.doi.org/10.1038/nmeth.1503.

GTEx Consortium (2013). The Genotype-Tissue Expression (GTEx) project. *Nature Genetics*, *45*(6), 580–585. http://dx.doi.org/10.1038/ng.2653.

Gusella, J. F., Wexler, N. S., Conneally, P. M., Naylor, S. L., Anderson, M. A., Tanzi, R. E., et al. (1983). A polymorphic DNA marker genetically linked to Huntington's disease. *Nature*, *306*(5940), 234–238.

Haley, C. S., & Knott, S. A. (1992). A simple regression method for mapping quantitative trait loci in line crosses using flanking markers. *Heredity*, *69*(4), 315–324.

Heinzen, E. L., Ge, D., Cronin, K. D., Maia, J. M., Shianna, K. V., Gabriel, W. N., et al. (2008). Tissue-specific genetic control of splicing: Implications for the study of complex traits. *PLoS Biology*, *6*(12), e1. http://dx.doi.org/10.1371/journal.pbio.1000001.

Hitzemann, R., Belknap, J. K., & McWeeney, S. K. (2008). Quantitative trait locus analysis: Multiple cross and heterogeneous stock mapping. *Alcohol Research & Health*, *31*(3), 261–265.

Hitzemann, R., Bottomly, D., Darakjian, P., Walter, N., Iancu, O., Searles, R., et al. (2013). Genes, behavior and next-generation RNA sequencing. *Genes, Brain, and Behavior*, *12*(1), 1–12. http://dx.doi.org/10.1111/gbb.12007.

Hitzemann, R., Bottomly, D., Iancu, O., Buck, K., Wilmot, B., Mooney, M., et al. (2014). The genetics of gene expression in complex mouse crosses as a tool to study the molecular underpinnings of behavior traits. *Mammalian Genome*, *25*(1–2), 12–22. http://dx.doi.org/10.1007/s00335-013-9495-6.

Houtkooper, R. H., Mouchiroud, L., Ryu, D., Moullan, N., Katsyuba, E., Knott, G., et al. (2013). Mitonuclear protein imbalance as a conserved longevity mechanism. *Nature*, *497*(7450), 451–457. http://dx.doi.org/10.1038/nature12188.

Hovatta, I., Zapala, M. A., Broide, R. S., Schadt, E. E., Libiger, O., Schork, N. J., et al. (2007). DNA variation and brain region-specific expression profiles exhibit different relationships between inbred mouse strains: Implications for eQTL mapping studies. *Genome Biology*, *8*(2), R25. http://dx.doi.org/10.1186/gb-2007-8-2-r25.

Huang, G. J., Shifman, S., Valdar, W., Johannesson, M., Yalcin, B., Taylor, M. S., et al. (2009). High resolution mapping of expression QTLs in heterogeneous stock mice in multiple tissues. *Genome Research*, *19*(6), 1133–1140. http://dx.doi.org/10.1101/gr.088120.108.

Hubner, N., Wallace, C. A., Zimdahl, H., Petretto, E., Schulz, H., Maciver, F., et al. (2005). Integrated transcriptional profiling and linkage analysis for identification of genes underlying disease. *Nature Genetics*, *37*(3), 243–253. http://dx.doi.org/10.1038/ng1522.

Islam, S., Zeisel, A., Joost, S., La Manno, G., Zajac, P., Kasper, M., et al. (2014). Quantitative single-cell RNA-seq with unique molecular identifiers. *Nature Methods*, *11*(2), 163–166. http://dx.doi.org/10.1038/nmeth.2772.

Jaitin, D. A., Kenigsberg, E., Keren-Shaul, H., Elefant, N., Paul, F., Zaretsky, I., et al. (2014). Massively parallel single-cell RNA-seq for marker-free decomposition of tissues into cell types. *Science*, *343*(6172), 776–779. http://dx.doi.org/10.1126/science.1247651.

Jansen, R. C., & Nap, J. P. (2001). Genetical genomics: The added value from segregation. *Trends in Genetics*, *17*(7), 388–391.

Jansen, R. C., & Stam, P. (1994). High resolution of quantitative traits into multiple loci via interval mapping. *Genetics*, *136*(4), 1447–1455.

Jiang, H., & Wong, W. H. (2009). Statistical inferences for isoform expression in RNA-Seq. *Bioinformatics*, *25*(8), 1026–1032. http://dx.doi.org/10.1093/bioinformatics/btp113.

Johnson, M. B., Kawasawa, Y. I., Mason, C. E., Krsnik, Z., Coppola, G., Bogdanovic, D., et al. (2009). Functional and evolutionary insights into human brain development through global transcriptome analysis. *Neuron*, *62*(4), 494–509. http://dx.doi.org/10.1016/j.neuron.2009.03.027.

Keane, T. M., Goodstadt, L., Danecek, P., White, M. A., Wong, K., Yalcin, B., et al. (2011). Mouse genomic variation and its effect on phenotypes and gene regulation. *Nature*, *477*(7364), 289–294. http://dx.doi.org/10.1038/nature10413.

Korir, P. K., & Seoighe, C. (2014). Inference of allele-specific expression from RNA-seq data. *Methods in Molecular Biology, 1112*, 49–69. http://dx.doi.org/10.1007/978-1-62703-773-0_4.

Laderas, T. G., Walter, N. A., Mooney, M., Vartanian, K., Darakjian, P., Buck, K., et al. (2011). Computational detection of alternative exon usage. *Frontiers in Neuroscience, 5*, 69. http://dx.doi.org/10.3389/fnins.2011.00069.

Lappalainen, T., Sammeth, M., Friedlander, M. R., t Hoen, P. A., Monlong, J., Rivas, M. A., et al. (2013). Transcriptome and genome sequencing uncovers functional variation in humans. *Nature, 501*(7468), 506–511. http://dx.doi.org/10.1038/nature12531.

Li, J., & Burmeister, M. (2005). Genetical genomics: Combining genetics with gene expression analysis. *Human Molecular Genetics, 14 Spec No. 2*, R163–R169. http://dx.doi.org/10.1093/hmg/ddi267.

Li, Z., Mulligan, M. K., Wang, X., Miles, M. F., Lu, L., & Williams, R. W. (2010). A transposon in Comt generates mRNA variants and causes widespread expression and behavioral differences among mice. *PLoS One, 5*(8), e12181. http://dx.doi.org/10.1371/journal.pone.0012181.

Lum, P. Y., Chen, Y., Zhu, J., Lamb, J., Melmed, S., Wang, S., et al. (2006). Elucidating the murine brain transcriptional network in a segregating mouse population to identify core functional modules for obesity and diabetes. *Journal of Neurochemistry, 97*(Suppl. 1), 50–62. http://dx.doi.org/10.1111/j.1471-4159.2006.03661.x.

MacDonald, M. E., Ambrose, C. M., Duyao, M. P., Myers, R. H., Lin, C., Srinidhi, L., et al. (1993). A novel gene containing a trinucleotide repeat that is expanded and unstable on Huntington's disease chromosomes. The Huntington's Disease Collaborative Research Group. *Cell, 72*(6), 971–983.

MacLaren, E. J., & Sikela, J. M. (2005). Cerebellar gene expression profiling and eQTL analysis in inbred mouse strains selected for ethanol sensitivity. *Alcoholism, Clinical and Experimental Research, 29*(9), 1568–1579.

Marc, R. E., Jones, B. W., Lauritzen, J. S., Watt, C. B., & Anderson, J. R. (2012). Building retinal connectomes. *Current Opinion in Neurobiology, 22*(4), 568–574. http://dx.doi.org/10.1016/j.conb.2012.03.011.

Massouras, A., Waszak, S. M., Albarca-Aguilera, M., Hens, K., Holcombe, W., Ayroles, J. F., et al. (2012). Genomic variation and its impact on gene expression in Drosophila melanogaster. *PLoS Genetics, 8*(11), e1003055. http://dx.doi.org/10.1371/journal.pgen.1003055.

McManus, C. J., Coolon, J. D., Duff, M. O., Eipper-Mains, J., Graveley, B. R., & Wittkopp, P. J. (2010). Regulatory divergence in Drosophila revealed by mRNA-seq. *Genome Research, 20*(6), 816–825. http://dx.doi.org/10.1101/gr.102491.109.

Miura, P., Shenker, S., Andreu-Agullo, C., Westholm, J. O., & Lai, E. C. (2013). Widespread and extensive lengthening of 3′ UTRs in the mammalian brain. *Genome Research, 23*(5), 812–825. http://dx.doi.org/10.1101/gr.146886.112.

Montgomery, S. B., Sammeth, M., Gutierrez-Arcelus, M., Lach, R. P., Ingle, C., Nisbett, J., et al. (2010). Transcriptome genetics using second generation sequencing in a Caucasian population. *Nature, 464*(7289), 773–777. http://dx.doi.org/10.1038/nature08903.

Morley, M., Molony, C. M., Weber, T. M., Devlin, J. L., Ewens, K. G., Spielman, R. S., et al. (2004). Genetic analysis of genome-wide variation in human gene expression. *Nature, 430*(7001), 743–747. http://dx.doi.org/10.1038/nature02797.

Mortazavi, A., Williams, B. A., McCue, K., Schaeffer, L., & Wold, B. (2008). Mapping and quantifying mammalian transcriptomes by RNA-Seq. *Nature Methods, 5*(7), 621–628. http://dx.doi.org/10.1038/nmeth.1226.

Mozhui, K., Hamre, K. M., Holmes, A., Lu, L., & Williams, R. W. (2007). Genetic and structural analysis of the basolateral amygdala complex in BXD recombinant inbred mice. *Behavior Genetics, 37*(1), 223–243. http://dx.doi.org/10.1007/s10519-006-9122-3.

Mozhui, K., Lu, L., Armstrong, W. E., & Williams, R. W. (2012). Sex-specific modulation of gene expression networks in murine hypothalamus. *Frontiers in Neuroscience*, *6*, 63. http://dx.doi.org/10.3389/fnins.2012.00063.

Mulligan, M. K., Wang, X., Adler, A. L., Mozhui, K., Lu, L., & Williams, R. W. (2012). Complex control of GABA(A) receptor subunit mRNA expression: Variation, covariation, and genetic regulation. *PLoS One*, 7(4), e34586. http://dx.doi.org/10.1371/journal.pone.0034586.

Myers, A. J., Gibbs, J. R., Webster, J. A., Rohrer, K., Zhao, A., Marlowe, L., et al. (2007). A survey of genetic human cortical gene expression. *Nature Genetics*, *39*(12), 1494–1499. http://dx.doi.org/10.1038/ng.2007.16.

Nembaware, V., Lupindo, B., Schouest, K., Spillane, C., Scheffler, K., & Seoighe, C. (2008). Genome-wide survey of allele-specific splicing in humans. *BMC Genomics*, *9*, 265. http://dx.doi.org/10.1186/1471-2164-9-265.

Nissim-Rafinia, M., & Kerem, B. (2005). The splicing machinery is a genetic modifier of disease severity. *Trends in Genetics*, *21*(9), 480–483. http://dx.doi.org/10.1016/j.tig.2005.07.005.

Overall, R. W., Kempermann, G., Peirce, J., Lu, L., Goldowitz, D., Gage, F. H., et al. (2009). Genetics of the hippocampal transcriptome in mouse: A systematic survey and online neurogenomics resource. *Frontiers in Neuroscience*, *3*, 55. http://dx.doi.org/10.3389/neuro.15.003.2009.

Pan, Q., Shai, O., Lee, L. J., Frey, B. J., & Blencowe, B. J. (2008). Deep surveying of alternative splicing complexity in the human transcriptome by high-throughput sequencing. *Nature Genetics*, *40*(12), 1413–1415. http://dx.doi.org/10.1038/ng.259.

Parsons, M. J., Grimm, C., Paya-Cano, J. L., Fernandes, C., Liu, L., Philip, V. M., et al. (2012). Genetic variation in hippocampal microRNA expression differences in C57BL/6 J X DBA/2 J (BXD) recombinant inbred mouse strains. *BMC Genomics*, *13*, 476. http://dx.doi.org/10.1186/1471-2164-13-476.

Parts, L., Hedman, A. K., Keildson, S., Knights, A. J., Abreu-Goodger, C., van de Bunt, M., et al. (2012). Extent, causes, and consequences of small RNA expression variation in human adipose tissue. *PLoS Genetics*, *8*(5), e1002704. http://dx.doi.org/10.1371/journal.pgen.1002704.

Peirce, J. L., Li, H., Wang, J., Manly, K. F., Hitzemann, R. J., Belknap, J. K., et al. (2006). How replicable are mRNA expression QTL? *Mammalian Genome*, *17*(6), 643–656. http://dx.doi.org/10.1007/s00335-005-0187-8.

Philip, V. M., Sokoloff, G., Ackert-Bicknell, C. L., Striz, M., Branstetter, L., Beckmann, M. A., et al. (2011). Genetic analysis in the Collaborative Cross breeding population. *Genome Research*, *21*(8), 1223–1238. http://dx.doi.org/10.1101/gr.113886.110.

Pickrell, J. K., Marioni, J. C., Pai, A. A., Degner, J. F., Engelhardt, B. E., Nkadori, E., et al. (2010). Understanding mechanisms underlying human gene expression variation with RNA sequencing. *Nature*, *464*(7289), 768–772. http://dx.doi.org/10.1038/nature08872.

Ramskold, D., Wang, E. T., Burge, C. B., & Sandberg, R. (2009). An abundance of ubiquitously expressed genes revealed by tissue transcriptome sequence data. *PLoS Computational Biology*, *5*(12), e1000598. http://dx.doi.org/10.1371/journal.pcbi.1000598.

Rapaport, F., Khanin, R., Liang, Y., Pirun, M., Krek, A., Zumbo, P., et al. (2013). Comprehensive evaluation of differential gene expression analysis methods for RNA-seq data. *Genome Biology*, *14*(9), R95. http://dx.doi.org/10.1186/gb-2013-14-9-r95.

Roberts, A., Pimentel, H., Trapnell, C., & Pachter, L. (2011). Identification of novel transcripts in annotated genomes using RNA-Seq. *Bioinformatics*, *27*(17), 2325–2329. http://dx.doi.org/10.1093/bioinformatics/btr355.

Robinson, M. D., McCarthy, D. J., & Smyth, G. K. (2010). edgeR: A bioconductor package for differential expression analysis of digital gene expression data. *Bioinformatics*, *26*(1), 139–140. http://dx.doi.org/10.1093/bioinformatics/btp616.

Rosen, G. D., Pung, C. J., Owens, C. B., Caplow, J., Kim, H., Mozhui, K., et al. (2009). Genetic modulation of striatal volume by loci on Chrs 6 and 17 in BXD recombinant inbred mice. *Genes, Brain, and Behavior, 8*(3), 296–308. http://dx.doi.org/10.1111/j.1601-183X.2009.00473.x.

Rozowsky, J., Abyzov, A., Wang, J., Alves, P., Raha, D., Harmanci, A., et al. (2011). AlleleSeq: Analysis of allele-specific expression and binding in a network framework. *Molecular Systems Biology*, 7, 522. http://dx.doi.org/10.1038/msb.2011.54.

Saarikangas, J., Hakanen, J., Mattila, P. K., Grumet, M., Salminen, M., & Lappalainen, P. (2008). ABBA regulates plasma-membrane and actin dynamics to promote radial glia extension. *Journal of Cell Science, 121*(Pt 9), 1444–1454. http://dx.doi.org/10.1242/jcs.027466.

Schadt, E. E., Monks, S. A., Drake, T. A., Lusis, A. J., Che, N., Colinayo, V., et al. (2003). Genetics of gene expression surveyed in maize, mouse and man. *Nature, 422*(6929), 297–302.

Shao, N. Y., Hu, H. Y., Yan, Z., Xu, Y., Hu, H., Menzel, C., et al. (2010). Comprehensive survey of human brain microRNA by deep sequencing. *BMC Genomics, 11*, 409. http://dx.doi.org/10.1186/1471-2164-11-409.

Skelly, D. A., Johansson, M., Madeoy, J., Wakefield, J., & Akey, J. M. (2011). A powerful and flexible statistical framework for testing hypotheses of allele-specific gene expression from RNA-seq data. *Genome Research, 21*(10), 1728–1737. http://dx.doi.org/10.1101/gr.119784.110.

Somel, M., Liu, X., Tang, L., Yan, Z., Hu, H., Guo, S., et al. (2011). MicroRNA-driven developmental remodeling in the brain distinguishes humans from other primates. *PLoS Biology, 9*(12), e1001214. http://dx.doi.org/10.1371/journal.pbio.1001214.

St George-Hyslop, P. H., Tanzi, R. E., Polinsky, R. J., Haines, J. L., Nee, L., Watkins, P. C., et al. (1987). The genetic defect causing familial Alzheimer's disease maps on chromosome 21. *Science, 235*(4791), 885–890.

Storey, J. D., Madeoy, J., Strout, J. L., Wurfel, M., Ronald, J., & Akey, J. M. (2007). Gene-expression variation within and among human populations. *American Journal of Human Genetics, 80*(3), 502–509. http://dx.doi.org/10.1086/512017.

Sun, W. (2012). A statistical framework for eQTL mapping using RNA-seq data. *Biometrics, 68*(1), 1–11. http://dx.doi.org/10.1111/j.1541-0420.2011.01654.x.

Sun, W., & Hu, Y. (2013). eQTL mapping using RNA-seq data. *Statistics in Biosciences, 5*(1), 198–219. http://dx.doi.org/10.1007/s12561-012-9068-3.

Svenson, K. L., Gatti, D. M., Valdar, W., Welsh, C. E., Cheng, R., Chesler, E. J., et al. (2012). High-resolution genetic mapping using the Mouse Diversity Outbred population. *Genetics, 190*(2), 437–447. http://dx.doi.org/10.1534/genetics.111.132597.

Taylor, B. A., Wnek, C., Kotlus, B. S., Roemer, N., MacTaggart, T., & Phillips, S. J. (1999). Genotyping new BXD recombinant inbred mouse strains and comparison of BXD and consensus maps. *Mammalian Genome, 10*(4), 335–348.

Templeton, J. P., Wang, X., Freeman, N. E., Ma, Z., Lu, A., Hejtmancik, F., et al. (2013). A crystallin gene network in the mouse retina. *Experimental Eye Research, 116*, 129–140. http://dx.doi.org/10.1016/j.exer.2013.08.001.

Threadgill, D. W., Miller, D. R., Churchill, G. A., & de Villena, F. P. (2011). The collaborative cross: A recombinant inbred mouse population for the systems genetic era. *ILAR Journal, 52*(1), 24–31.

Trapnell, C., Williams, B. A., Pertea, G., Mortazavi, A., Kwan, G., van Baren, M. J., et al. (2010). Transcript assembly and quantification by RNA-Seq reveals unannotated transcripts and isoform switching during cell differentiation. *Nature Biotechnology, 28*(5), 511–515. http://dx.doi.org/10.1038/nbt.1621.

Turk, R., 't Hoen, P. A., Sterrenburg, E., de Menezes, R. X., de Meijer, E. J., Boer, J. M., et al. (2004). Gene expression variation between mouse inbred strains. *BMC Genomics, 5*(1), 57.

Ule, J., Ule, A., Spencer, J., Williams, A., Hu, J. S., Cline, M., et al. (2005). Nova regulates brain-specific splicing to shape the synapse. *Nature Genetics*, *37*(8), 844–852. http://dx.doi.org/10.1038/ng1610.

Valdar, W. S., Flint, J., & Mott, R. (2003). QTL fine-mapping with recombinant-inbred heterogeneous stocks and in vitro heterogeneous stocks. *Mammalian Genome*, *14*(12), 830–838. http://dx.doi.org/10.1007/s00335-003-3021-1.

Valdar, W., Solberg, L. C., Gauguier, D., Burnett, S., Klenerman, P., Cookson, W. O., et al. (2006). Genome-wide genetic association of complex traits in heterogeneous stock mice. *Nature Genetics*, *38*(8), 879–887. http://dx.doi.org/10.1038/ng1840.

Van Gelder, R. N., von Zastrow, M. E., Yool, A., Dement, W. C., Barchas, J. D., & Eberwine, J. H. (1990). Amplified RNA synthesized from limited quantities of heterogeneous cDNA. *Proceedings of the National Academy of Sciences of the United States of America*, *87*(5), 1663–1667.

Vanderlinden, L. A., Saba, L. M., Kechris, K., Miles, M. F., Hoffman, P. L., & Tabakoff, B. (2013). Whole brain and brain regional coexpression network interactions associated with predisposition to alcohol consumption. *PLoS One*, *8*(7), e68878. http://dx.doi.org/10.1371/journal.pone.0068878.

Waterston, R. H., Lindblad-Toh, K., Birney, E., Rogers, J., Abril, J. F., Agarwal, P., et al. (2002). Initial sequencing and comparative analysis of the mouse genome. *Nature*, *420*(6915), 520–562. http://dx.doi.org/10.1038/nature01262.

Webster, J. A., Gibbs, J. R., Clarke, J., Ray, M., Zhang, W., Holmans, P., et al. (2009). Genetic control of human brain transcript expression in Alzheimer disease. *American Journal of Human Genetics*, *84*(4), 445–458. http://dx.doi.org/10.1016/j.ajhg.2009.03.011.

Williams, R. W., Bennett, B., Lu, L., Gu, J., DeFries, J. C., Carosone-Link, P. J., et al. (2004). Genetic structure of the LXS panel of recombinant inbred mouse strains: A powerful resource for complex trait analysis. *Mammalian Genome*, *15*(8), 637–647. http://dx.doi.org/10.1007/s00335-004-2380-6.

Williams, R. W., Gu, J., Qi, S., & Lu, L. (2001). The genetic structure of recombinant inbred mice: High-resolution consensus maps for complex trait analysis. *Genome Biology*, *2*(11) RESEARCH0046.

Williams, R. t., Lim, J. E., Harr, B., Wing, C., Walters, R., Distler, M. G., et al. (2009). A common and unstable copy number variant is associated with differences in Glo1 expression and anxiety-like behavior. *PLoS One*, *4*(3), e4649.

Williams, R. W., & Mulligan, M. K. (2012). Genetic and molecular network analysis of behavior. *International Review of Neurobiology*, *104*, 135–157. http://dx.doi.org/10.1016/B978-0-12-398323-7.00006-9.

Wolen, A. R., Phillips, C. A., Langston, M. A., Putman, A. H., Vorster, P. J., Bruce, N. A., et al. (2012). Genetic dissection of acute ethanol responsive gene networks in prefrontal cortex: Functional and mechanistic implications. *PLoS One*, 7(4), e33575. http://dx.doi.org/10.1371/journal.pone.0033575.

Woo, S. L., Lidsky, A. S., Guttler, F., Thirumalachary, C., & Robson, K. J. (1984). Prenatal diagnosis of classical phenylketonuria by gene mapping. *The Journal of the American Medical Association*, *251*(15), 1998–2002.

Xu, Q., Modrek, B., & Lee, C. (2002). Genome-wide detection of tissue-specific alternative splicing in the human transcriptome. *Nucleic Acids Research*, *30*(17), 3754–3766.

Yang, X., Schadt, E. E., Wang, S., Wang, H., Arnold, A. P., Ingram-Drake, L., et al. (2006). Tissue-specific expression and regulation of sexually dimorphic genes in mice. *Genome Research*, *16*(8), 995–1004. http://dx.doi.org/10.1101/gr.5217506.

Ye, R., Carneiro, A. M., Airey, D., Sanders-Bush, E., Williams, R. W., Lu, L., et al. (2014). Evaluation of heritable determinants of blood and brain serotonin homeostasis using

recombinant inbred mice. *Genes, Brain, and Behavior, 13*(3), 247–260. http://dx.doi.org/10.1111/gbb.12092.
Zeng, Z. B. (1994). Precision mapping of quantitative trait loci. *Genetics, 136*(4), 1457–1468.
Ziats, M. N., & Rennert, O. M. (2013). Identification of differentially expressed microRNAs across the developing human brain. *Molecular Psychiatry, 19*, 848–852. http://dx.doi.org/10.1038/mp.2013.93.
Zou, F., Chai, H. S., Younkin, C. S., Allen, M., Crook, J., Pankratz, V. S., et al. (2012). Brain expression genome-wide association study (eGWAS) identifies human disease-associated variants. *PLoS Genetics, 8*(6), e1002707. http://dx.doi.org/10.1371/journal.pgen.1002707.

CHAPTER NINE

Transcriptomic Changes in Brain Development

Allissa A. Dillman, Mark R. Cookson[1]
Cell Biology and Gene Expression Section, Laboratory of Neurogenetics, National Institute on Aging, Bethesda, Maryland, USA
[1]Corresponding author: e-mail address: cookson@mail.nih.gov

Contents

Abstract

The transcriptome changes hugely during development of the brain. Whole genes, alternate exons, and single base pair changes related to RNA editing all show differences between embryonic and mature brain. Collectively, these changes control proteomic diversity as the brain develops. Additionally, there are many changes in noncoding RNAs (miRNA and lncRNA) that interact with mRNA to influence the overall transcriptional landscape. Here, we will discuss what is known about such changes in brain development, particularly focusing on high-throughput approaches and how those can be used to infer mechanisms by which gene expression is controlled in the brain as it matures.

1. INTRODUCTION

The cellular and molecular complexity of the mature adult brain is influenced both by processes in development and by the experience-dependent formation of neuronal circuits (Innocenti & Price, 2005; Sur & Rubenstein, 2005). Brain development occurs throughout embryonic growth and, in most species, continues after birth with a wide variety of developmental programs between species (Borrell & Calegari, 2014). To

International Review of Neurobiology, Volume 116
ISSN 0074-7742
http://dx.doi.org/10.1016/B978-0-12-801105-8.00009-6

form the mature brain, many different cell types need to differentiate from basal progenitors, migrate to their anatomical positions, and, for neurons, form synapses. As such, brain development is a highly regulated process.

Part of the regulation of brain development includes the coordinated expression of many different genes in a spatially and temporally appropriate context. Virtually all levels of gene expression, from whole genes to splicing and RNA editing, show evidence of regulation during brain development. The purpose of this review is to discuss all levels of gene expression in the development of the brain with a particular emphasis on genome-wide techniques that have allowed for an overall view of the generality of expression changes in this organ.

2. GENE EXPRESSION

A number of studies have employed microarray technology in an attempt to study the molecular changes occurring during brain development including many reports at a genome-wide scale. Although these could be organized by species or brain region, here we will use the order of publication as, in general, the depth of coverage in genome-wide techniques has increased over time.

In 2001, two papers were published using arrays to look at development in the mouse hippocampus (Mody et al., 2001) and cerebellum (Kagami & Furuichi, 2001) using Affymetrix arrays. These two brain regions have slightly different trajectories of development in the mouse and the two studies used slightly different choices for developmental time point. Mody et al. examined the hippocampus at embryonic day 16 (E16) and postnatal days 1 (P1), P7, P16, and P30, whereas E18, P7, P14, P21, and P56 were used in the cerebellum by Kagami and Furuichi. Despite the differences between the two regions and slight differences in the choices of time points, there are some substantial overlap in the genes and, more importantly, types of genes identified in both studies. In both studies, there was a decrement in expression of gene related to neuronal proliferation or cell division from embryonic to postnatal stages (Kagami & Furuichi, 2001; Mody et al., 2001). This presumably represents the maturation from dividing neuronal precursor cells to mature, postmitotic neurons. One example of gene expression that tended to increase with brain development noted by Mody et al. was the upregulation of genes involved in glycolysis, consistent with a shift from ketone to glucose metabolism with brain maturation (Mody et al., 2001). Similarly, gene expression profiles consistent with synaptic maturation and associated signal transduction were clearly seen in both studies (Kagami & Furuichi, 2001; Mody et al., 2001). A summary of some of the key types of genes that are developmentally regulated in different brain regions is shown in Fig. 9.1.

A similar study of the developing cerebellum focused particularly on granule cells which largely develop during the postnatal period in the mouse (Díaz et al., 2002). As well as examining expression patterns in the cerebellum, Diaz et al. used the same technology to look at cultured granule cells developing *in vitro*, mutant mice where granule cells are lost in the postnatal period, as well as the pontine nucleus to which the granule cells project. Although the data series are therefore complex to interpret, they indicate that gene expression in development involves both cell autonomous and non-cell autonomous regulation.

Further studies extended these observations using whole brains using additional embryonic stages (Matsuki, Hori, & Furuichi, 2005), focusing on the prefrontal region of the cerebral cortex in postnatal development (Semeralul et al., 2006), or examining the cortex from embryonic to postnatal development (Pramparo et al., 2011). In general, these studies confirmed earlier results in that there were consistent decreases in cell

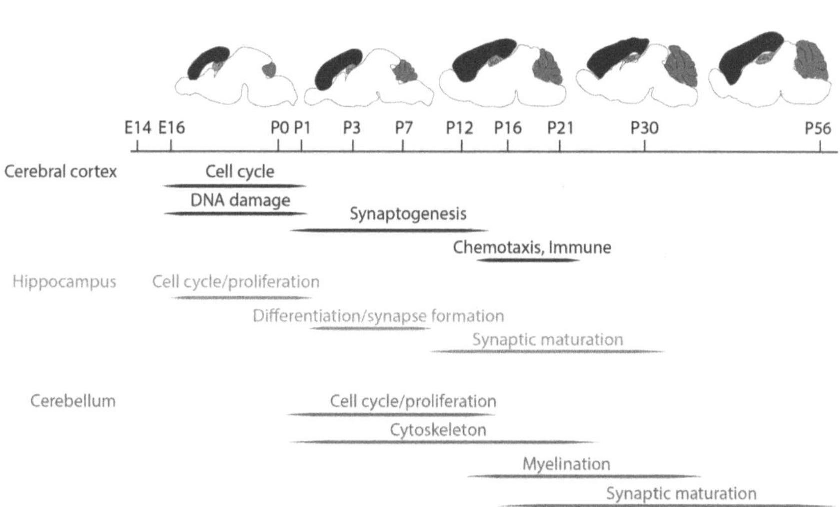

Figure 9.1 Categories of gene expression in the developing mouse brain. The timeline from embryonic (E) to postnatal (P) gene expression is given in days and above the timeline are schematics of the brain approximately equivalent to their positions. Three brain regions where gene expression has been studied are colored in blue (dark gray in the print version) (cerebral cortex), red (light gray in the print version) (hippocampus), and green (gray in the print version) (cerebellum). Below the time line are groups of genes that are prominently expressed at given developmental ranges in each region.

division proteins and acquisition of genes that encode for synaptic proteins. Another interesting observations across several of these experiments are the sheer numbers of genes that show some evidence of regulation during brain development. For example, Matsuki et al. reported that 1413 genes (~11%) showed altered expression in the prenatal period, while Semeralul et al. reported 366 differentially expressed probe sets in the postnatal period. Although these two studies are not strictly comparable, they suggest that effects on gene expression are greater in the prenatal compared to postnatal period.

A similar picture of gene expression during development emerges from studies of the human prefontal cortex (Colantuoni et al., 2011). The absolute number of genes that show differential expression in prenatal development is higher than those in postnatal development, which is higher still than in aging. Furthermore, the magnitude of changes is higher in fetal development than in later stages. Also consistent with mouse data were the types of genes that showed changes during development with, for example, diminishment of cell cycle genes and increases in synaptic components. One complication with human gene expression studies is the genetic diversity of humans compared to inbred mouse strains where genetic variation is minimal. Because gene expression is under genetic control, including in the brain (Gibbs et al., 2010), it is potentially important to dissect out genotypic and developmental effects on gene expression. However, single nucleotide polymorphisms (SNPs) that influence gene expression appear to influence overall expression levels rather than rates of change, that is to say that most genes that are developmentally regulated retain that regulation irrespective of genotype (Colantuoni et al., 2011).

One of the attractions of using genome-wide approaches is that they might be mined to look for unexpected associations that, in turn, might be used to predict mechanisms. One useful approach is to generate self-organizing clusters of genes with similar trajectories of change with development. For example, several transcription factors (SMBP2, FE65, and Sox-M) show correlated expression with genes that increase in the postnatal time period in the hippocampus (Mody et al., 2001). As specific transcription factors are involved in neuronal specification (Thompson & Ziman, 2011), the increase in transcription factor expression in the postnatal period suggests that such proteins might also be important in the maintenance of neuronal phenotype through adulthood.

Genome-wide approaches can further be used to understand how specific transcription factors work during development. For example, knockout of the transcription factor aristaless-related homeobox (*Arx*) results in altered migration of interneurons and abnormal neuronal differentiation. Using Affymetrix arrays in a conditional knockout where *Arx* was depleted in the subpallium, Fulp et al. were able to reconstruct a genetic network where *Arx* normally represses additional transcription factors including *Lmo1*, *Ebf3*, and *Shox2* (Fulp et al., 2008).

As another example of the use of mutant mice, Prampano et al. compared gene expression in several different mutations that are associated with deficits in neuronal migrations (*Lis1*, *Dcx*, and *Ywhae*) and found alterations in classes of genes expressed (Pramparo et al., 2011). Specifically, disruption in neuronal migration genes caused alterations in cell cycle and cytoskeleton categories, but for some mutants, there were also differences in genes encoding synaptic proteins. These observations suggest that there is a dependency of later gene expression events on earlier ones, i.e., that neuronal migration is a required step for synaptic maturation.

The above datasets were generated using microarrays, but there are many other ways to look at gene expression. *In situ* hybridization (ISH) has been applied at a genome-wide level to both mouse (Liscovitch & Chechik, 2013) and human (Miller et al., 2014) brain development. The principle advantage of ISH over arrays is that gene expression can be addressed at the level of brain regions, layers, and even single cells. Such analyses have generally found similar categories of genes as in the array studies but indicate that there are specific genes that are developmentally regulated in germinal zones, for example (Miller et al., 2014).

A deeper view of gene expression changes in the developing mouse cerebral cortex has been achieved using transcriptome sequencing or RNA-Seq (Dillman et al., 2013; Han et al., 2009). Such techniques have been increasingly chosen for gene expression studies as it has been reported that RNA-Seq has greater linearity and reliability compared to microarrays ('t Hoen et al., 2008). Consistent with previous array studies, genes with higher expression in the embryonic brain included many genes involved in cell division, while those that were more highly expressed in the adult brain were related to neurotransmission and ion homeostasis (Dillman et al., 2013). However, probably due to the improved dynamic range of RNA-Seq, such approaches tend to nominate many more genes as being regulated during development compared to array-based studies. In our hands, from ~24,000 genes identified, about 4000 were differentially expressed comparing E17 mice with adult (3–4 months old) females. There was also good quantitative agreement between RNA-Seq and qRT-PCR for a subset of genes chosen for validation, again supporting the idea that RNA-Seq reliably estimates fold differences between conditions. A similar estimate (4000/16,000 detected genes) of differentially expressed genes was reported by Han et al., comparing E18 and P7 mouse cortex (Han et al., 2009). RNA-Seq data are also available for the human brain at various stages of development (e.g., http://www.brainspan.org). Analysis of this dataset again shows a distinct developmental trajectory for expression of a large number of genes including many involved in synaptic function (Parikshak et al., 2013).

An important additional utility of RNA-Seq is that as well as estimating overall gene expression, we examine more complex aspects of gene expression, which will be discussed later in the review, namely splicing and RNA editing. However, it is perhaps interesting to discuss some genetic events that may underly some of the changes in gene expression in brain development.

3. DNA SEQUENCE VARIATION AND EPIGENETIC MODIFICATION IN BRAIN DEVELOPMENT

A discussion of gene expression at the RNA level should also consider the architecture of DNA itself. Within a given species, DNA is highly polymorphic and some of that variation can manifest itself as differences in expression levels rather than coding sequence of genes. Mapping the relationship between DNA variation and gene expression levels identifies expression quantitative trait loci (eQTLs). Conceptually, eQTLs are regions of the genome where polymorphic variants are statistically associated with differences in mRNA expression levels. Studies in human (Gibbs et al., 2010) and mouse (van Nas et al., 2010) brain have identified a large number of such eQTLs for many genes. Such polymorphic loci might, therefore, influence brain development. In one study overlaying genetic data with gene expression across brain development, there were very few examples of genetic polymorphisms that altered the trajectory of gene expression changes throughout development (Colantuoni et al., 2011). Nonetheless, genetic background effects do need to be considered in gene expression profiling experiments, including brain development.

As discussed above, alterations in expression levels of transcription factors may be important in the control of gene expression during brain development. However, the underlying interaction between transcription factors and DNA is dynamic as DNA is subject to a number of regulatory modifications, including methylation. DNA methylation generally occurs at cytosine bases to form 5-methylcytosine in the promoter region of genes. Cytosine methylation generally represses gene expression (Tate & Bird, 1993), although it may also be a mechanism involved in generation of alternate splicing events (Zhou, Luo, Wise, & Lou, 2014).

Methylation seems to play a particularly important role in cell differentiation in the brain. For instance, astrocyte differentiation is dependent on the transcription factor STAT3, but expression levels alone are not enough to trigger differentiation as the promoter of the astrocyte marker GFAP is methylated to prevent STAT3 binding. Once this site is demethylated, the cells can respond to the presence of STAT3 and differentiation can occur (Takizawa et al., 2001). In this way, transcription factors and DNA modifications work together to control gene expression.

DNA methylation can also be involved in genomic imprinting, in which a gene is expressed in a parent-of-origin-specific manner. A substantial proportion of imprinted genes are highly expressed in the brain with unique spatial and temporal expression. For example, *UBE3A* has maternally based expression in specific subpopulations of neurons in the hippocampus and cerebellum but is biallelically expressed in the rest of the brain and body (Albrecht et al., 1997; Rougeulle, Glatt, & Lalande, 1997). In chimeric mice embryos, duplicated maternal genomes contributed to the development of the hypothalamic but not to the cerebral cortex, while a duplicated paternal genome contributed to cortical but not to hypothalamic structures indicating unique differential roles for parent of origin genomes (Keverne, Fundele, Narasimha, Barton, & Surani, 1996). An example of temporal regulation is the gene *Murr1*, which has biallelic expression in embryonic and neonatal mice but only the maternal allele is expressed in adult brain (Wang et al., 2004).

Collectively, these examples show that the transcriptome of the brain is regulated at multiple levels in a manner that depends on epigenetic modification. As might therefore be expected, these single examples likely generalize across the genome. It has been demonstrated recently that widespread DNA methylation changes occur in development in both the mouse and human brain (Lister et al., 2013; Numata et al., 2012). As seen with expression changes, the most dramatic differences in DNA methylation occur during prenatal development with a slowing of progression after birth and even more modest changes in aging (Numata et al., 2012). In some cases, there are DNA methylation events that reverse course after initial development, i.e. where a sequence may undergo demethylation before birth then becoming methylated after birth. This is generally consistent with previous data using smaller sets of methylation events that showed a general increase in methylation in the human brain with age that was also confirmed using isolated neurons (Siegmund et al., 2007).

4. ALTERNATIVE SPLICING

Many of the above approaches generally considered each "gene" as a single unit. However, many tissues, including the brain, show a large number of splicing events with perhaps half of all genes showing some evidence of alternate exons being incorporated into mature mRNA (Lee & Irizarry, 2003).

As might therefore be expected, there are many examples of regulated alternative splicing in neuronal development. In mice, fetal *Mapt* has only minor incorporation exon 10 but by postnatal day 24 all tau contains this exon (McMillan et al., 2008). Interestingly, human *MAPT* retains exon 10 throughout adulthood (Liu & Gong, 2008), perhaps related to the larger size of human neurons compared to neurons leading to a higher requirement for axonal stability. The glutamate receptor gene *Gria2* has a pair of exons that can be spliced in or out leading to two different protein isoforms, flip and flop, that have different electrophysiological characteristics (Sommer et al., 1990). In rats, flip is expressed at stable levels throughout development, while flop expression is low until postnatal day 8 (Monyer, Seeburg, & Wisden, 1991). Although not comprehensive, these examples show how alternate splicing in brain-expressed genes can be functionally important in different species.

Another level of regulation related to splicing is intron retention, where sequences that would normally be spliced out are included in the mature mRNA. In general, retention of introns is high in the brain than other tissues and is developmentally regulated, with levels of retention higher in the fetal brain than in the adult (Ameur et al., 2011). One example of intron retention during development is in the axon guidance molecule Robo3 (Colak, Ji, Porse, & Jaffrey, 2013). A Robo3 isoform containing an intronic sequence (Robo3.2) is expressed but translationally repressed and allows for neuronal attraction to the spinal cord midline. Once the axon crosses the spinal cord midline, it receives signals from the floor plate to translate Robo3.2 allowing nonsense-mediated decay to occur, causing repellence to the midline. Whether other examples of intron retention are similarly functionally important in brain development is not known, but given that intron retention is frequent in the embryonic brain (Ameur et al., 2011), it is likely that this is an important mechanism of gene regulation relevant for neuronal maturation.

The brain also has been found to have longer 3′-UTR regions (Miura, Shenker, Andreu-Agullo, Westholm, & Lai, 2013; Ramsköld, Wang, Burge, & Sandberg, 2009) than other tissues, and this lengthening of UTRs occurs during development (Ji, Lee, Pan, Jiang, & Tian, 2009). This may be related to stability of mRNA transcripts, as 3′-UTR regions contain polyadenylation signals that control the turnover of mRNA; interestingly, the brain has more alternate poly-adenylation than other tissues (Hu, Liu, & Yan, 2014). Alternate 3′-UTR signals may also be important in targeting mRNA to neuronal processes as there are signals that direct mRNAs to axons and dendrites (Mohr, 1999).

There have been several studies looking at alternative exon usage in brain development in different species including humans and other primates using genome-wide exon arrays (Johnson et al., 2009; Mazin et al., 2013). What is impressive about these studies is that, like measures of overall gene expression, they estimate that a large proportion of genes show alternative splicing. In our own work using RNA-Seq in the mouse brain, we found almost 400 exons that were differentially expressed with examples where exon inclusion were higher or lower in the adult compared to embryonic brain and these included well-characterized examples such as *Mapt*. We also found that there were many types of alternative exon usage, including 5′- and 3′-UTR sequences (Dillman et al., 2013), showing that some of the specific examples discussed above may generalize to many genes.

An obvious mechanism for alterations in splicing during development is that splicing factors might themselves be differentially expressed. There is some support for this from large-scale experiments, which have found age-dependent changes in expression of *PTBP1*, *PTBP2*, *hnRNPA1*, *hnRNPF*, *hnRNPH1*, and *hnRNPH3* in the developing human cerebral cortex (Mazin et al., 2013). Differential expression of RNA binding proteins also occurs in mouse development. One of the genes with the largest differences in gene expression in our own dataset (Dillman et al., 2013) was *Igf2bp1*, which is associated with translational repression of a subset of mRNA (Bell et al., 2013). Nova2, a neuron-specific RNA binding protein, is required for the development of the spinal cord and brain stem. Using high-throughput sequencing of RNA isolated by crosslinking, it was discovered that the binding of Nova2 affects alternative splicing (Licatalosi et al., 2008). These observations demonstrate that alternative splicing is therefore required for normal brain development.

It is likely that there are additional levels of complexity in transcript generation that would also be relevant to brain development. One of the limitations of that RNA-Seq we used is that sequences were limited to ~200 bp, although there are technologies that allow for longer reads and hence to recover a greater depth of information about full-length transcripts (Au et al., 2013). Applying a similar approach to the developing brain would be of particular interest in the future.

5. RNA EDITING

An additional source of transcriptome diversity is generated at the single base level via RNA editing. Although there were some early claims of a huge diversity of RNA editing events in the mammalian genome (Li et al., 2011), many of the observed events were shown to be sequencing errors and other technical artifacts (Pickrell, Gilad, & Pritchard, 2012). Instead, it is generally accepted that in many species, RNA editing events are limited to adenosine to inosine and cytosine to uracil, both of which have a well-defined enzymatic basis.

Adenosine-to-inosine substitutions in mammalian RNA are carried out by adenosine deaminases (ADARs), of which there are three isoforms. ADAR1 and ADAR2 are ubiquitously expressed, with expression levels are highest in the brain while ADAR3 is exclusively expressed in the brain (Hogg, Paro, Keegan, & O'Connell, 2011). ADARs act on double-stranded RNA and may require dimerization to be enzymatically active (Cho et al., 2003; Gallo, Keegan, Ring, & O'Connell, 2003). ADARs are localized primarily in the nucleolus and are bound to ribosomal RNA (Sansam, Wells, & Emeson, 2003) but can translocate to the nucleus upon expression of specific ADAR substrates (Desterro et al., 2003).

Inosine is recognized as guanosine in translation and, as such, editing in the coding region of a gene can result in a change in the amino acid sequence (Sommer, Köhler, Sprengel, & Seeburg, 1991). Editing may be particularly important in the brain as there are multiple isoforms of neurotransmitter receptors that are targeted by ADARs (Seeburg, 2000). The majority of A-to-I editing sites are highly conserved across genetically divergent mouse strains (Danecek et al., 2012), supporting the idea that editing is biologically important. In mice, editing of Gria2 leads to a lower permeability of this glutamate channel to calcium ions. If only the unedited isoform of Gria2 is present, mice die within 2 weeks of birth due to seizures (Higuchi et al., 2000). In octopuses, RNA editing plays a role in the temperature adaptability of potassium channels (Garrett & Rosenthal, 2012), again showing that RNA editing may influence neuronal excitability. There are examples of genes that are both edited and that undergo alternative splicing during development (Barresi et al., 2014).

The other major editing enzyme in mammals, APOBEC1, deaminates cytidine to produce uracil (Koito & Ikeda, 2012). Genome-wide surveys suggest that C-to-U editing is far less numerous than A-to-I editing (Kleinman, Adoue, & Majewski, 2012). APOBEC1 is expressed in neurons and is thought primarily to have an antiviral role (Gee et al., 2011) and, perhaps because of this, at the time of this review any potential role in development has not been well studied.

There have been attempts to look at editing in a genome-wide manner in brain development. For example, Wahlstedt et al. described examined 28 known A-to-I editing sites and found that many showed an increase in editing as the brain develops (Wahlstedt, Daniel, Ensterö, & Ohman, 2009). In our own analysis of A-to-I editing, we discovered 176 sites in the mouse brain. Although some coding edits were found, the majority of sites were in the 3′-UTR of genes. We also confirmed that there was a tendency for increase in the proportion of edited transcripts with development (Dillman et al., 2013). The tendency of RNA to become more completely edited during development suggests that protein diversity is less tolerated in the mature CNS than during development.

The mechanism underlying increased completion of A-to-I editing may be partly related to increased expression of Adar enzymes through the postnatal period (Dillman et al., 2013). However, why there is variation in the level of editing between different sites, varying from less than 20% to nearly 100% edited as for *Gria2*, is not at all clear. One possible future experiment would be to examine RNA editing in mice lacking specific *Adar* isoenzymes, which might allow for estimation of the redundancy in editing between the different *Adar* genes.

6. NONCODING RNA

In recent years, it has become clear that the genome contains many types of RNA distinct from protein-coding mRNA species. Some small noncoding RNA (ncRNA), including microRNA (miRNA), have important roles in regulating stability and translation of mRNAs (Dogini et al., 2014). Others such as long noncoding RNAs (lncRNA) influence epigenetic regulation by structural mechanisms (Peschansky & Wahlestedt, 2014) and are highly conserved across species (Chodroff et al., 2010). Human accelerated regions (HARs) are noncoding regions of the genome that are conserved throughout vertebrate evolution but have significant substitution rates in humans. There is significant enrichment adjacent to genes known to play a role in neuronal development (17082449, 16915236). One particular HAR, HAR1, had the most genomic changes in humans with 18 substitutions compared to chimpanzees, while there were only 2 base differences comparing chimpanzees to chickens. HAR1 overlaps with two ncRNAs HAR1F and HAR1R. Interestingly HAR1F is specifically expressed in the fetal brain in Cajal–Retzius neurons along with reelin a gene critical in the specification of layering in the cortex (16915236). As might be expected, there are many single examples of changes in expression of ncRNAs as the brain develops (Barry, 2014; Iyengar et al., 2014; Nowak & Michlewski, 2013), including instances where a neuronal specific function is impacted by miRNA expression levels during development (Schratt et al., 2006).

Several studies have attempted to use genome-wide approaches to look at miRNA in brain development. In the developing rat forebrain from E2 to P5, about 20% of mature miRNA species were shown to have altered expression patterns using a custom array (Krichevsky, King, Donahue, Khrapko, & Kosik, 2003). All of the proposed differences were validated by Northern blots, suggesting that such changes are methodologically robust. Similarly, in studies using several different microarray platforms, a large proportion of miRNAs were found to show changes in expression during development of the mouse brain (Miska et al., 2004; Sempere et al., 2004), in the pig cortex and cerebellum (Podolska et al., 2011) and in many regions of the human brain (Moreau, Bruse, Jornsten, Liu, & Brzustowicz, 2013). ncRNA can also be quantified by RNA-Seq, usually by making libraries that are enriched for small RNA species. As for conventional mRNA, sequencing tends to identify a greater number of genes than arrays in brain (Juhila et al., 2011). These types of methods have been applied to the developing pig hypothalamus and pituitary and again reported a large number of differences (Zhang et al., 2013). Collectively, these results show that many miRNA are regulated during development.

Alterations in the expression levels of miRNA are an additional mechanism that might contribute to some of the changes in mRNA expression and splicing discussed above. Because miRNA generally bind multiple mature mRNA species, it has been predicted that they might be important for coordinated control of gene expression in multiple species (Favre, Banta Lavenex, & Lavenex, 2012). It has been suggested that miRNA:mRNA interactions are particularly important in allowing for maintaining the overall stability in gene expression levels while still allowing for fine-tuning in response to developmental stimuli (Follert, Cremer, & Béclin, 2014). The lncRNA Evf2 binds to intergenic regions and influences expression of proximal genes *Dlx5/6* and *Gad1* (Bond et al., 2009). There is some evidence that a relationship between expression of lncRNAs and nearby protein-coding genes generalizes across many examples (Mercer, Dinger, Sunkin, Mehler, & Mattick, 2008). Small RNAs may also contribute to the generation of transcript diversity. For example, a miRNA expressed selectively in the nervous system can influence splicing via the factors PTBP1 and PTBP2 (Makeyev, Zhang, Carrasco, & Maniatis, 2007).

7. SUMMARY

The examples above show that the brain transcriptome undergoes a number of significant changes throughout development. Importantly, there are many levels of regulation including at the levels of whole gene, single exons, and single base pairs in the case of RNA editing sites. The mechanisms underlying all of these changes are not always understood, but many are likely to be important in the functional specification of the brain. Future challenges include developing additional ways to look at the whole transcriptome in an unbiased manner.

REFERENCES

Albrecht, U., Sutcliffe, J. S., Cattanach, B. M., Beechey, C. V., Armstrong, D., Eichele, G., et al. (1997). Imprinted expression of the murine Angelman syndrome gene, Ube3a, in hippocampal and Purkinje neurons. *Nature Genetics*, *17*(1), 75–78.

Ameur, A., Zaghlool, A., Halvardson, J., Wetterbom, A., Gyllensten, U., Cavelier, L., et al. (2011). Total RNA sequencing reveals nascent transcription and widespread co-transcriptional splicing in the human brain. *Nature Structural & Molecular Biology*, *18*(12), 1435–1440.

Au, K. F., Sebastiano, V., Afshar, P. T., Durruthy, J. D., Lee, L., Williams, B. A., et al. (2013). Characterization of the human ESC transcriptome by hybrid sequencing. *Proceedings of the National Academy of Sciences of the United States of America*, *110*(50), E4821–E4830.

Barresi, S., Tomaselli, S., Athanasiadis, A., Galeano, F., Locatelli, F., Bertini, E., et al. (2014). Oligophrenin-1 (OPHN1), a gene involved in X-linked intellectual disability, undergoes RNA editing and alternative splicing during human brain development. *PLoS One*, *9*(3), e91351.

Barry, G. (2014). Integrating the roles of long and small non-coding RNA in brain function and disease. *Molecular Psychiatry*, *19*(4), 410–416.

Bell, J. L., Wächter, K., Mühleck, B., Pazaitis, N., Köhn, M., Lederer, M., et al. (2013). Insulin-like growth factor 2 mRNA-binding proteins (IGF2BPs): Post-transcriptional drivers of cancer progression? *Cellular and Molecular Life Sciences: CMLS*, *70*(15), 2657–2675.

Bond, A. M., Vangompel, M. J. W., Sametsky, E. A., Clark, M. F., Savage, J. C., Disterhoft, J. F., et al. (2009). Balanced gene regulation by an embryonic brain ncRNA is critical for adult hippocampal GABA circuitry. *Nature Neuroscience*, *12*(8), 1020–1027.

Borrell, V., & Calegari, F. (2014 Apr 28). Mechanisms of brain evolution: Regulation of neural progenitor cell diversity and cell cycle length. *Neuroscience Research*. http://www.ncbi.nlm.nih.gov/pubmed/24786671#, pii: S0168-0102(14)00069-8. http://dx.doi.org/10.1016/j.neures.2014.04.004. [Epub ahead of print].

Cho, D.-S. C., Yang, W., Lee, J. T., Shiekhattar, R., Murray, J. M., & Nishikura, K. (2003). Requirement of dimerization for RNA editing activity of adenosine deaminases acting on RNA. *The Journal of Biological Chemistry*, *278*(19), 17093–17102.

Chodroff, R. A., Goodstadt, L., Sirey, T. M., Oliver, P. L., Davies, K. E., Green, E. D., et al. (2010). Long noncoding RNA genes: Conservation of sequence and brain expression among diverse amniotes. *Genome Biology*, *11*(7), R72.

Colak, D., Ji, S.-J., Porse, B. T., & Jaffrey, S. R. (2013). Regulation of axon guidance by compartmentalized nonsense-mediated mRNA decay. *Cell*, *153*(6), 1252–1265.

Colantuoni, C., Lipska, B. K., Ye, T., Hyde, T. M., Tao, R., Leek, J. T., et al. (2011). Temporal dynamics and genetic control of transcription in the human prefrontal cortex. *Nature*, *478*(7370), 519–523.

Danecek, P., Nellåker, C., McIntyre, R. E., Buendia-Buendia, J. E., Bumpstead, S., Ponting, C. P., et al. (2012). High levels of RNA-editing site conservation amongst 15 laboratory mouse strains. *Genome Biology*, *13*(4), 26.

Desterro, J. M. P., Keegan, L. P., Lafarga, M., Berciano, M. T., O'Connell, M., & Carmo-Fonseca, M. (2003). Dynamic association of RNA-editing enzymes with the nucleolus. *Journal of Cell Science*, *116*(Pt 9), 1805–1818.

Díaz, E., Ge, Y., Yang, Y. H., Loh, K. C., Serafini, T. A., Okazaki, Y., et al. (2002). Molecular analysis of gene expression in the developing pontocerebellar projection system. *Neuron*, *36*(3), 417–434.

Dillman, A. A., Hauser, D. N., Gibbs, J. R., Nalls, M. A., McCoy, M. K., Rudenko, I. N., et al. (2013). mRNA expression, splicing and editing in the embryonic and adult mouse cerebral cortex. *Nature Neuroscience*, *16*(4), 499–506.

Dogini, D. B., Pascoal, V. D. B., Avansini, S. H., Vieira, A. S., Pereira, T. C., & Lopes-Cendes, I. (2014). The new world of RNAs. *Genetics and Molecular Biology*, *37*(1 Suppl.), 285–293.

Favre, G., Banta Lavenex, P., & Lavenex, P. (2012). miRNA regulation of gene expression: A predictive bioinformatics analysis in the postnatally developing monkey hippocampus. *PLoS One*, 7(8), e43435.

Follert, P., Cremer, H., & Béclin, C. (2014). MicroRNAs in brain development and function: A matter of flexibility and stability. *Frontiers in Molecular Neuroscience*, 7, 5.

Fulp, C. T., Cho, G., Marsh, E. D., Nasrallah, I. M., Labosky, P. A., & Golden, J. A. (2008). Identification of Arx transcriptional targets in the developing basal forebrain. *Human Molecular Genetics*, *17*(23), 3740–3760.

Gallo, A., Keegan, L. P., Ring, G. M., & O'Connell, M. A. (2003). An ADAR that edits transcripts encoding ion channel subunits functions as a dimer. *The EMBO Journal*, *22*(13), 3421–3430.

Garrett, S., & Rosenthal, J. J. C. (2012). RNA editing underlies temperature adaptation in K + channels from polar octopuses. *Science (New York, N.Y.)*, *335*(6070), 848–851.

Gee, P., Ando, Y., Kitayama, H., Yamamoto, S. P., Kanemura, Y., Ebina, H., et al. (2011). APOBEC1-mediated editing and attenuation of herpes simplex virus 1 DNA indicate that neurons have an antiviral role during herpes simplex encephalitis. *Journal of Virology*, *85*(19), 9726–9736.

Gibbs, J. R., van der Brug, M. P., Hernandez, D. G., Traynor, B. J., Nalls, M. A., Lai, S.-L., et al. (2010). Abundant quantitative trait loci exist for DNA methylation and gene expression in human brain. *PLoS Genetics*, *6*(5), e1000952.

Han, X., Wu, X., Chung, W.-Y., Li, T., Nekrutenko, A., Altman, N. S., et al. (2009). Transcriptome of embryonic and neonatal mouse cortex by high-throughput RNA sequencing. *Proceedings of the National Academy of Sciences of the United States of America*, *106*(31), 12741–12746.

Higuchi, M., Maas, S., Single, F. N., Hartner, J., Rozov, A., Burnashev, N., et al. (2000). Point mutation in an AMPA receptor gene rescues lethality in mice deficient in the RNA-editing enzyme ADAR2. *Nature*, *406*(6791), 78–81.

Hogg, M., Paro, S., Keegan, L. P., & O'Connell, M. A. (2011). RNA editing by mammalian ADARs. *Advances in Genetics*, *73*, 87–120.

Hu, W., Liu, Y., & Yan, J. (2014). Microarray meta-analysis of RNA-binding protein functions in alternative polyadenylation. *PLoS One*, *9*(3), e90774.

Innocenti, G. M., & Price, D. J. (2005). Exuberance in the development of cortical networks. *Nature Reviews. Neuroscience*, *6*(12), 955–965.

Iyengar, B. R., Choudhary, A., Sarangdhar, M. A., Venkatesh, K. V., Gadgil, C. J., & Pillai, B. (2014). Non-coding RNA interact to regulate neuronal development and function. *Frontiers in Cellular Neuroscience*, *8*, 47.

Ji, Z., Lee, J. Y., Pan, Z., Jiang, B., & Tian, B. (2009). Progressive lengthening of 3' untranslated regions of mRNAs by alternative polyadenylation during mouse embryonic development. *Proceedings of the National Academy of Sciences of the United States of America*, *106*(17), 7028–7033.

Johnson, M. B., Kawasawa, Y. I., Mason, C. E., Krsnik, Z., Coppola, G., Bogdanović, D., et al. (2009). Functional and evolutionary insights into human brain development through global transcriptome analysis. *Neuron*, *62*(4), 494–509.

Juhila, J., Sipilä, T., Icay, K., Nicorici, D., Ellonen, P., Kallio, A., et al. (2011). MicroRNA expression profiling reveals miRNA families regulating specific biological pathways in mouse frontal cortex and hippocampus. *PLoS One*, *6*(6), e21495.

Kagami, Y., & Furuichi, T. (2001). Investigation of differentially expressed genes during the development of mouse cerebellum. *Brain Research. Gene Expression Patterns*, *1*(1), 39–59.

Keverne, E. B., Fundele, R., Narasimha, M., Barton, S. C., & Surani, M. A. (1996). Genomic imprinting and the differential roles of parental genomes in brain development. *Brain Research. Developmental Brain Research*, *92*(1), 91–100.

Kleinman, C. L., Adoue, V., & Majewski, J. (2012). RNA editing of protein sequences: A rare event in human transcriptomes. *RNA (New York, N.Y.)*, *18*(9), 1586–1596.

Koito, A., & Ikeda, T. (2012). Apolipoprotein B mRNA-editing, catalytic polypeptide cytidine deaminases and retroviral restriction. *Wiley Interdisciplinary Reviews. RNA*, *3*(4), 529–541.

Krichevsky, A. M., King, K. S., Donahue, C. P., Khrapko, K., & Kosik, K. S. (2003). A microRNA array reveals extensive regulation of microRNAs during brain development. *RNA (New York, N.Y.)*, *9*(10), 1274–1281.

Lee, C. J., & Irizarry, K. (2003). Alternative splicing in the nervous system: An emerging source of diversity and regulation. *Biological Psychiatry*, *54*(8), 771–776.

Li, M., Wang, I. X., Li, Y., Bruzel, A., Richards, A. L., Toung, J. M., et al. (2011). Widespread RNA and DNA sequence differences in the human transcriptome. *Science (New York, N.Y.)*, *333*(6038), 53–58.

Licatalosi, D. D., Mele, A., Fak, J. J., Ule, J., Kayikci, M., Chi, S. W., et al. (2008). HITS-CLIP yields genome-wide insights into brain alternative RNA processing. *Nature*, *456*(7221), 464–469.

Liscovitch, N., & Chechik, G. (2013). Specialization of gene expression during mouse brain development. *PLoS Computational Biology*, *9*(9), e1003185.

Lister, R., Mukamel, E. A., Nery, J. R., Urich, M., Puddifoot, C. A., Johnson, N. D., et al. (2013). Global epigenomic reconfiguration during mammalian brain development. *Science (New York, N.Y.)*, *341*(6146), 1237905.

Liu, F., & Gong, C.-X. (2008). Tau exon 10 alternative splicing and tauopathies. *Molecular Neurodegeneration, 3*, 8.

Makeyev, E. V., Zhang, J., Carrasco, M. A., & Maniatis, T. (2007). The microRNA miR-124 promotes neuronal differentiation by triggering brain-specific alternative pre-mRNA splicing. *Molecular Cell, 27*(3), 435–448.

Matsuki, T., Hori, G., & Furuichi, T. (2005). Gene expression profiling during the embryonic development of mouse brain using an oligonucleotide-based microarray system. *Brain Research. Molecular Brain Research, 136*(1–2), 231–254.

Mazin, P., Xiong, J., Liu, X., Yan, Z., Zhang, X., Li, M., et al. (2013). Widespread splicing changes in human brain development and aging. *Molecular Systems Biology, 9*, 633.

McMillan, P., Korvatska, E., Poorkaj, P., Evstafjeva, Z., Robinson, L., Greenup, L., et al. (2008). Tau isoform regulation is region- and cell-specific in mouse brain. *The Journal of Comparative Neurology, 511*(6), 788–803.

Mercer, T. R., Dinger, M. E., Sunkin, S. M., Mehler, M. F., & Mattick, J. S. (2008). Specific expression of long noncoding RNAs in the mouse brain. *Proceedings of the National Academy of Sciences of the United States of America, 105*(2), 716–721.

Miller, J. A., Ding, S.-L., Sunkin, S. M., Smith, K. A., Ng, L., Szafer, A., et al. (2014). Transcriptional landscape of the prenatal human brain. *Nature, 508*(7495), 199–206.

Miska, E. A., Alvarez-Saavedra, E., Townsend, M., Yoshii, A., Sestan, N., Rakic, P., et al. (2004). Microarray analysis of microRNA expression in the developing mammalian brain. *Genome Biology, 5*(9), R68.

Miura, P., Shenker, S., Andreu-Agullo, C., Westholm, J. O., & Lai, E. C. (2013). Widespread and extensive lengthening of 3' UTRs in the mammalian brain. *Genome Research, 23*(5), 812–825.

Mody, M., Cao, Y., Cui, Z., Tay, K. Y., Shyong, A., Shimizu, E., et al. (2001). Genome-wide gene expression profiles of the developing mouse hippocampus. *Proceedings of the National Academy of Sciences of the United States of America, 98*(15), 8862–8867.

Mohr, E. (1999). Subcellular RNA compartmentalization. *Progress in Neurobiology, 57*(5), 507–525.

Monyer, H., Seeburg, P. H., & Wisden, W. (1991). Glutamate-operated channels: Developmentally early and mature forms arise by alternative splicing. *Neuron, 6*(5), 799–810.

Moreau, M. P., Bruse, S. E., Jornsten, R., Liu, Y., & Brzustowicz, L. M. (2013). Chronological changes in microRNA expression in the developing human brain. *PLoS One, 8*(4), e60480.

Nowak, J. S., & Michlewski, G. (2013). miRNAs in development and pathogenesis of the nervous system. *Biochemical Society Transactions, 41*(4), 815–820.

Numata, S., Ye, T., Hyde, T. M., Guitart-Navarro, X., Tao, R., Wininger, M., et al. (2012). DNA methylation signatures in development and aging of the human prefrontal cortex. *American Journal of Human Genetics, 90*(2), 260–272.

Parikshak, N. N., Luo, R., Zhang, A., Won, H., Lowe, J. K., Chandran, V., et al. (2013). Integrative functional genomic analyses implicate specific molecular pathways and circuits in autism. *Cell, 155*(5), 1008–1021.

Peschansky, V. J., & Wahlestedt, C. (2014). Non-coding RNAs as direct and indirect modulators of epigenetic regulation. *Epigenetics: Official Journal of the DNA Methylation Society, 9*(1), 3–12.

Pickrell, J. K., Gilad, Y., & Pritchard, J. K. (2012). Comment on "Widespread RNA and DNA sequence differences in the human transcriptome" *Science (New York, N.Y.), 335*(6074), 1302.

Podolska, A., Kaczkowski, B., Kamp Busk, P., Søkilde, R., Litman, T., Fredholm, M., et al. (2011). MicroRNA expression profiling of the porcine developing brain. *PLoS One, 6*(1), e14494.

Pramparo, T., Libiger, O., Jain, S., Li, H., Youn, Y. H., Hirotsune, S., et al. (2011). Global developmental gene expression and pathway analysis of normal brain development and mouse models of human neuronal migration defects. *PLoS Genetics*, 7(3), e1001331.

Ramsköld, D., Wang, E. T., Burge, C. B., & Sandberg, R. (2009). An abundance of ubiquitously expressed genes revealed by tissue transcriptome sequence data. *PLoS Computational Biology*, *5*(12), e1000598.

Rougeulle, C., Glatt, H., & Lalande, M. (1997). The Angelman syndrome candidate gene, UBE3A/E6-AP, is imprinted in brain. *Nature Genetics*, *17*(1), 14–15.

Sansam, C. L., Wells, K. S., & Emeson, R. B. (2003). Modulation of RNA editing by functional nucleolar sequestration of ADAR2. *Proceedings of the National Academy of Sciences of the United States of America*, *100*(24), 14018–14023.

Schratt, G. M., Tuebing, F., Nigh, E. A., Kane, C. G., Sabatini, M. E., Kiebler, M., et al. (2006). A brain-specific microRNA regulates dendritic spine development. *Nature*, *439*(7074), 283–289.

Seeburg, P. H. (2000). RNA helicase participates in the editing game. *Neuron*, *25*(2), 261–263.

Semeralul, M. O., Boutros, P. C., Likhodi, O., Okey, A. B., Van Tol, H. H. M., & Wong, A. H. C. (2006). Microarray analysis of the developing cortex. *Journal of Neurobiology*, *66*(14), 1646–1658.

Sempere, L. F., Freemantle, S., Pitha-Rowe, I., Moss, E., Dmitrovsky, E., & Ambros, V. (2004). Expression profiling of mammalian microRNAs uncovers a subset of brain-expressed microRNAs with possible roles in murine and human neuronal differentiation. *Genome Biology*, *5*(3), R13.

Siegmund, K. D., Connor, C. M., Campan, M., Long, T. I., Weisenberger, D. J., Biniszkiewicz, D., et al. (2007). DNA methylation in the human cerebral cortex is dynamically regulated throughout the life span and involves differentiated neurons. *PLoS One*, *2*(9), e895.

Sommer, B., Keinänen, K., Verdoorn, T. A., Wisden, W., Burnashev, N., Herb, A., et al. (1990). Flip and flop: A cell-specific functional switch in glutamate-operated channels of the CNS. *Science (New York, N.Y.)*, *249*(4976), 1580–1585.

Sommer, B., Köhler, M., Sprengel, R., & Seeburg, P. H. (1991). RNA editing in brain controls a determinant of ion flow in glutamate-gated channels. *Cell*, *67*(1), 11–19.

Sur, M., & Rubenstein, J. L. R. (2005). Patterning and plasticity of the cerebral cortex. *Science (New York, N.Y.)*, *310*(5749), 805–810.

Takizawa, T., Nakashima, K., Namihira, M., Ochiai, W., Uemura, A., Yanagisawa, M., et al. (2001). DNA methylation is a critical cell-intrinsic determinant of astrocyte differentiation in the fetal brain. *Developmental Cell*, *1*(6), 749–758.

Tate, P. H., & Bird, A. P. (1993). Effects of DNA methylation on DNA-binding proteins and gene expression. *Current Opinion in Genetics & Development*, *3*(2), 226–231.

't Hoen, P. A. C., Ariyurek, Y., Thygesen, H. H., Vreugdenhil, E., Vossen, R. H. A. M., de Menezes, R. X., et al. (2008). Deep sequencing-based expression analysis shows major advances in robustness, resolution and inter-lab portability over five microarray platforms. *Nucleic Acids Research*, *36*(21), e141.

Thompson, J. A., & Ziman, M. (2011). Pax genes during neural development and their potential role in neuroregeneration. *Progress in Neurobiology*, *95*(3), 334–351.

van Nas, A., Ingram-Drake, L., Sinsheimer, J. S., Wang, S. S., Schadt, E. E., Drake, T., et al. (2010). Expression quantitative trait loci: Replication, tissue- and sex-specificity in mice. *Genetics*, *185*(3), 1059–1068.

Wahlstedt, H., Daniel, C., Enserö, M., & Ohman, M. (2009). Large-scale mRNA sequencing determines global regulation of RNA editing during brain development. *Genome Research*, *19*(6), 978–986.

Wang, Y., Joh, K., Masuko, S., Yatsuki, H., Soejima, H., Nabetani, A., et al. (2004). The mouse Murr1 gene is imprinted in the adult brain, presumably due to transcriptional interference by the antisense-oriented U2af1-rs1 gene. *Molecular and Cellular Biology*, *24*(1), 270–279.

Zhang, L., Cai, Z., Wei, S., Zhou, H., Zhou, H., Jiang, X., et al. (2013). MicroRNA expression profiling of the porcine developing hypothalamus and pituitary tissue. *International Journal of Molecular Sciences*, *14*(10), 20326–20339.

Zhou, H.-L., Luo, G., Wise, J. A., & Lou, H. (2014). Regulation of alternative splicing by local histone modifications: Potential roles for RNA-guided mechanisms. *Nucleic Acids Research*, *42*(2), 701–713.

CHAPTER TEN

Gene Expression in the Addicted Brain

Zhifeng Zhou[1], Mary-Anne Enoch, David Goldman

Laboratory of Neurogenetics, National Institute on Alcohol Abuse and Alcoholism, National Institutes of Health, Bethesda, Maryland, USA

[1]Corresponding author: e-mail address: zhouz@mail.nih.gov

Contents

Abstract

Addiction is due to changes in the structure and function of the brain, including neuronal networks and the cells that comprise them. Within cells, gene expression changes can track and help explain their altered function. Transcriptional changes induced by addictive agents are dynamic and divergent and range from signal pathway-specific perturbations to widespread molecular and cellular dysregulation that can be measured by "omic" methods and that can be used to identify new pathways. The molecular effects of addiction depend on timing of exposure or withdrawal, the stage of adaptation, the brain region, and the behavioral model, there being many models of addiction. However, the molecular neural adaptations across different drug exposures, conditions, and regions are to some extent shared and can reflect common actions on pathways relevant to addiction. Epigenetic studies of DNA methylation and histone modifications and studies of regulatory RNA networks have been informative for elucidating the mechanisms of transcriptional change in the addicted brain.

International Review of Neurobiology, Volume 116
ISSN 0074-7742
http://dx.doi.org/10.1016/B978-0-12-801105-8.00010-2

1. INTRODUCTION

Substance abuse and addiction to drugs and other addictive agents lead to cellular and molecular changes and are also caused in part by adaptations in epigenetic regulation and gene expression that can be measured in cells. Addictive behaviors are the outcome of allostatic maladaptation of neural circuitries (Goldman, Oroszi, & Ducci, 2005; Koob & Le Moal, 2001). Although great efforts have been made to understand the molecular basis of addiction, the mechanisms are elusive, in part because they are likely to be multiple. However, the study of gene expression in the addicted brain has already yielded valuable insights to the molecular mechanisms of maladaption. In model organisms and cellular models, several important pathway-related changes induced by acute and chronic drug exposure have been discovered. Human studies enabled by the availability of postmortem brain tissues from addicted individuals (Albertson et al., 2004; Albertson, Schmidt, Kapatos, & Bannon, 2006; Bannon, Kapatos, & Albertson, 2005; Kristiansen, Bannon, & Meador-Woodruff, 2009; Lehrmann et al., 2003; Lewohl et al., 2011; Liu, Chen, Lerner, Brackett, & Matsumoto, 2005; Mash et al., 2007; Ponomarev, Wang, Zhang, Harris, & Mayfield, 2012; Tang, Fasulo, Mash, & Hemby, 2003; Zhou, Yuan, Mash, & Goldman, 2011) have also provided critical, although somewhat divergent results for the understanding of addiction. Overall, there appear to be many common neuronal changes in gene expression among individuals addicted to various agents (Lehrmann et al., 2006; Marie-Claire et al., 2007; Zhou et al., 2011) and some commonalities with observations from model organisms, reflecting impact on shared molecular pathways involved in neuronal adaptation as well as drug-specific changes (Albertson et al., 2006; Celentano et al., 2009; Zhou et al., 2011).

It is clear that several differences in type of exposure alter the pattern of altered gene expression. One such factor is course of the exposure. Specific changes in early response genes and signal transduction pathways are more visible in the early stages of drug-induced neural adaptive processes (Celentano et al., 2009; Marie-Claire et al., 2007; Zhou et al., 2011), whereas prolonged exposure leads to widespread transcriptional changes of genes involved in diverse cellular functions such as ion transport, chromosome remodeling, stress and immune response, cell adhesion, cell cycle, apoptosis, protein and lipid metabolism, and mitochondrial functions (Albertson et al., 2004; Bannon et al., 2005; Mash et al., 2007; Renthal et al., 2007; Zhou et al., 2011). The impact of drug exposure on transcription is also brain region specific. In two components of the mesolimbic system, the dorsal striatum and nucleus accumbens (NAc), the expression of genes involved in dopaminergic, glutamatergic, and GABAergic transmission (Ghasemzadeh, Mueller, & Vasudevan, 2009; Hyman & Malenka, 2001; McClung et al., 2005; Schumann & Yaka, 2009) and that play key roles in drug-reward and drug-seeking behavior is strongly altered. In the hippocampus, a brain region critical for associative learning and memory, addiction alters the expression of genes involved in long-term potentiation (LTP) (Zhou et al., 2011). Genetic studies, especially ones using genomic sequencing of animal models selectively bred for addiction phenotypes, have uncovered functional variants of genes involved in neural adaptation that are directly responsible for genetic differences in the propensity to use addictive agents and in response (Zhou et al., 2013). Using "omic" approaches, it has also become possible to analyze the whole transcriptome and epigenetic patterning of the genome, and new molecular adaptive processes that contribute to addiction have recently been revealed by applying these methods both in humans and in model organisms.

2. MOLECULAR ADAPTATIONS ACCOMPANYING EARLY RESPONSE AND LONG-TERM ADAPTATIONS IN THE ADDICTED BRAIN

Knowledge of early gene expression changes in response to drug exposure has largely derived from animal studies, many of which have focused on preselected candidate genes and pathways. The molecular targets are often drug specific, for example, the dopamine transporter for cocaine and amphetamine exposure (Calipari, Ferris, Salahpour, Caron, & Jones, 2013; Peraile et al., 2010), opioid receptors and propeptide genes for opioid exposure (Diaz, Barros, Antonelli, Rubio, & Balerio, 2006), and GABA and glutamate receptors for other drug and alcohol exposure (Enoch et al., 2012; Meinhardt et al., 2013; Nona, Li, & Nobrega, 2013; Schumann & Yaka, 2009; Swanson, Baker, Carson, Worley, & Kalivas, 2001; Zhang et al., 2009; Zhou et al., 2013). Certain aspects of cell signaling, early transcriptional response, and learning have been obvious, and fruitful, targets for study in the addictions. Acute exposure to cocaine induces expression of immediate-early genes such as *Jun* and *Fos*, which encode transcription factors. The transcripts of these gene return to control levels, and following repeated administration of the drugs, desentization is seen (Hope, Kosofsky,

Hyman, & Nestler, 1992). The transcription factors FosB (Hope et al., 1992; Nestler, 2008) and CREB (Carlezon, Duman, & Nestler, 2005) have also been well documented as key components targeted by multiple signal transduction pathways and are involved in regulating expression of drug response genes. Binding of the Fos/Jun heterodimer to AP-1 sites and CREB to cAMP-response elements (CREs) in gene promoters activates transcription of the targeted genes. Another group of well-studied immediate-early gene products is the Nur transcription factors that bind to Nur-responsive elements. These are widely present in the hypothalamus–pituitary–adrenal axis and show rapid and transient increases in expression during acute exposure to addictive drugs (Campos-Melo, Galleguillos, Sanchez, Gysling, & Andres, 2013).

In the past decade, global analysis of gene expression using high-throughput microarrays and, more recently, the use of genomic sequencing, have been frequently applied to the problem of addiction (Hitzemann et al., 2013) and have shed new light on molecular pathways that are altered in the addicted brain. These studies have been conducted in diverse contexts including rodents, nonhuman primates, and postmortem human brain samples and have revealed some important divergences. The differences in what is observed appear to be mainly due to timing and exposure: each study is a snapshot of the addicted brain in dynamic processes. In rodents, some studies have profiled gene expression during drug-self administration, whereas others during withdrawal. An important distinction between the rodent models and human postmortem brain is that in rodents, "chronic exposure" usually refers to a few days or weeks, whereas in humans, it usually denotes many years of heavy use. This one fact appears to explain most of the differences observed in studies of rodents versus those on people. More widespread and divergent molecular and cellular changes have been observed in the chronically addicted human postmortem brain. In a study of postmortem prefrontal cortex from chronic cocaine abusers, Lehrmann and colleagues found expression alterations in multiple cellular functional domains, including energy metabolism, mitochondrial oxidative phosphorylation, oligodendrocyte function, cytoskeleton and related signaling, and neuronal plasticity (Lehrmann et al., 2003). Interestingly, they also noted two distinctive states of transcription regulation, an elevated gene expression profile in the recent active cocaine abusers and decreased expression state in the non-active abusers. Altered expression in cocaine addicts has also been shown in myelin-related genes. In a study by Albertson and colleagues (Albertson et al., 2004; Bannon et al., 2005) on human postmortem NAc, the most prominent changes were decreases of myelin basic protein (MBP), proteolipid protein, and myelin-associated oligodendrocyte basic protein. The expression changes were also consistent with a decrease in the number of MBP-immunoactive oligodendrocytes. A study by Mash and colleagues also found cocaine-induced expression changes in genes involved in regulating extracellular matrix integrity and angiogenesis (Mash et al., 2007). At the top of the list of affected genes was *RECK*, encoding a membrane-anchored glycoprotein serving as an inhibitor for matrix metalloproteinase-9. In addition, they also observed altered expression of genes involved in apoptosis and cell death, neurogenesis and axon guidance, signal transduction, transcriptional and translational regulation, and ion transport (Mash et al., 2007).

Our study (Zhou et al., 2011) with genomic sequencing directly examined mRNA-based transcriptome (RNA-Seq) in human postmortem hippocampal tissue from 24 men who were either cocaine addicts or alcoholics, or age-, ethnicity-, and postmortem interval-matched drug-free controls. Expression of 16,008 Refseq genes was detected. Among these, at an uncorrected $P < 0.05$, we observed a total of 1994 differentially expressed genes in cocaine addicts, and 1275 differentially expressed genes in the alcoholics. After genome-wide multiple testing correction using a relatively stringent FDR cutoff of less than 0.2, there were 394 differentially expressed genes in the cocaine addicts and 48 in the alcoholics. At FDR < 0.05, there were 80 differentially expressed genes in the cocaine addicts and 11 in the alcoholics (Fig. 10.1A and B).

These genes differentially expressed in the human chronic cocaine brains we studied are involved in diverse cellular functions, but there were patterns that strongly implicated certain cellular functions. Cocaine depressed the transcript levels for all five members of the BEX gene family (*BEX 1–5*), which encodes brain expressed, X-linked proteins that are thought to mediate neurotrophin signaling and neuronal differentiation (Vilar et al., 2006). There were also significant expression changes for some histone protein genes. Genes involved in regulation of transcription, gene silencing, and chromatin modification were also affected. Several of these genes had been previously implicated in cocaine addiction, including *DNMT3a*, a DNA methyltransferase which was reported to play an important role in regulating cocaine response and spine plasticity in the NAc in the rat (LaPlant et al., 2010) and *HDAC2*, a histone deacetylase found to be involved in cocaine-induced transcription changes in rat NAc and cocaine seeking behavior (Chandrasekar & Dreyer, 2010). In addition, there was also convergent

A

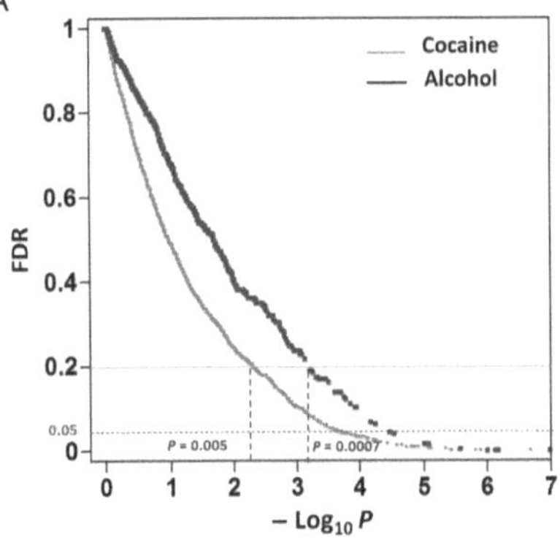

B

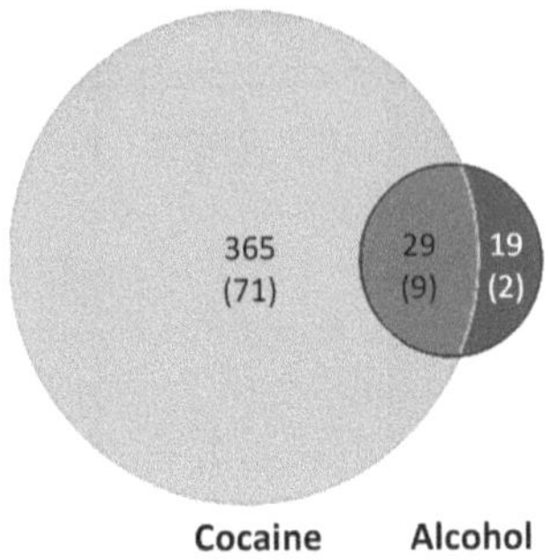

C

Genes with significant differential expression (FDR < 0.2) observed in both chronic cocaine- and alcohol-addicted individuals

Gene symbol	Gene name	Cocaine				Alcohol			
		log FC	*P* Value	FDR	Rank	log FC	*P* Value	FDR	Rank
HIST1H4E	Histone cluster 1, h4e	1.57	1.6E−10	0.000	*1*	1.70	2.5E−11	0.000	*1*
RN7SK	RNA, 7SK small nuclear	2.42	5.3E−10	0.000	*2*	1.31	3.6E−05	0.044	*11*
CDR1	Cerebellar degeneration-related protein 1, 34kda	1.81	5.2E−09	0.000	*3*	1.36	1.1E−06	0.004	*6*
SNORD89	Small nucleolar RNA, C/D box 89	1.53	5.6E−08	0.000	*5*	1.30	1.1E−06	0.004	*5*
SNORA73A	Small nucleolar RNA, H/ACA box 73A	1.41	2.2E−07	0.001	*6*	0.81	6.5E−04	0.190	*48*
SCARNA17	Small Cajal body-specific RNA 17	1.30	1.6E−06	0.004	*8*	1.08	3.0E−05	0.042	*10*
PAPD1	Mitochondrial poly(A) polymerase	0.95	4.1E−06	0.004	*14*	1.17	1.1E−07	0.001	*2*
CACNB2	Calcium channel, voltage-dependent, beta 2 subunit	0.87	4.8E−06	0.005	*15*	0.61	4.8E−04	0.171	*38*
LRCH4	Leucine-rich repeats and calponin homology (CH) domain containing 4	0.80	8.8E−06	0.008	*18*	0.65	1.5E−04	0.112	*16*
SNORD42A	Small nucleolar RNA, C/D box 42A	−1.47	1.2E−05	0.010	*19*	−1.48	1.1E−05	0.018	*9*
SNORA47	Small nucleolar RNA, H/ACA box 47	1.82	1.5E−05	0.010	*22*	1.48	2.3E−04	0.140	*21*
LENG8	Leukocyte receptor cluster (LRC) member 8	0.77	2.3E−05	0.014	*26*	0.72	6.2E−05	0.066	*12*
FAM123A	Family with sequence similarity 123A	0.72	3.9E−05	0.019	*30*	0.92	8.0E−07	0.004	*4*
HIVEP3	Human immunodeficiency virus type I enhancer binding protein 3	0.70	3.9E−05	0.019	*31*	0.59	3.3E−04	0.163	*26*
HNRPH1	Heterogeneous nuclear ribonucleoprotein H1	0.81	6.5E 05	0.026	*38*	0.72	2.7E−04	0.141	*25*
ZGPAT	Zinc finger, CCCH-type with G patch domain	0.55	9.2E 05	0.033	*43*	0.51	2.7E−04	0.141	*24*
ERF	Ets2 repressor factor	0.65	2.2E 04	0.046	*73*	0.82	8.7E−06	0.018	*8*
SNORD116-29	Small nucleolar RNA, C/D box 116-29	−1.68	3.0E 04	0.053	*89*	−1.82	1.2E−04	0.103	*14*
C9orf139	Chromosome 9 open reading frame 139	0.90	4.4E 04	0.068	*103*	0.89	5.2E−04	0.176	*41*
C9orf3	Aminopeptidase O	0.70	6.5E 04	0.084	*119*	0.73	4.0E−04	0.171	*29*
KCNA2	Potassium voltage-gated channel, shaker-related subfamily, member 2	0.67	1.1E 03	0.106	*153*	0.74	3.7E−04	0.171	*27*
EXOC6B	Exocyst complex component 6B	0.51	1.4E−03	0.124	*180*	0.55	6.4E−04	0.190	*45*
CENTB5	Arfgap with coiled-coil, ankyrin repeat and PH domains 3	0.54	1.8E−03	0.139	*198*	0.61	4.7E−04	0.171	*37*
TAOK2	TAO kinase 2	0.56	1.8E−03	0.139	*201*	0.68	2.4E−04	0.141	*22*
TNRC6C	Trinucleotide repeat containing 6C	0.59	2.8E−03	0.170	*255*	0.69	6.5E−04	0.190	*47*
ADAMTS4	ADAM metallopeptidase with thrombospondin type 1 motif, 4	0.65	2.9E−03	0.173	*263*	0.79	4.3E−04	0.171	*33*
MSH4	Muts homolog 4 (E. Coli)	0.79	3.5E−03	0.180	*302*	0.99	4.2E−04	0.171	*31*
C16orf72	Chromosome 16 open reading frame 72	0.54	4.3E−03	0.189	*358*	0.66	6.0E−04	0.186	*44*
CCR5	Chemokine (C-C motif) receptor 5	−0.79	4.7E−03	0.194	*386*	−1.19	7.4E−05	0.070	*13*

Figure 10.1 Differentially expressed genes in the hippocampus of cocaine- and alcohol-addicted individuals, as detected by RNA-Seq. (A) Scatter plot of $-\log_{10} P$ (uncorrected *P* value) versus FDR *q* value for all 16,008 expressed genes. FDR thresholds of 0.2 and 0.05 are marked, as well as corresponding uncorrected *P* values at an FDR of 0.2. (B) Genes differentially expressed (FDR < 0.2) in cocaine only, alcohol only, and common to both. (C) Genes with significant differential expression (FDR < 0.2) observed in both chronic cocaine- and alcohol-addicted individuals.

evidence that chronic cocaine exposure alters expression of genes involved in RNA processing, including significant alteration in the expression of genes encoding RNA-binding and processing proteins and enrichment of differentially expressed small nucleolar (sno) RNA genes, which are involved in both ribosomal RNA and mRNA processing (Kishore & Stamm, 2006).

A particularly salient effect of long-term cocaine exposure in postmortem brains we studied was alteration in the expression of genes involved in mitochondrial inner membrane functions and oxidative phosphorylation (Zhou et al., 2011). Interestingly, these genes integral to cellular energy production have also been implicated in neurodegenerative diseases (Cho, Nakamura, & Lipton, 2010). Among the 90 genes encoding components of oxidative phosphorylation whose expression could be reliably evaluated by RNA-Seq of hippocampal mRNA, 32 were differentially expressed (uncorrected $P < 0.05$), and all were downregulated. Furthermore, 74 of the 90 genes (including the 32 genes that were significantly downregulated) displayed reduced expression levels in cocaine addicts. These findings were also highly consistent with previous brain imaging studies that have revealed negative effects of cocaine on brain glucose metabolism (London et al., 1990; Lyons, Friedman, Nader, & Porrino, 1996; Macey, Rice, Freedland, Whitlow, & Porrino, 2004; Thanos, Michaelides, Benveniste, Wang, & Volkow, 2008). Furthermore, alteration of certain genes encoding for mitochondrial components induced by cocaine (Lehrmann et al., 2003) and nicotine (Wang, Kim, Donovan, Becker, & Li, 2009) exposure had also been reported previously.

3. SUBSTANCE-SPECIFIC AND SHARED GENE EXPRESSION CHANGES IN ADDICTED BRAIN

Clearly, cellular and molecular changes of neural adaptation in addiction occur in a substance-specific fashion as well as through common transduction in neurotransmission pathways. Substance-specific effects are not only due to the direct action of certain drugs of abuse on specific receptors, but also by distinctive molecular and cellular changes related to the drugs that may be related to distinctive signaling mechanisms or differences in the modulation of specific neurons and circuits. Using microarray gene expression profiling, Bannon and colleagues (Albertson et al., 2006) observed decreased expression of many genes involved in presynaptic release of neurotransmitter in the NAc of chronic heroin abusers, but not in chronic cocaine abusers. Similarly, the prominent depressed expression of myelin-related genes found in cocaine abusers was not observed in heroin abusers. Their results suggested the divergent effects of cocaine and heroin on gene expression in the NAc, despite their common effects on dopaminergic transmission. Another study by Marie-Claire et al. (2007) also reported differential effects of cocaine and 3,4-methylenedioxymethamphetamine on expression of the *Rnd* gene family involved in actin cytoskeleton regulation in mouse striatum and noted that the two drugs might act through distinctive pathways to regulate these genes.

Comparing cocaine exposure with alcohol exposure, our analysis (Zhou et al., 2011) of transcriptomes revealed a stronger shift in hippocampal mRNA expression in cocaine-addicted brains. This was manifested by both the larger number of differentially expressed genes (Fig. 10.1B) and changes in molecular and cellular functions defined by gene ontology. For example, the unidirectional depression of expression for the genes encoding for mitochondrial inner membrane and oxidative phosphorylation was only observed in the cocaine-addicted brain (Zhou et al., 2011). This strongly suggests that the inhibition is a specific effect of chronic cocaine exposure, with potential negative implications for brain energy metabolism and diverse brain functions that depend on it and with different metabolic consequences in alcoholism.

Drug-induced neuroplasticity involves some common molecular and cellular changes of the neurocircuitries (Hyman & Malenka, 2001; Kauer & Malenka, 2007; Koob & Volkow, 2010), such as dopaminergic transmission in mesolimbic system, and corticotropin-releasing factor and norepinephrine systems in the extended amygdale. We have also observed significant overlap of gene expression and pathway alteration in both cocaine- and alcohol-addicted brains. Among the 48 differentially expressed genes ($FDR < 0.2$) in the alcoholics, 29 were common to cocaine addicts and in each case the change was in the same direction. More strikingly, for the 11 most significantly differentially expressed genes ($FDR < 0.05$) in alcoholics, 9 were also altered to the same degree in the cocaine addicts, suggesting shared pathways impacted by both cocaine and alcohol in neuronal adaptation. It is apparent that these commonly affected protein-coding genes (Fig. 10.1C) play important roles in neuronal functions. They include *CDR1*, a cerebellar degeneration-related protein; *LRCH4*, a leucine-rich repeat-containing neuronal protein; *CACNB2*, a subunit of voltage-gated calcium channel and involved in neuronal functions; and *FAM123A* (*AMER2*), a member of the gene family involved in neurogenesis

(Comai, Boutet, Neirijnck, & Schedl, 2010). Other commonly and most significantly affected genes also encode for proteins critical in cellular functions such as histone (*HIST1H4E*), transcription regulations (*ZGPAT*, *ERF*, and *HIVEP3*) (Li et al., 2007), and mitochrondrial poly(A) polymerase (*PAPD1*). A subsequent pathway-targeted analysis of GABAergic genes also revealed common expression changes in the cocaine addicts and alcoholics from our study, such as the downregulation of *GABBR1*, *GABRG2*, and *GPHN*, a gene encoding the associated scaffolding protein gephyrin (Enoch et al., 2012).

4. REGION-SPECIFIC GENE EXPRESSION CHANGES IN ADDICTED BRAIN

It is apparent that many neuronal gene expression changes in the drug-induced adaptive process are region specific and cell specific. The mesolimbic system is critically involved in drug-reward and drug-seeking behavior and has been a focus for studies of addiction. In the dorsal striatum and NAc, medium-sized spiny neurons mediate dopaminergic, glutamatergic, and GABAergic neurotransmission (Hyman & Malenka, 2001) and rodents exposed to cocaine or during withdrawal show significant changes in dopaminergic, glutamatergic, and GABAergic neurotransmission (Ghasemzadeh et al., 2009; Hyman & Malenka, 2001; Nestler, 2001). Expression changes of genes targeted by dopaminergic and glutamatergic transmissions or genes involved in mediating transmission were also initiated during adaptation to drug exposure. Some of these genes have been relatively well analyzed such as *CART* (Douglass, McKinzie, & Couceyro, 1995), the *Fos* family (Hope et al., 1992; Nestler, 2008), *CREB* (Carlezon et al., 2005), *Arc* (Fosnaugh, Bhat, Yamagata, Worley, & Baraban, 1995), *EGR1* (O'Donovan, Tourtellotte, Millbrandt, & Baraban, 1999), *Homer-1* (Swanson et al., 2001), *MKP-1* (Ujike, Takaki, Kodama, & Kuroda, 2002), *Narp* (Hyman & Malenka, 2001), *NFκB* (Ang et al., 2001), and *CdK5* (Bibb et al., 2001). During the adaptive process, changes in the striatum take place in synergy with changes in other brain regions, particularly with changes in dopaminergic neurons in the midbrain ventral tegmental area where cocaine-induced glutamate release activates calcium–calmodulin-dependent protein kinases such as CaMKII which are involved in the process of behavioral sensitization (Fernandez-Espejo, Ramiro-Fuentes, Portavella, & Moreno-Paublete, 2008).

In our study of the hippocampal transcriptome (Zhou et al., 2011) of cocaine addicts and alcoholics, we did not observe significant expression changes for some of the genes that have previously been shown to be altered in striatum, such as *CART*, *FOSB*, *CdK5*, *NFκB*, and *HOMER*. These differences may be a manifestation of brain region-specific changes or may also be the result of stage-specific alterations in response to drug exposure because the rodent studies were performed following relatively short-term drug exposure. However, in the cocaine addicts, we did observe expression changes in genes important for hippocampal functions, such as LTP. Hippocampal functions related to short- and long-term memory processes involve synaptic plasticity, and drug-associated learning and memories are important in craving. The hippocampus also directly projects excitatory efferents to the NAc and can also activate dopaminergic neurons of the ventral tegmental area, further implicating its involvement in drug-induced changes of neural plasticity. The genes involved in LTP include specific ionotropic and metabotropic glutamatergic receptors, calcium signaling-related proteins such as calmodulin, calcium/calmodulin-dependent protein kinase, protein phosphotase, adenylate cyclase, protein kinase A and C, mitogen-activated protein kinase, and cAMP-response element binding protein (CREB). Among these, the most significantly affected genes are the *N*-methyl D-aspartate (NMDA) receptor 2B (*GRIN2B*), a subunit of the ionotropic glutamate receptor; protein phosphatase 3 catalytic subunit α isoform (*PPP3CA*), a part of the calcium-dependent phosphatase calcineurin; and calcium/calmodulin-dependent protein kinase type II Δ chain subunit (*CAMK2D*). In addition, the list of genes relevant to LTP whose expression is altered by long-term cocaine exposure includes protein phosphatase 1 catalytic subunit β and γ isoforms (*PPP1CB* and *PPP1CC*), calmodulin 2 (*CALM2*), CREB (*CREB1*), adenylate cyclase 1 (*ADCY1*), protein kinase C β1 (*PRKCB1*), and an N-ras oncogene with intrinsic GTPase activity (*NRAS*). Although we did not observe significant changes of the LTP pathway in the alcoholics, the phosphatidylinositol signaling system, which is closely related to the LTP pathway, was significantly altered by chronic exposure to both cocaine and alcohol. These findings of gene expression changes, together with other studies that have shown the effects of cocaine on LTP (del Olmo et al., 2006; Dunwiddie, Proctor, & Tyma, 1988; Guan, Zhang, Xu, & Li, 2009; Huang, Lin, & Hsu, 2007; Smith, Browning, & Dunwiddie, 1993; Thompson, Gosnell, & Wagner, 2002; Thompson, Swant, & Wagner, 2005), provide evidence that chronic exposure to cocaine, and possibly alcohol, leads to long-term changes in the plasticity of the hippocampus and underlines the importance in addiction of molecular mechanisms for learning.

5. PERTURBATION OF THE GLUTAMATERGIC SYSTEM IN ADDICTED BRAIN

The glutamatergic system, the major excitatory system in the central nervous system, is of particular relevance to addiction through the network of interactions with dopaminergic and GABAergic transmission that underlie alcohol and drug craving and relapse. Glutamate receptors work in synergy with dopamine receptors in dendritic spines of medium-sized spiny neurons in the striatum (Cahill, Salery, Vanhoutte, & Caboche, 2014). Epistatic interactions of glutamatergic and dopaminergic genes have been claimed in alcoholics (Puls et al., 2008). Acute and chronic exposure to alcohol affects glutamate transmission (Ding, Engleman, Rodd, & McBride, 2012) and hyperfunctioning of glutamate transmission has been observed during ethanol or drug withdrawal (Hermann et al., 2012; Prior & Galduroz, 2011). Conditional knockout of the NMDA receptor GluN2B subunit in mice eliminates LTP in the bed nucleus of the stria terminalis (Wills et al., 2012) and makes the animals more sensitive to the locomotor effects of ethanol (Badanich et al., 2011). Pharmacological manipulations have demonstrated that activation of group II metabotropic glutamate receptors decreases alcohol (Rodd et al., 2006; Zhao et al., 2006) and cocaine (Jin et al., 2010) seeking and decreases alcohol-induced neurodegeneration (Cippitelli et al., 2010) in rats.

Alteration of gene expression has been linked to persistent behavioral changes in alcohol- or drug-dependent individuals in both animal and human studies (Edenberg et al., 2005; Heilig & Koob, 2007; Hwang, Stewart, Zhang, Lumeng, & Li, 2004; Liang et al., 2010; Zhou et al., 2011). In the "post-dependent" rats generated by intermittent alcohol vapor intoxication and withdrawal, Meinhardt et al. (2013) identified a pronounced deficit of the metabotropic glutamate receptor II (mGluR2) in the pyramidal neurons of the infralimbic cortex. Among a group of glutamatergic genes that showed enriched downregulation of expression, *Grm2*, which encodes for mGluR2, was one of the genes that were most significantly affected, although the expression of *Grm3*, which encodes mGluR3, the other member of the group II metabotropic glutamate receptors, was not altered in this region. Reduction of extracellular glutamate levels in the NAc, which was readily observed in control rats upon systemic injection of mGluR2/mGluR3 agonists, was also absent in the post-dependent rats, consistent with the lack of mGluR2 function as a presynaptic receptor to downregulate glutamate release upon activation. The role of mGluR2 was further demonstrated by restoring the receptor through bilateral injection of a lentiviral vector expressing mGluR2 into infralimbic cortex. Expression of the receptor significantly reduced alcohol seeking in the post-dependent rats during the cue-induced reinstatement tests.

Using genomic sequencing, genetic linkage, functional validation, and transcriptome analysis, we found that a *Grm2* stop codon functions as a genetic determinant for alcohol preference in selectively bred alcohol preferring (P) and nonpreferring (NP) rats (Zhou et al., 2013). In contrast, genetic studies in humans to identify genes and variants underlying complex disorders and addiction have achieved only limited success, largely due to genetic heterogeneity and the limited effect size of individual loci. Animals selectively bred for alcohol and drug dependence provide potentially powerful models for the identification of genetic variants influencing addiction behaviors both because the artificial selection may collect to high frequencies variants that are rare or uncommon in the ancestral population and because of the ability to control environmental exposures and test animals under the same conditions. By exome sequencing, we uncovered a *Grm2* C407* variant from 25,715 SNPs that homozygously segregates between P and NP rats. All P rats were homozygous for this stop codon in the mGluR2 receptor ligand-binding domain, whereas none of the NP rats carried this allele. The levels of *Grm2* transcript in both striatum and hippocampus were significantly lower, and expression of mGluR2 protein was undetectable in P rats (Fig. 10.2A and B). The loss of the mGluR2 receptor was also consistent with the observation of uncompensated impairment in synaptic depression in P rats, measured as field excitatory postsynaptic potential or population spike in dentate gyrus/hippocampal and striatal slices, upon activation of the receptor by the group II mGluR agonist LY379268.

The causal role of the mGluR2 stop codon in altered alcohol preference was supported by multiple layers of evidence (Zhou et al., 2013). Genetic linkage analysis in the F2 rats derived from intercrossed inbred P and NP rats showed that homozygous stop codon carriers had significantly increased alcohol consumption and preference. Pharmacological blockade of mGluR2 receptor by mGluR2/mGluR3 antagonist LY341495 also significantly escalated alcohol self-administration in Wistar rats trained in an operant self-reinforcement paradigm. To further validate whether the loss

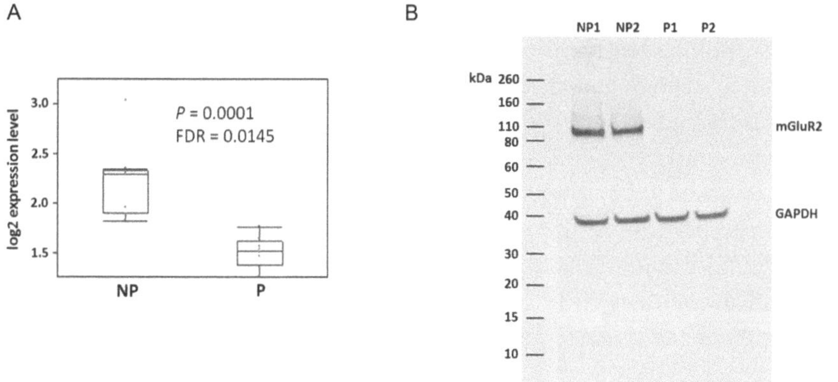

C

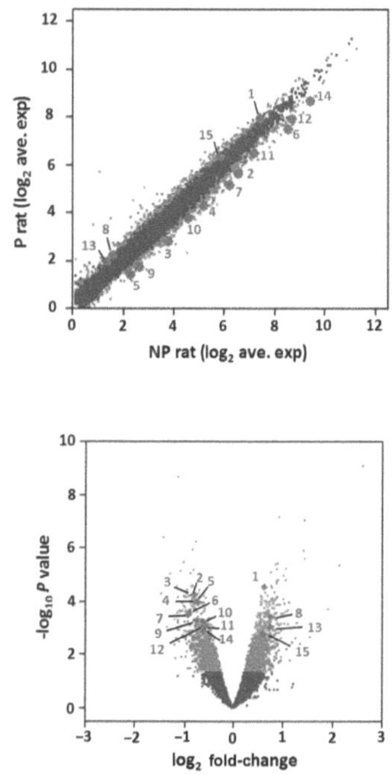

	Gene symbol	Gene name	logFC (P/NP)	*P* Value	FDR
1	*Camkv*	CaM kinase-like vesicle-associated	0.628	2.8E-05	9.7E-03
2	*Syn1*	Synapsin I	-0.795	5.7E-05	1.3E-02
3	*Synpo*	Synaptopodin	-0.812	5.6E-05	1.3E-02
4	*Grin2b*	Glutamate receptor, ionotropic, N-methyl D-aspartate 2B	-0.738	1.1E-04	1.5E-02
5	*Grm2*	Glutamate receptor, metabotropic 2	-0.729	1.1E-04	1.5E-02
6	*Gabbr1*	Gamma-aminobutyric acid (GABA) B receptor 1	-0.905	2.6E-04	2.0E-02
7	*Adrbk1*	Adrenergic, beta, receptor kinase 1	-0.923	3.2E-04	2.2E-02
8	*Oprm1*	Opioid receptor, mu 1	0.759	3.7E-04	2.4E-02
9	*Chrm4*	Cholinergic receptor, muscarinic 4	-0.774	5.1E-04	2.9E-02
10	*Camk4*	Calcium/calmodulin-dependent protein kinase IV	-0.610	5.6E-04	3.0E-02
11	*Snph*	Syntaphilin	-0.582	8.6E-04	3.4E-02
12	*Camk2b*	Calcium/calmodulin-dependent protein kinase II beta	-0.599	9.4E-04	3.5E-02
13	*Gabrg3*	Gamma-aminobutyric acid (GABA) A receptor, gamma 3	0.755	9.7E-04	3.5E-02
14	*Slc1a2*	Solute carrier family 1 (glial high affinity glutamate transporter), member 2	-0.599	1.3E-03	4.1E-02
15	*Gria4*	Glutamate receptor, ionotropic, AMPA4	0.606	1.6E-03	4.5E-02

Figure 10.2 Differential gene expression in hippocampus of P and NP rats. (A) Hippocampal *Grm2* mRNA levels in P and NP rats. (B) Western blot of hippocampal mGluR2 protein in NP and P rats (2 individuals in each group). (C) Average expression levels (normalized $\log_2$ read counts, upper left) of 11,406 genes measured by RNA-Seq are compared between P and NP rats. The fold change (P vs. NP) and nominal *P* values ($-\log_{10}$, *t* test) for each gene are also plotted (lower left). Significance of difference between P and NP for each gene is color coded: blue (black in the print version), not significant; orange (dark gray in the print version), $P < 0.05$; green (light gray in the print version), false discovery rate < 0.05. Fifteen differentially expressed genes among those in the overrepresented functional domains are highlighted and details of their expression differences are listed in the table (right).

of mGluR2 causally contributes to excessive alcohol intake, we examined alcohol-drinking behavior in *Grm2* knockout mice. Tested by a two-bottle free choice scheme and an escalation procedure, *Grm2*-null mice showed significantly higher levels of alcohol consumption and preference than the wild-type control.

To examine gene expression changes related to alcohol-drinking behavior, we also performed hippocampal transcriptome analysis in P and NP rats (Fig. 10.2C) using RNA-sequencing (Zhou et al., 2013). The results indicated an overall pattern of altered expression of genes involved in neural development and synaptic functions. A total of 485 genes were differentially expressed at FDR < 0.05 following correction for genome-wide testing. Differentially expressed genes were significantly enriched with segregating SNPs located in the coding regions and UTRs, indicating the potential involvement of *cis*-regulatory elements in these genes. Using functional annotation analysis with twofold enrichment as a cutoff, we identified several functional domains among the 485 differentially expressed genes, including calmodulin binding, synapse, and neuronal projection. Of particular interest was overrepresentation of the genes that function in glutamate, GABA, opioid, cholinergic, and adrenergic transmission (Fig. 10.2C). This pattern was consistent with the loss of mGluR2 receptor in P rats, but also more readily points to overall neuronal differences between P and NP rats that influence alcohol-drinking behaviors. The gene expression differences between P and NP rats are thus consistent and convergent with their genetic and phenotypic differences and are likely to be influenced by their overall genetic differences or the interaction of *Grm2* C407* with other loci.

6. EPIGENETIC REGULATION OF GENE EXPRESSION IN ADDICTED BRAIN

Epigenetics plays a key role in regulating gene expression. Studies have shown that drug exposure causes changes in DNA methylation that lead to alterations in transcription. Acute cocaine treatment was reported to increase the expression of DNA methyltransferase genes, *Dnmt3a* and *Dnmt 3b*, in mouse NAc, resulting in DNA hypermethylation and the increased binding of methyl CpG binding protein 2 (MeCP2) at the promoter of protein phosphatase-1 catalytic subunit gene (*Pp1c*). As a result, *Pp1c* expression was decreased (Anier, Malinovskaja, Aonurm-Helm, Zharkovsky, & Kalda, 2010). In contrast, chronic cocaine administration was found to decrease expression of *Dnmt3a*. The attenuation of DNA methylation led to potentiated cocaine reward (LaPlant et al., 2010). Acute and repeated cocaine administration was also shown to cause hypomethylation at the FosB promoter, leading to upregulation of FosB expression (Anier et al., 2010). It has also been shown that in heroin addicts, there was elevated methylation at several CpG sites in the promoter of a μ-opioid receptor gene, *OPRM1*, in lymphocytes, which might result in reduced expression of that gene (Nielsen et al., 2009). Differential DNA methylation at the promoter of the pro-opiomelanocortin gene (*POMC*) was found to be associated with alcoholism in a human study (Muschler et al., 2010). DNA hypomethylation was associated with activation of endogenous retroviruses in alcoholic brain (Ponomarev et al., 2012).

Chronic drug exposure also causes significant changes of histone modification. Histone acetylation is known to be associated with activated gene expression. Chronic cocaine exposure was shown to inhibit the function of *Hdac5*, a histone deacetylase, in mouse NAc (Renthal et al., 2007). Activation of dopamine D1 receptor induced upregulation of histone acetylation at the promoters of tyrosine hydroxylase (*Th*) and brain-derived neurotrophic factor (*Bdnf*) genes in mouse NAc and the expression of the two genes (Schroeder et al., 2008). There was a reported association in mice between histone H3 acetylation-activated transcription of addiction-related genes, such as CamkII-α and the motivation for cocaine (L. Wang et al., 2010). In mice chronically administered amphetamine, the Δ-FosB-mediated responses were also found to involve recruiting *Hdac1* to its target gene promoters (Renthal et al., 2008). Inhibition of histone deacetylase reduced behavioral sensitization to morphine in mice (Jing et al., 2011). Alteration of histone methylation also plays important roles in neuronal adaptation of addicted brain. Repeated cocaine administration in mice was shown to repress the expression of lysine dimethyltransferase G9a, resulting in decrease of histone lysine 9 dimethylation (H3K9me2) in NAc (Maze et al., 2010).

Our chromatin immunoprecipitation and genomic sequencing (ChIP-Seq)-based analysis in the postmortem hippocampus of cocaine addicts and alcoholics revealed significant changes in histone H3 lysine 4 trimethylation (H3K4me3) (Zhou et al., 2011), a histone mark known to be associated with activation of gene expression. Similar to the changes observed in gene expression, there was a more widespread and greater impact in response to chronic cocaine exposure than to alcohol exposure. There were also concordant changes between H3K4me3 and gene expression at some loci. In cocaine addicts, these included components of the mitochondrial oxidative phosphorylation pathway or regulators of cellular energy metabolism such as *NDUFS2*, *NDUFA12L*, *UQCRB*, *INSR*, and *IGF1R*; genes involved in LTP and other neuronal functions such as calmodulin 2 (*CALM2*), Synaptophysin-like protein 2 (*SYPL2*), sodium/chloride-dependent neurotransmitter transporter (*SLC6A15*), and nociceptin (*PNOC*). In alcoholics, there were concordant changes between H3K4me3 and gene expression of Protocadherin alpha-7 (*PCDHA7*), Aquaporin-11 (*AQP11*), and potassium inwardly-rectifying channel, subfamily J, member 5 (*CIR*), all of which are involved in critical neuronal and cellular functions. Globally, among all 13,113 histone H3K4me3 peaks mapped to the promoters of hippocampal expressed genes, there was a trend of correlation between H3K4me3 and expression changes in cocaine addicts. However, this trend was not observed in alcoholics. Overall, there was no significant overlap between the genes with either significant H3K4me3 changes or expression changes in both cocaine addicts and alcoholics. This may reflect the fact that epigenetic regulation of gene expression through chromatin remodeling involves many different types of histone modifications at many different histone residues (Barski et al., 2007; Wang et al., 2008), and cocaine- and alcohol-induced expression changes are very likely the results of alterations of those many different histone modifications.

Gene expression changes caused by chronic drug exposure may also be mediated by regulatory RNAs, such as microRNAs (miRNAs) and long non-coding RNAs (lncRNAs). Regulatory RNAs modify gene expression through multiple means, such as altering mRNA stability, basal transcription machinery, translational efficiency, and chromosome modification. miRNA array analysis in human prefrontal cortex revealed upregulation of approximately 35 miRNAs in alcoholics relative to controls with predicted target genes implicated in apoptosis, cell cycle, cell adhesion, nervous system development, and cell–cell signaling (Lewohl et al., 2011). Exposure of zebrafish embryos to cocaine reduced the expression of miR-133b in the CNS, and this difference in miR signaling might in turn modulate expression of dopamine receptors, the dopamine transporter, and tyrosine hydroxylase (Barreto-Valer, Lopez-Bellido, Macho Sanchez-Simon, & Rodriguez, 2012). The let-7 miR family may also interact with the 3′-untranslated region of μ-opioid receptor mRNA to regulate opioid tolerance (He & Wang, 2012). miRNAs are both synaptically enriched and depleted by drug exposure. Cocaine modulates levels of the miR-8 family which is enriched at postsynaptic densities and regulates expression of cell adhesion molecules (Eipper-Mains, Eipper, & Mains, 2012). Differential expression of multiple lncRNAs was identified in the NAc of cocaine-conditioned mice and those lncRNAs were reported to regulate their target loci through both *cis*- and *trans*-actions (Bu et al., 2012).

7. CONCLUSION

Profiling gene expression in the addicted brain has revealed both agent-specific and common drug-induced neural adaptations, providing valuable insights for the understanding of the relevant molecular and cellular mechanisms. The development of transcriptome-based sequencing analysis has equipped us with potent tools that can be combined with neuroscience tools and approaches including the isolation of particular regions, circuits, and cells involved in addiction, genetic models including artificially selected strains and humans varying in vulnerability and response, and interventional models including pharmacological challenges and gene-based manipulations of pathway function and response. These approaches will further enable us to deconstruct the transcription machinery and epigenetic regulation in the addicted brain.

REFERENCES

Albertson, D. N., Pruetz, B., Schmidt, C. J., Kuhn, D. M., Kapatos, G., & Bannon, M. J. (2004). Gene expression profile of the nucleus accumbens of human cocaine abusers: Evidence for dysregulation of myelin. *Journal of Neurochemistry*, *88*(5), 1211–1219.

Albertson, D. N., Schmidt, C. J., Kapatos, G., & Bannon, M. J. (2006). Distinctive profiles of gene expression in the human nucleus accumbens associated with cocaine and heroin abuse. *Neuropsychopharmacology*, *31*(10), 2304–2312.

Ang, E., Chen, J., Zagouras, P., Magna, H., Holland, J., Schaeffer, E., et al. (2001). Induction of nuclear factor-kappaB in nucleus accumbens by chronic cocaine administration. *Journal of Neurochemistry*, *79*(1), 221–224.

Anier, K., Malinovskaja, K., Aonurm-Helm, A., Zharkovsky, A., & Kalda, A. (2010). DNA methylation regulates cocaine-induced behavioral sensitization in mice. *Neuropsychopharmacology*, *35*(12), 2450–2461.

Badanich, K. A., Doremus-Fitzwater, T. L., Mulholland, P. J., Randall, P. K., Delpire, E., & Becker, H. C. (2011). NR2B-deficient mice are more sensitive to the locomotor stimulant and depressant effects of ethanol. *Genes, Brain, and Behavior*, *10*(7), 805–816.

Bannon, M., Kapatos, G., & Albertson, D. (2005). Gene expression profiling in the brains of human cocaine abusers. *Addiction Biology*, *10*(1), 119–126.

Barreto-Valer, K., Lopez-Bellido, R., Macho Sanchez-Simon, F., & Rodriguez, R. E. (2012). Modulation by cocaine of dopamine receptors through miRNA-133b in zebrafish embryos. *PLoS One*, 7(12), e52701.

Barski, A., Cuddapah, S., Cui, K., Roh, T. Y., Schones, D. E., Wang, Z., et al. (2007). High-resolution profiling of histone methylations in the human genome. *Cell*, *129*(4), 823–837.

Bibb, J. A., Chen, J., Taylor, J. R., Svenningsson, P., Nishi, A., Snyder, G. L., et al. (2001). Effects of chronic exposure to cocaine are regulated by the neuronal protein Cdk5. *Nature*, *410*(6826), 376–380.

Bu, Q., Hu, Z., Chen, F., Zhu, R., Deng, Y., Shao, X., et al. (2012). Transcriptome analysis of long non-coding RNAs of the nucleus accumbens in cocaine-conditioned mice. *Journal of Neurochemistry*, *123*(5), 790–799.

Cahill, E., Salery, M., Vanhoutte, P., & Caboche, J. (2014). Convergence of dopamine and glutamate signaling onto striatal ERK activation in response to drugs of abuse. *Frontiers in Pharmacology*, *4*, 172.

Calipari, E. S., Ferris, M. J., Salahpour, A., Caron, M. G., & Jones, S. R. (2013). Methylphenidate amplifies the potency and reinforcing effects of amphetamines by increasing dopamine transporter expression. *Nature Communications*, *4*, 2720.

Campos-Melo, D., Galleguillos, D., Sanchez, N., Gysling, K., & Andres, M. E. (2013). Nur transcription factors in stress and addiction. *Frontiers in Molecular Neuroscience*, *6*, 44.

Carlezon, W. A., Jr., Duman, R. S., & Nestler, E. J. (2005). The many faces of CREB. *Trends in Neurosciences*, *28*(8), 436–445.

Celentano, M., Caprioli, D., Dipasquale, P., Cardillo, V., Nencini, P., Gaetani, S., et al. (2009). Drug context differently regulates cocaine versus heroin self-administration and cocaine- versus heroin-induced Fos mRNA expression in the rat. *Psychopharmacology*, *204*(2), 349–360.

Chandrasekar, V., & Dreyer, J. L. (2010). The brain-specific neural zinc finger transcription factor 2b (NZF-2b/7ZFMyt1) suppresses cocaine self-administration in rats. *Frontiers in Behavioral Neuroscience*, *4*, 14.

Cho, D. H., Nakamura, T., & Lipton, S. A. (2010). Mitochondrial dynamics in cell death and neurodegeneration. *Cellular and Molecular Life Sciences*, *67*(20), 3435–3447.

Cippitelli, A., Damadzic, R., Frankola, K., Goldstein, A., Thorsell, A., Singley, E., et al. (2010). Alcohol-induced neurodegeneration, suppression of transforming growth factor-beta, and cognitive impairment in rats: Prevention by group II metabotropic glutamate receptor activation. *Biological Psychiatry*, *67*(9), 823–830.

Comai, G., Boutet, A., Neirijnck, Y., & Schedl, A. (2010). Expression patterns of the Wtx/Amer gene family during mouse embryonic development. *Developmental Dynamics*, *239*(6), 1867–1878.

del Olmo, N., Miguens, M., Higuera-Matas, A., Torres, I., Garcia-Lecumberri, C., Solis, J. M., et al. (2006). Enhancement of hippocampal long-term potentiation induced by cocaine self-administration is maintained during the extinction of this behavior. *Brain Research*, *1116*(1), 120–126.

Diaz, S. L., Barros, V. G., Antonelli, M. C., Rubio, M. C., & Balerio, G. N. (2006). Morphine withdrawal syndrome and its prevention with baclofen: Autoradiographic study of mu-opioid receptors in prepubertal male and female mice. *Synapse*, *60*(2), 132–140.

Ding, Z. M., Engleman, E. A., Rodd, Z. A., & McBride, W. J. (2012). Ethanol increases glutamate neurotransmission in the posterior ventral tegmental area of female wistar rats. *Alcoholism, Clinical and Experimental Research*, *36*(4), 633–640.

Douglass, J., McKinzie, A. A., & Couceyro, P. (1995). PCR differential display identifies a rat brain mRNA that is transcriptionally regulated by cocaine and amphetamine. *The Journal of Neuroscience*, *15*(3 Pt 2), 2471–2481.

Dunwiddie, T. V., Proctor, W. R., & Tyma, J. (1988). Local anaesthetic actions of cocaine: Effects on excitatory and inhibitory synaptic responses in the hippocampus in vitro. *British Journal of Pharmacology*, *95*(4), 1117–1124.

Edenberg, H. J., Strother, W. N., McClintick, J. N., Tian, H., Stephens, M., Jerome, R. E., et al. (2005). Gene expression in the hippocampus of inbred alcohol-preferring and -nonpreferring rats. *Genes, Brain, and Behavior*, *4*(1), 20–30.

Eipper-Mains, J. E., Eipper, B. A., & Mains, R. E. (2012). Global approaches to the role of miRNAs in drug-induced changes in gene expression. *Frontiers in Genetics*, *3*, 109.

Enoch, M. A., Zhou, Z., Kimura, M., Mash, D. C., Yuan, Q., & Goldman, D. (2012). GABAergic gene expression in postmortem hippocampus from alcoholics and cocaine addicts; corresponding findings in alcohol-naive P and NP rats. *PLoS One*, 7(1), e29369.

Fernandez-Espejo, E., Ramiro-Fuentes, S., Portavella, M., & Moreno-Paublete, R. (2008). Role for D-serine within the ventral tegmental area in the development of cocaine's sensitization. *Neuropsychopharmacology*, *33*(5), 995–1003.

Fosnaugh, J. S., Bhat, R. V., Yamagata, K., Worley, P. F., & Baraban, J. M. (1995). Activation of arc, a putative "effector" immediate early gene, by cocaine in rat brain. *Journal of Neurochemistry*, *64*(5), 2377–2380.

Ghasemzadeh, M. B., Mueller, C., & Vasudevan, P. (2009). Behavioral sensitization to cocaine is associated with increased glutamate receptor trafficking to the postsynaptic density after extended withdrawal period. *Neuroscience*, *159*(1), 414–426.

Goldman, D., Oroszi, G., & Ducci, F. (2005). The genetics of addictions: Uncovering the genes. *Nature Reviews. Genetics*, *6*(7), 521–532.

Guan, X., Zhang, R., Xu, Y., & Li, S. (2009). Cocaine withdrawal enhances long-term potentiation in rat hippocampus via changing the activity of corticotropin-releasing factor receptor subtype 2. *Neuroscience*, *161*(3), 665–670.

He, Y., & Wang, Z. J. (2012). Let-7 microRNAs and opioid tolerance. *Frontiers in Genetics*, *3*, 110.

Heilig, M., & Koob, G. F. (2007). A key role for corticotropin-releasing factor in alcohol dependence. *Trends in Neurosciences*, *30*(8), 399–406.

Hermann, D., Weber-Fahr, W., Sartorius, A., Hoerst, M., Frischknecht, U., Tunc-Skarka, N., et al. (2012). Translational magnetic resonance spectroscopy reveals excessive central glutamate levels during alcohol withdrawal in humans and rats. *Biological Psychiatry*, *71*(11), 1015–1021.

Hitzemann, R., Bottomly, D., Darakjian, P., Walter, N., Iancu, O., Searles, R., et al. (2013). Genes, behavior and next-generation RNA sequencing. *Genes, Brain, and Behavior*, *12*(1), 1–12.

Hope, B., Kosofsky, B., Hyman, S. E., & Nestler, E. J. (1992). Regulation of immediate early gene expression and AP-1 binding in the rat nucleus accumbens by chronic cocaine. *Proceedings of the National Academy of Sciences of the United States of America*, *89*(13), 5764–5768.

Huang, C. C., Lin, H. J., & Hsu, K. S. (2007). Repeated cocaine administration promotes long-term potentiation induction in rat medial prefrontal cortex. *Cerebral Cortex*, *17*(8), 1877–1888.

Hwang, B. H., Stewart, R., Zhang, J. K., Lumeng, L., & Li, T. K. (2004). Corticotropin-releasing factor gene expression is down-regulated in the central nucleus of the amygdala of alcohol-preferring rats which exhibit high anxiety: A comparison between rat lines selectively bred for high and low alcohol preference. *Brain Research*, *1026*(1), 143–150.

Hyman, S. E., & Malenka, R. C. (2001). Addiction and the brain: The neurobiology of compulsion and its persistence. *Nature Reviews. Neuroscience*, *2*(10), 695–703.

Jin, X., Semenova, S., Yang, L., Ardecky, R., Sheffler, D. J., Dahl, R., et al. (2010). The mGluR2 positive allosteric modulator BINA decreases cocaine self-administration and cue-induced cocaine-seeking and counteracts cocaine-induced enhancement of brain reward function in rats. *Neuropsychopharmacology*, *35*(10), 2021–2036.

Jing, L., Luo, J., Zhang, M., Qin, W. J., Li, Y. L., Liu, Q., et al. (2011). Effect of the histone deacetylase inhibitors on behavioural sensitization to a single morphine exposure in mice. *Neuroscience Letters*, *494*, 169–173.

Kauer, J. A., & Malenka, R. C. (2007). Synaptic plasticity and addiction. *Nature Reviews. Neuroscience*, *8*(11), 844–858.

Kishore, S., & Stamm, S. (2006). Regulation of alternative splicing by snoRNAs. *Cold Spring Harbor Symposia on Quantitative Biology*, *71*, 329–334.

Koob, G. F., & Le Moal, M. (2001). Drug addiction, dysregulation of reward, and allostasis. *Neuropsychopharmacology*, *24*(2), 97–129.

Koob, G. F., & Volkow, N. D. (2010). Neurocircuitry of addiction. *Neuropsychopharmacology*, *35*(1), 217–238.

Kristiansen, L. V., Bannon, M. J., & Meador-Woodruff, J. H. (2009). Expression of transcripts for myelin related genes in postmortem brain from cocaine abusers. *Neurochemical Research*, *34*(1), 46–54.

LaPlant, Q., Vialou, V., Covington, H. E., 3rd, Dumitriu, D., Feng, J., Warren, B. L., et al. (2010). Dnmt3a regulates emotional behavior and spine plasticity in the nucleus accumbens. *Nature Neuroscience*, *13*(9), 1137–1143.

Lehrmann, E., Colantuoni, C., Deep-Soboslay, A., Becker, K. G., Lowe, R., Huestis, M. A., et al. (2006). Transcriptional changes common to human cocaine, cannabis and phencyclidine abuse. *PLoS One*, *1*, e114.

Lehrmann, E., Oyler, J., Vawter, M. P., Hyde, T. M., Kolachana, B., Kleinman, J. E., et al. (2003). Transcriptional profiling in the human prefrontal cortex: Evidence for two activational states associated with cocaine abuse. *The Pharmacogenomics Journal*, *3*(1), 27–40.

Lewohl, J. M., Nunez, Y. O., Dodd, P. R., Tiwari, G. R., Harris, R. A., & Mayfield, R. D. (2011). Up-regulation of microRNAs in brain of human alcoholics. *Alcoholism, Clinical and Experimental Research*, *35*(11), 1928–1937.

Li, Y. J., Deng, J., Mayhew, G. M., Grimsley, J. W., Huo, X., & Vance, J. M. (2007). Investigation of the PARK10 gene in Parkinson disease. *Annals of Human Genetics*, *71*(Pt 5), 639–647.

Liang, T., Kimpel, M. W., McClintick, J. N., Skillman, A. R., McCall, K., Edenberg, H. J., et al. (2010). Candidate genes for alcohol preference identified by expression profiling in alcohol-preferring and -nonpreferring reciprocal congenic rats. *Genome Biology*, *11*(2), R11.

Liu, Y., Chen, G. D., Lerner, M. R., Brackett, D. J., & Matsumoto, R. R. (2005). Cocaine up-regulates Fra-2 and sigma-1 receptor gene and protein expression in brain regions involved in addiction and reward. *The Journal of Pharmacology and Experimental Therapeutics*, *314*(2), 770–779.

London, E. D., Cascella, N. G., Wong, D. F., Phillips, R. L., Dannals, R. F., Links, J. M., et al. (1990). Cocaine-induced reduction of glucose utilization in human brain. A study using positron emission tomography and [fluorine 18]-fluorodeoxyglucose. *Archives of General Psychiatry*, *47*(6), 567–574.

Lyons, D., Friedman, D. P., Nader, M. A., & Porrino, L. J. (1996). Cocaine alters cerebral metabolism within the ventral striatum and limbic cortex of monkeys. *The Journal of Neuroscience*, *16*(3), 1230–1238.

Macey, D. J., Rice, W. N., Freedland, C. S., Whitlow, C. T., & Porrino, L. J. (2004). Patterns of functional activity associated with cocaine self-administration in the rat change over time. *Psychopharmacology*, *172*(4), 384–392.

Marie-Claire, C., Salzmann, J., David, A., Courtin, C., Canestrelli, C., & Noble, F. (2007). Rnd family genes are differentially regulated by 3,4-methylenedioxymethamphetamine and cocaine acute treatment in mice brain. *Brain Research*, *1134*(1), 12–17.

Mash, D. C., ffrench-Mullen, J., Adi, N., Qin, Y., Buck, A., & Pablo, J. (2007). Gene expression in human hippocampus from cocaine abusers identifies genes which regulate extracellular matrix remodeling. *PLoS One*, *2*(11), e1187.

Maze, I., Covington, H. E., 3rd, Dietz, D. M., LaPlant, Q., Renthal, W., Russo, S. J., et al. (2010). Essential role of the histone methyltransferase G9a in cocaine-induced plasticity. *Science*, *327*(5962), 213–216.

McClung, C. A., Sidiropoulou, K., Vitaterna, M., Takahashi, J. S., White, F. J., Cooper, D. C., et al. (2005). Regulation of dopaminergic transmission and cocaine reward by the Clock gene. *Proceedings of the National Academy of Sciences of the United States of America*, *102*(26), 9377–9381.

Meinhardt, M. W., Hansson, A. C., Perreau-Lenz, S., Bauder-Wenz, C., Stahlin, O., Heilig, M., et al. (2013). Rescue of infralimbic mGluR2 deficit restores control over drug-seeking behavior in alcohol dependence. *The Journal of Neuroscience*, *33*(7), 2794–2806.

Muschler, M. A., Hillemacher, T., Kraus, C., Kornhuber, J., Bleich, S., & Frieling, H. (2010). DNA methylation of the POMC gene promoter is associated with craving in alcohol dependence. *Journal of Neural Transmission*, *117*(4), 513–519.

Nestler, E. J. (2001). Molecular basis of long-term plasticity underlying addiction. *Nature Reviews. Neuroscience*, *2*(2), 119–128.

Nestler, E. J. (2008). Review. Transcriptional mechanisms of addiction: Role of DeltaFosB. *Philosophical Transactions of the Royal Society of London. Series B, Biological Sciences*, *363*(1507), 3245–3255.

Nielsen, D. A., Yuferov, V., Hamon, S., Jackson, C., Ho, A., Ott, J., et al. (2009). Increased OPRM1 DNA methylation in lymphocytes of methadone-maintained former heroin addicts. *Neuropsychopharmacology*, *34*(4), 867–873.

Nona, C. N., Li, R., & Nobrega, J. N. (2013). Altered NMDA receptor subunit gene expression in brains of mice showing high vs. low sensitization to ethanol. *Behavioural Brain Research*, *260C*, 58–66.

O'Donovan, K. J., Tourtellotte, W. G., Millbrandt, J., & Baraban, J. M. (1999). The EGR family of transcription-regulatory factors: Progress at the interface of molecular and systems neuroscience. *Trends in Neurosciences*, *22*(4), 167–173.

Peraile, I., Torres, E., Mayado, A., Izco, M., Lopez-Jimenez, A., Lopez-Moreno, J. A., et al. (2010). Dopamine transporter down-regulation following repeated cocaine: Implications for 3,4-methylenedioxymethamphetamine-induced acute effects and long-term neurotoxicity in mice. *British Journal of Pharmacology*, *159*(1), 201–211.

Ponomarev, I., Wang, S., Zhang, L., Harris, R. A., & Mayfield, R. D. (2012). Gene coexpression networks in human brain identify epigenetic modifications in alcohol dependence. *The Journal of Neuroscience*, *32*(5), 1884–1897.

Prior, P. L., & Galduroz, J. C. (2011). Glutamatergic hyperfunctioning during alcohol withdrawal syndrome: Therapeutic perspective with zinc and magnesium. *Medical Hypotheses*, *77*(3), 368–370.

Puls, I., Mohr, J., Wrase, J., Priller, J., Behr, J., Kitzrow, W., et al. (2008). Synergistic effects of the dopaminergic and glutamatergic system on hippocampal volume in alcohol-dependent patients. *Biological Psychology*, *79*(1), 126–136.

Renthal, W., Carle, T. L., Maze, I., Covington, H. E., 3rd, Truong, H. T., Alibhai, I., et al. (2008). Delta FosB mediates epigenetic desensitization of the c-fos gene after chronic amphetamine exposure. *The Journal of Neuroscience*, *28*(29), 7344–7349.

Renthal, W., Maze, I., Krishnan, V., Covington, H. E., 3rd, Xiao, G., Kumar, A., et al. (2007). Histone deacetylase 5 epigenetically controls behavioral adaptations to chronic emotional stimuli. *Neuron*, *56*(3), 517–529.

Rodd, Z. A., McKinzie, D. L., Bell, R. L., McQueen, V. K., Murphy, J. M., Schoepp, D. D., et al. (2006). The metabotropic glutamate 2/3 receptor agonist LY404039 reduces alcohol-seeking but not alcohol self-administration in alcohol-preferring (P) rats. *Behavioural Brain Research*, *171*(2), 207–215.

Schroeder, F. A., Penta, K. L., Matevossian, A., Jones, S. R., Konradi, C., Tapper, A. R., et al. (2008). Drug-induced activation of dopamine D(1) receptor signaling and inhibition of class I/II histone deacetylase induce chromatin remodeling in reward circuitry and modulate cocaine-related behaviors. *Neuropsychopharmacology*, *33*(12), 2981–2992.

Schumann, J., & Yaka, R. (2009). Prolonged withdrawal from repeated noncontingent cocaine exposure increases NMDA receptor expression and ERK activity in the nucleus accumbens. *The Journal of Neuroscience*, *29*(21), 6955–6963.

Smith, D. A., Browning, M., & Dunwiddie, T. V. (1993). Cocaine inhibits hippocampal long-term potentiation. *Brain Research*, *608*(2), 259–265.

Swanson, C. J., Baker, D. A., Carson, D., Worley, P. F., & Kalivas, P. W. (2001). Repeated cocaine administration attenuates group I metabotropic glutamate receptor-mediated glutamate release and behavioral activation: A potential role for Homer. *The Journal of Neuroscience*, *21*(22), 9043–9052.

Tang, W. X., Fasulo, W. H., Mash, D. C., & Hemby, S. E. (2003). Molecular profiling of midbrain dopamine regions in cocaine overdose victims. *Journal of Neurochemistry*, *85*(4), 911–924.

Thanos, P. K., Michaelides, M., Benveniste, H., Wang, G. J., & Volkow, N. D. (2008). The effects of cocaine on regional brain glucose metabolism is attenuated in dopamine transporter knockout mice. *Synapse*, *62*(5), 319–324.

Thompson, A. M., Gosnell, B. A., & Wagner, J. J. (2002). Enhancement of long-term potentiation in the rat hippocampus following cocaine exposure. *Neuropharmacology*, *42*(8), 1039–1042.

Thompson, A. M., Swant, J., & Wagner, J. J. (2005). Cocaine-induced modulation of long-term potentiation in the CA1 region of rat hippocampus. *Neuropharmacology*, *49*(2), 185–194.

Ujike, H., Takaki, M., Kodama, M., & Kuroda, S. (2002). Gene expression related to synaptogenesis, neuritogenesis, and MAP kinase in behavioral sensitization to psychostimulants. *The Annals of the New York Academy of Sciences*, *965*, 55–67.

Vilar, M., Murillo-Carretero, M., Mira, H., Magnusson, K., Besset, V., & Ibanez, C. F. (2006). Bex1, a novel interactor of the p75 neurotrophin receptor, links neurotrophin signaling to the cell cycle. *The EMBO Journal*, *25*(6), 1219–1230.

Wang, J., Kim, J. M., Donovan, D. M., Becker, K. G., & Li, M. D. (2009). Significant modulation of mitochondrial electron transport system by nicotine in various rat brain regions. *Mitochondrion*, *9*(3), 186–195.

Wang, L., Lv, Z., Hu, Z., Sheng, J., Hui, B., Sun, J., et al. (2010). Chronic cocaine-induced H3 acetylation and transcriptional activation of CaMKIIalpha in the nucleus accumbens is critical for motivation for drug reinforcement. *Neuropsychopharmacology*, *35*(4), 913–928.

Wang, Z., Zang, C., Rosenfeld, J. A., Schones, D. E., Barski, A., Cuddapah, S., et al. (2008). Combinatorial patterns of histone acetylations and methylations in the human genome. *Nature Genetics*, *40*(7), 897–903.

Wills, T. A., Klug, J. R., Silberman, Y., Baucum, A. J., Weitlauf, C., Colbran, R. J., et al. (2012). GluN2B subunit deletion reveals key role in acute and chronic ethanol sensitivity of glutamate synapses in bed nucleus of the stria terminalis. *Proceedings of the National Academy of Sciences of the United States of America*, *109*(5), E278–E287.

Zhang, G. C., Vu, K., Parelkar, N. K., Mao, L. M., Stanford, I. M., Fibuch, E. E., et al. (2009). Acute administration of cocaine reduces metabotropic glutamate receptor 8 protein expression in the rat striatum in vivo. *Neuroscience Letters*, *449*(3), 224–227.

Zhao, Y., Dayas, C. V., Aujla, H., Baptista, M. A., Martin-Fardon, R., & Weiss, F. (2006). Activation of group II metabotropic glutamate receptors attenuates both stress and cue-induced ethanol-seeking and modulates c-fos expression in the hippocampus and amygdala. *The Journal of Neuroscience, 26*(39), 9967–9974.

Zhou, Z., Karlsson, C., Liang, T., Xiong, W., Kimura, M., Tapocik, J. D., et al. (2013). Loss of metabotropic glutamate receptor 2 escalates alcohol consumption. *Proceedings of the National Academy of Sciences of the United States of America, 110*(42), 16963–16968.

Zhou, Z., Yuan, Q., Mash, D. C., & Goldman, D. (2011). Substance-specific and shared transcription and epigenetic changes in the human hippocampus chronically exposed to cocaine and alcohol. *Proceedings of the National Academy of Sciences of the United States of America, 108*(16), 6626–6631.

CHAPTER ELEVEN

RNA-Seq Reveals Novel Transcriptional Reorganization in Human Alcoholic Brain

Sean P. Farris, R. Dayne Mayfield[1]

Waggoner Center for Alcohol and Addiction Research, The University of Texas at Austin, Austin, TX 78712
[1]Corresponding author: e-mail address: dayne.mayfield@austin.utexas.edu

Contents

Abstract

DNA microarrays have been used for over a decade to profile gene expression on a genomic scale. While this technology has advanced our understanding of complex cellular function, the reliance of microarrays on hybridization kinetics results in several technical limitations. For example, knowledge of the sequences being probed is required, distinguishing similar sequences is difficult because of cross-hybridization, and the relatively narrow dynamic range of the signal limits sensitivity. Recently, new technologies have been introduced that are based on novel sequencing methodologies. These next-generation sequencing methods do not have the limitations inherent to microarrays. Next-generation sequencing is unique since it allows the detection of all known and novel RNAs present in biological samples without bias toward known transcripts. In addition, the expression of coding and noncoding RNAs, alternative splicing events, and expressed single nucleotide polymorphisms (SNPs) can be identified in a single experiment. Furthermore, this technology allows for remarkably higher throughput while lowering sequencing costs. This significant shift in throughput and pricing makes low-cost access to whole genomes possible and more importantly expands sequencing applications far beyond traditional uses (Morozova & Marra, 2008) to include sequencing the transcriptome (RNA-Seq), providing detail on gene structure, alternative

International Review of Neurobiology, Volume 116
ISSN 0074-7742
http://dx.doi.org/10.1016/B978-0-12-801105-8.00011-4

splicing events, expressed SNPs, and transcript size (Mane et al., 2009; Tang et al., 2009; Walter et al., 2009), in a single experiment, while also quantifying the absolute abundance of genes, all with greater sensitivity and dynamic range than the competing cDNA microarray technology (Mortazavi, Williams, McCue, Schaeffer, & Wold, 2008).

1. OVERVIEW

RNA-Seq utilizes highly efficient sequencing techniques and subsequent mapping of short sequence reads to a reference genome, making it possible to identify exons and introns by mapping their boundaries of genes, which in turn allows investigation of the complexity of transcriptomes in unparalleled detail. Moreover, RNA-Seq enables identification of transcription initiation sites and new splicing variants and permits quantitative determination of exon and splicing isoform expression. This innovative technology facilitates detailed examination of individual expression differences in human brain and makes it possible to dissect the genetic complexities of alcoholism and a variety of physiological conditions (Wang, Gerstein, & Snyder, 2009).

This review addresses three critical barriers to progress in alcohol research: (1) Regulation of cell function often occurs at the level of alternative splicing of mRNAs (Hartmann & Valcárcel, 2009; Tazi et al., 2009), and emerging evidence indicates that this can be important for alcohol tolerance (Pietrzykowski et al., 2008), yet we have little information about splicing changes in human alcoholism. This can now be examined using next-generation sequencing of brain RNA from alcoholics and controls. (2) We do not know if our rodent and nonhuman primate models of alcohol consumption or dependence contain any of the molecular signatures found in human alcoholic brain. Because these animal models must serve as the basis for future medication development, it is essential to determine which, if any, display genomic convergence with human alcoholics. (3) Noncoding RNAs (ncRNAs) are emerging as "master regulators" of gene expression and may underlie many of the widespread genomic changes produced by chronic alcohol consumption, yet we have limited knowledge of changes in brain miRNA levels in human alcoholics or animal models and even less is understood regarding the behavioral significance of changes in ncRNAs.

2. RNA-SEQ OF POSTMORTEM BRAIN TISSUE

Transcriptome profiling of postmortem brain tissue from alcoholics and matched controls has revealed novel and detailed gene expression changes, generating new avenues for addiction research. Although there are certain difficulties inherent with using postmortem brain tissue, such as difficulty in obtaining samples and accounting for variable patterns of alcohol use and other human variables, postmortem brain tissue remains the gold standard against which all other model systems should be evaluated. Next-generation sequencing provides a more comprehensive and accurate tool for transcriptome analysis of this limited, valuable resource.

A first-pass examination of the transcriptome of alcoholics and matched controls identified a number of molecular constituents within a specific brain region (Fig. 11.1). The type of RNA molecules uncovered depends on the initial experimental design, but novel biological features may also be revealed. By design, RNA-Seq of the prefrontal cortex primarily identified protein-coding transcripts and also discerned an appreciable number of pseudogenes and small nucleolar RNAs (snoRNAs) (Fig. 11.1). In addition to having recognized roles in RNA processing and ribosomal RNA modification (Eddy, 2001; Kiss, 2002), snoRNAs are implicated in regulating CNS function (Cao, Yeo, Muotri, Kuwabara, & Gage, 2006; Rogelj, Hartmann, Yeo, Hunt, & Giese, 2003). The expression of snoRNAs and other ncRNAs may have important roles in alcoholism and other diseases.

Comparing the overall expression of detected biological RNA categories within individual samples illustrates consistency among nonalcoholics and

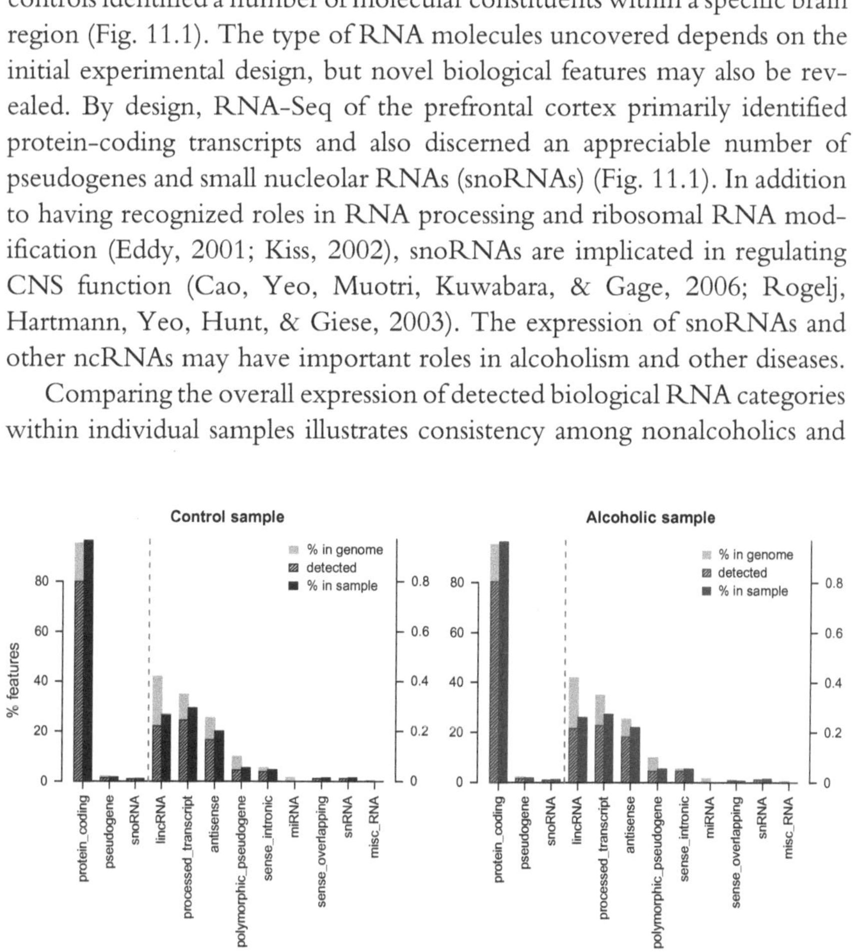

Figure 11.1 RNA-Seq detection of biological features for alcoholics and matched controls. Bar plot demonstrates the percentage of features detected in a representative control (blue (black in the print version)) and alcoholic (red (dark gray in the print version)) sample from a cohort of the prefrontal cortex. The left axis shows percentage of features for the top three biotypes with the right axis showing percentage of remaining biotypes (separated by dotted green (light gray in the print version) vertical line). Protein-coding transcripts were the predominant feature detected in both groups.

alcoholics (Fig. 11.2). Alternatively, any possible discrepancies that may need special consideration in downstream analyses may also be revealed through this type of comparison. The global expression level of transcriptome elements is stable among individuals from alcoholics and matched controls (Fig. 11.3A), and is also similar in terms of sensitivity, determined through the number of counts per million (CPM) mapped reads over varying levels of stringency (Fig. 11.3B). General agreement across the samples indicates the absence of a potential batch effect or outliers within the examined cohort. Sequencing depth, based on reliably and unambiguously mapped reads, is nearly uniform across controls and alcoholics for all of the biological units (Fig. 11.4A), as well as for only protein-coding transcripts (Fig. 11.4B). Although increased sequencing depth could improve expression estimates, continuity between specimens suggests reliable biotype measurements for comparing alcoholic and nonalcoholic subjects. Importantly, lack of overall expression differences in proportion to disease state does not exclude finding potential differences in discrete RNA molecules, which may be important players in the development of alcohol use disorder.

3. DETECTION OF TECHNICAL BIASES IN RNA-SEQ DATA

Obtaining an accurate assessment of RNA molecules that correspond with disease is not a trivial undertaking and should include a comprehensive evaluation of expression estimates for potential areas of artificial biases (Ozsolak & Milos, 2011). Transcript length and guanine–cytosine content (GC content) are two particular characteristics that may influence the quantification of RNA-Seq data (Oshlack & Wakefield, 2009; Pickrell et al., 2010). Nonnormalized expression counts follow a similar trend for alcoholics and matched controls with respect to the length (Fig. 11.5A) and percentage of GC content of identified transcripts (Fig. 11.6A). The length and GC content for mapped features, without normalization, are significantly associated with expression for both groups (Figs. 11.5B and 11.6B). Correcting expression estimates based on the number of collected reads per kilobase per million (RPKM) mapped reads, one method accounting for molar concentration and transcript length (Mortazavi, Williams, McCue, Schaeffer, & Wold, 2008), effectively alleviated the significant bias introduced by transcript length within controls and alcoholics (Fig. 11.5D). Utilizing RPKM values also blunted the relationship between GC content and computed expression values (Fig. 11.6D), although not to the same degree as the length of expressed biotypes. The effect of GC content on

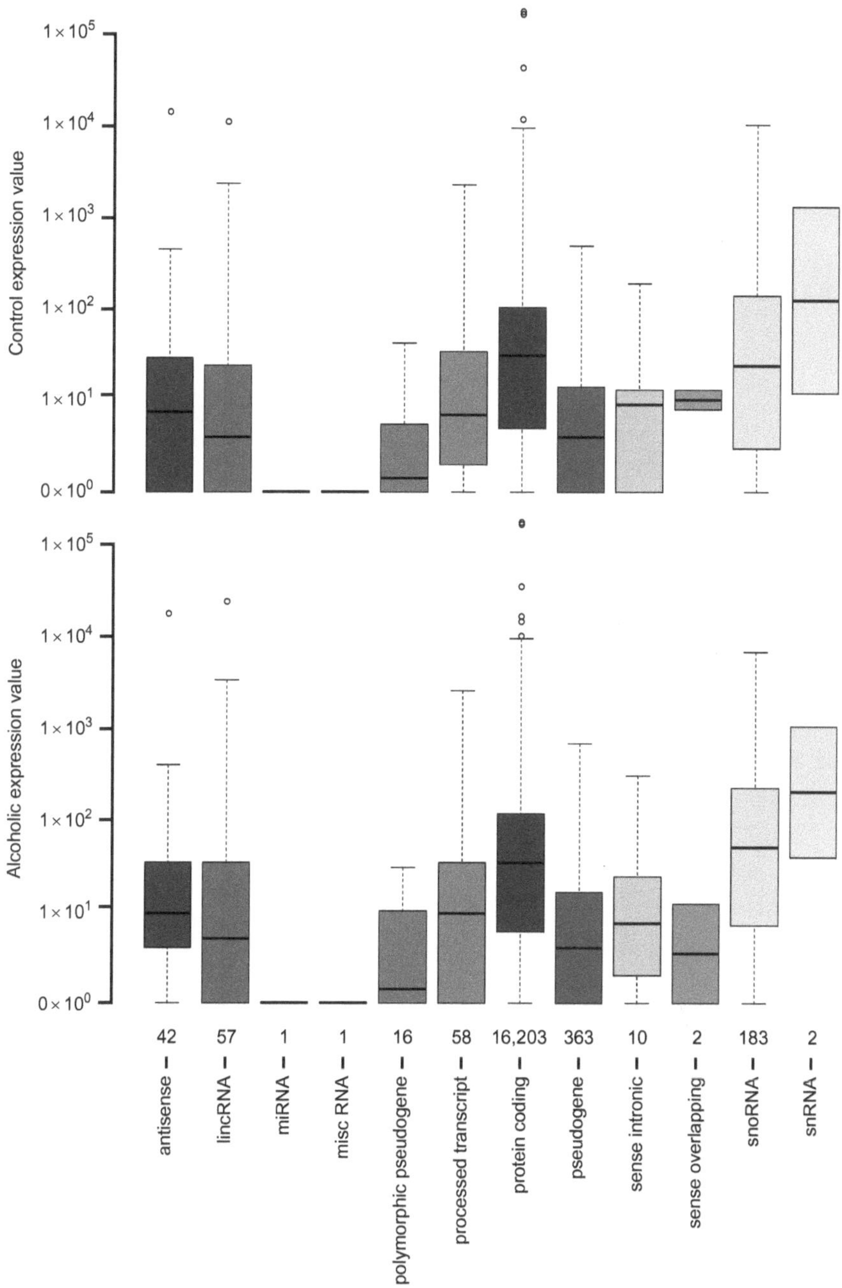

Figure 11.2 Expression values of the detected biological features for alcoholics and matched controls. Box and whisker plots for expression of biological features in representative controls (A) and alcoholics (B) from the prefrontal cortex. Shown along the *x*-axis is the number of corresponding biotypes determined for all samples having greater than zero counts. The two groups have similar overall expression values for biological features.

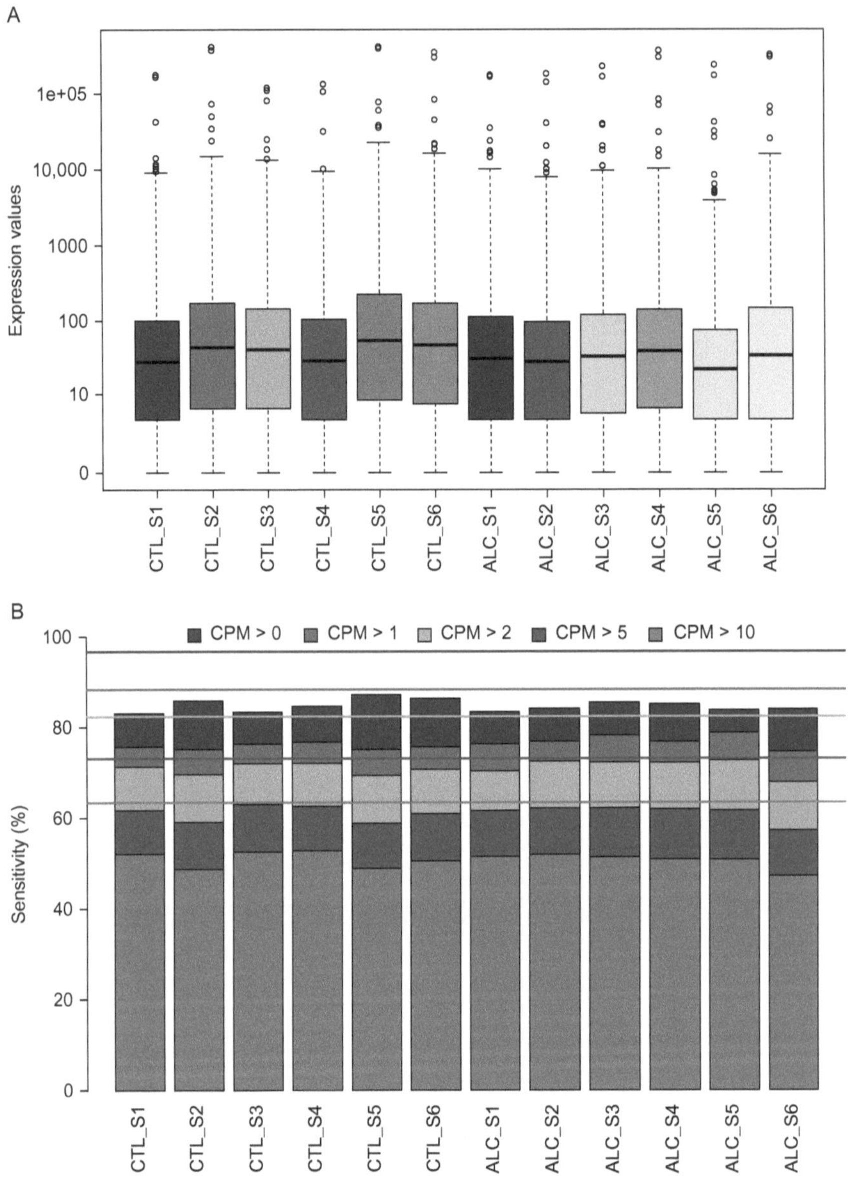

Figure 11.3 Comparison of individual samples for overall expression and sensitivity. (A) Box and whisker plots for overall expression within individual samples and (B) stacked bar plot of binned expression based on counts per million (CPM) mapped reads across individuals to determine intersample consistency and the percentage of low expression values that may interfere with downstream analyses. Horizontal lines depict the percentage of CPM expressed in at least one specimen to help determine an appropriate range of sensitivity.

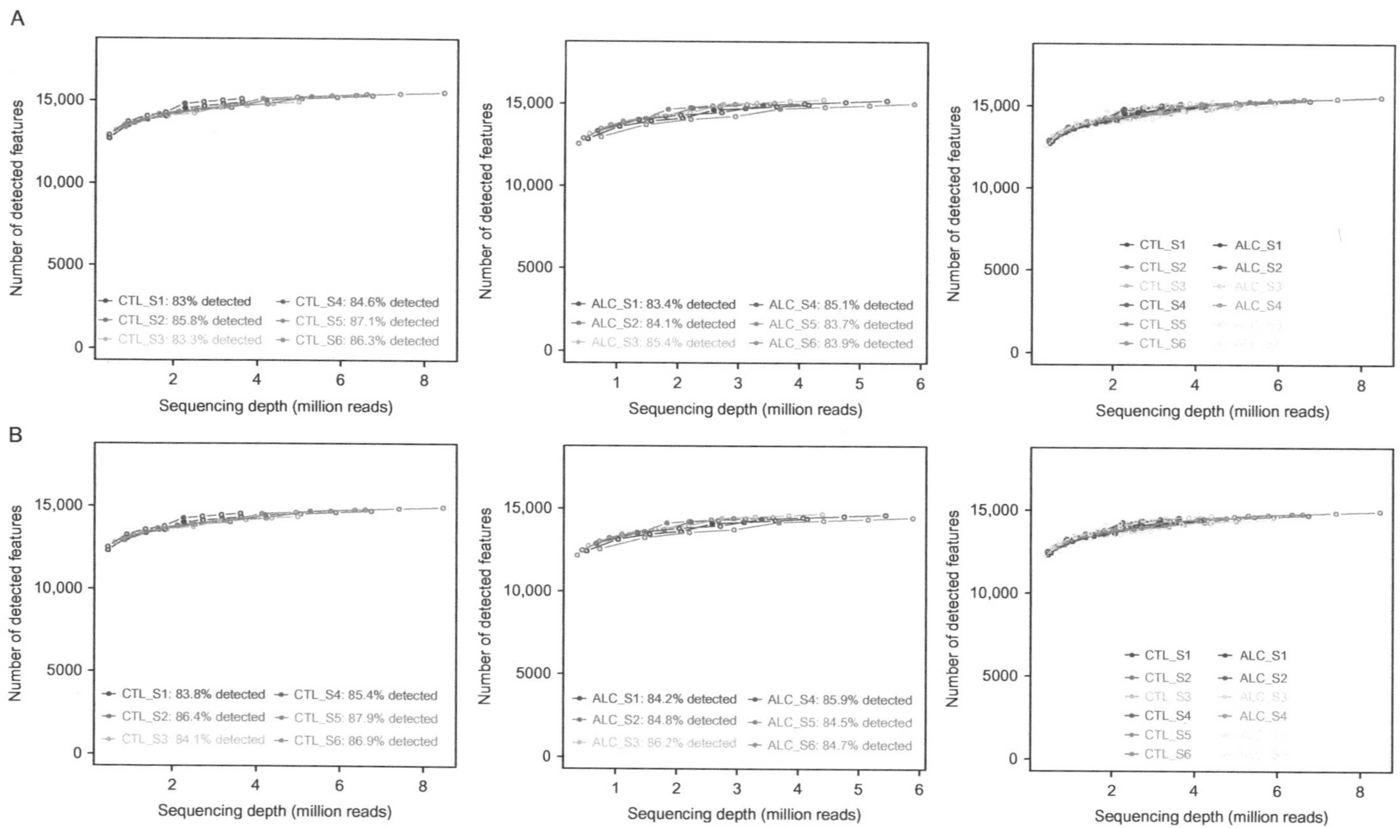

Figure 11.4 Saturation plots for assessing quality control across individual samples and disease groups. Number of detected features compared with sequencing depth in million mapped reads for controls (left), alcoholics (middle), and all samples (right) for all biological features detected (A) and "protein-coding" transcripts only (B). Figures demonstrate all samples have comparable saturation slopes and can be included in downstream analysis.

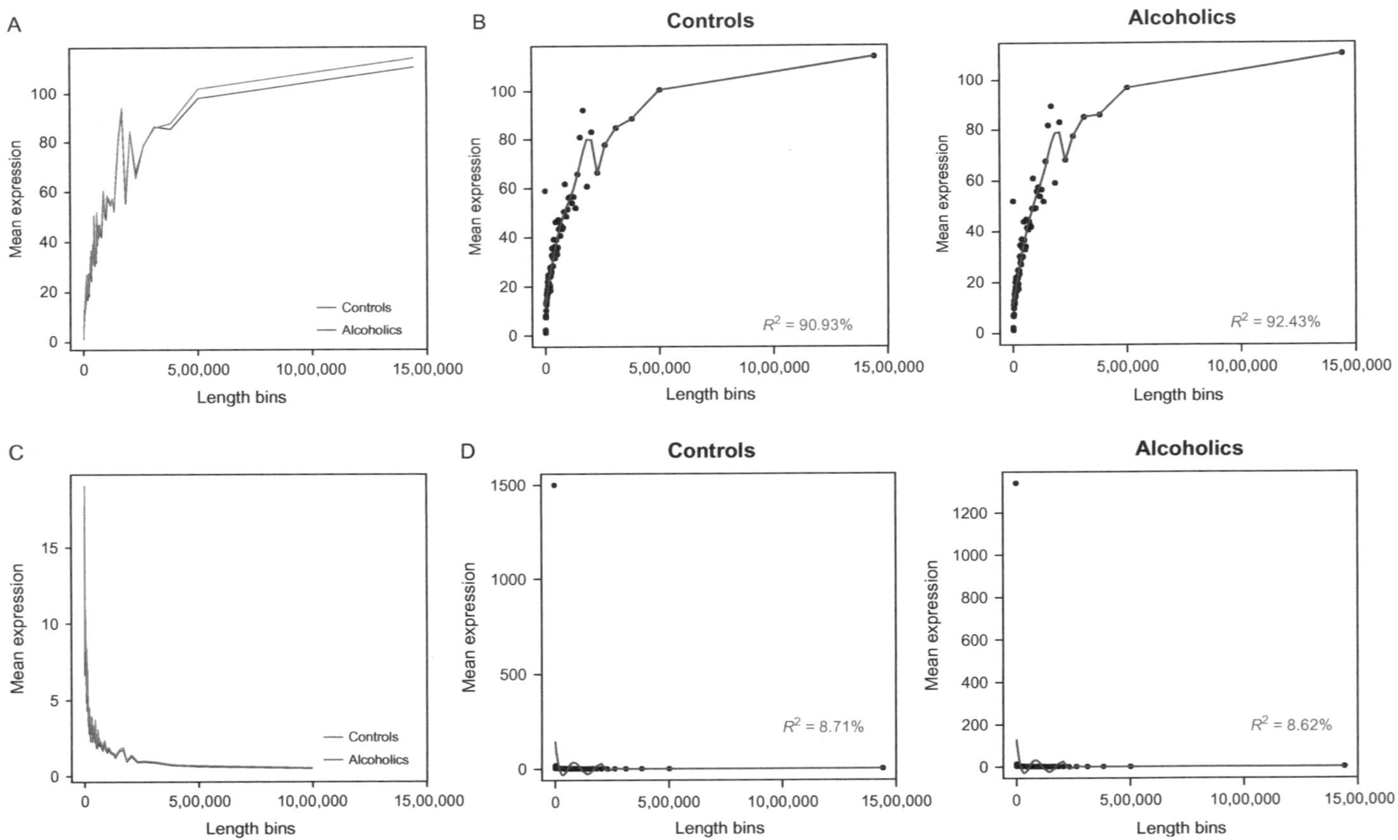

Figure 11.5 Assessment of gene expression for bias in sequencing length. The size of detected biotypes is binned along the *x*-axis and compared to mean expression values of raw counts (top) and normalized expression by reads per kilobase per million (RPKM) mapped reads (bottom). Alcoholic and control samples follow similar trends in raw mean expression values (A) and RPKM values (C). Mean raw count values in controls and alcoholics are highly associated with feature length (B); however, feature length is not strongly correlated with mean RPKM (D).

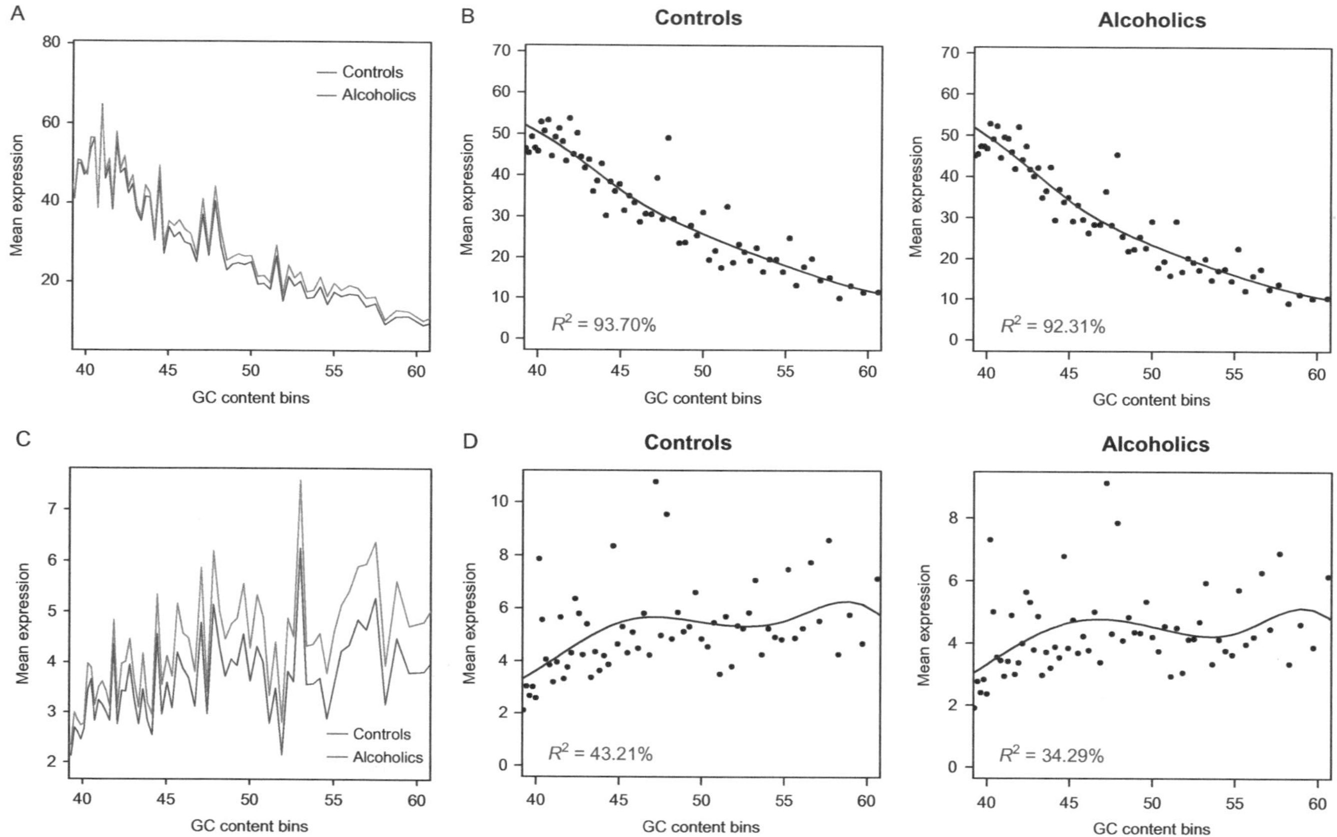

Figure 11.6 Assessment of gene expression for bias in GC content. The GC% of detected biotypes is binned along the *x*-axis and compared to mean expression values of raw counts (top) and normalized expression by reads per kilobase per million (RPKM) mapped reads (bottom). Alcoholic and control samples follow similar trends in raw mean expression values (A) and RPKM values (C) in relation to GC%. Mean raw count values in controls and alcoholics are persistently correlated with % of GC content (B); however, GC% shows lower correlation with normalized expression (D).

expression may be minimized with additional processing/normalization strategies (Hansen, Irizarry, & Wu, 2012; Risso, Schwartz, Sherlock, & Dudoit, 2011). At first glance, expression may appear nearly indistinguishable between alcoholics and controls (Figs. 11.5 and 11.6); however, detailed examination of RNA-Seq can expose biases that may affect expression estimates, and an appropriate normalization strategy is therefore crucial.

4. NORMALIZATION OF RNA-SEQ DATA

No single procedure has yet emerged as a gold standard for RNA-Seq analyses. Differing methodologies for profiling expression can reveal discrepant findings in the identification of differentially expressed genes from the same experimental dataset (Rapaport et al., 2013; Soneson & Delorenzi, 2013; Tarazona, García-Alcalde, Dopazo, Ferrer, & Conesa, 2011). In order to adequately manage bioinformatics pipelines, multiple *in silico* experimental designs, rather than a one size fits all approach, may initially need to be explored before selecting a suitable model of normalization. RNA-Seq expression data from the prefrontal cortex are illustrated using different representative methods of normalization. The intersample correlations among controls and alcoholics fluctuate according to the normalization strategy and impact the extent of within group variation (Fig. 11.7). RPKM values have the highest proportion of variability, which may impede the identification of differentially expressed features between alcoholics and controls. A comprehensive evaluation of normalization techniques for RNA-Seq data has previously suggested the RPKM approach is ineffective and should cease to be used for evaluating differential expression (Dillies et al., 2013). Additionally, RPKM data may fail to adequately account for RNA composition bias (Robinson & Oshlack, 2010) or gene length (Bullard, Purdom, Hansen, & Dudoit, 2010) in the detection of differentially expressed features. Practical recommendations are available for generating fairly robust datasets (Dillies et al., 2013) and will continue to evolve as RNA-Seq is adopted in a larger number of laboratories. Selecting the appropriate statistical method for minimizing the effects of technical error will also depend upon additional known sources of systematic variation.

5. ALTERNATIVE SPLICING AND DIFFERENTIAL EXPRESSION

Once an acceptable normalization method is determined, summarized read counts can be evaluated for divergent expression profiles between two

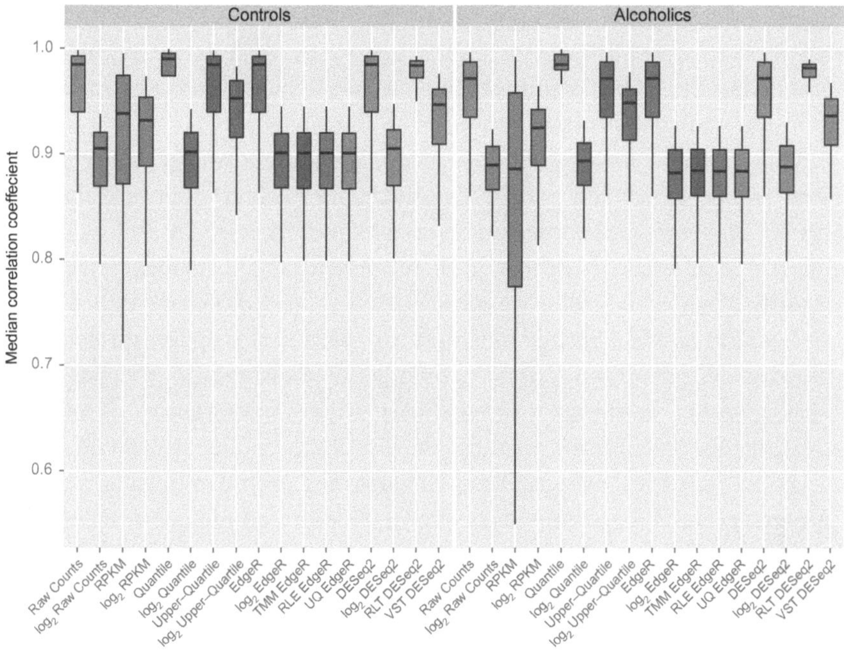

Figure 11.7 Comparison of normalization strategies for assessing gene expression in controls and alcoholics. Box and whisker plots for intersample Pearson correlation coefficients of control and alcoholic prefrontal cortex gene expression across multiple strategies for normalizing RNA-Seq count data. Differing methods of normalization exhibit differing median intersample consistency and within group variation, which may affect experimental outcomes. The appropriate method should be based on the quality control measures and hypothesis in question.

or more conditions. RNA-Seq is a powerful tool for the detection of differentially expressed features, capable of capturing weakly expressed genes and alternatively splices transcripts within a single experiment (Bottomly et al., 2011; Marioni, Mason, Mane, Stephens, & Gilad, 2008). Although it is challenging to use short-read sequencers to quantify splice variants having identical exons, several algorithms exist for computing the expression of full-length isoforms (Garber, Grabherr, Guttman, & Trapnell, 2011; Trapnell et al., 2012; Xing et al., 2006). Recognition of alternatively spliced transcripts, and their individual exons, is an important aspect for interpreting the neurobiology of disease. The human transcriptome is able to generate a tremendous degree of biodiversity, with ~95% of all multiexon genes undergoing alternative splicing (Pan, Kaiguo, Razak, Westwood, & Gerlai, 2011). Humans, and closely related primates, exhibit the greatest degree of complexity in splicing, with the human brain being the most diverse among several tissue types (Barbosa-Morais et al., 2012). The higher rate of alternative splicing in human brain may underscore evolutionary remodeling for higher cognitive function while generating greater susceptibility to neuropsychiatric diseases.

Differential expression between alcoholics and controls is able to distinguish ~1000 genes and ~1200 significant alternatively spliced transcripts (Fig. 11.8). Genes and corresponding spliced isoforms tend to follow similar patterns of differential expression in relation to alcohol dependence. The majority of genes and splice variants with a p-value ≤ 0.05 have only a modest, less than twofold change in expression. Statistically significant genes, or alternatively spliced transcripts, with larger fold changes in expression are usually weaker in overall expression. In some circumstances, it may be advantageous to remove missing or low-level counts; however, some methods of statistical inference may account for extreme or missing variables (Anders et al., 2013). Individual exons are more abundant in RNA-Seq counts, making no assumption of its interconnection with other units to form a functional RNA molecule. Approximately, 11,000 individual exons are differentially expressed in the prefrontal cortex of alcoholics (Fig. 11.8). Additionally, a greater number of exon features have fold-change values >2, which may suggest some gene or isoform reconstructions underestimate some of the differences occurring with alcohol dependence. Although individual exons cannot function solely on their own accord, these molecular units might serve as surrogate markers for differences in the functional

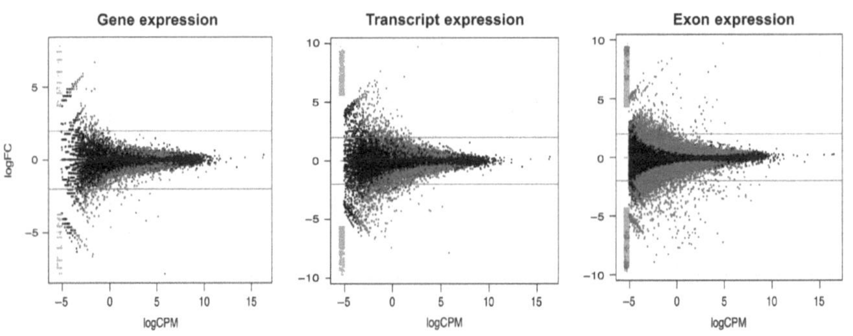

Figure 11.8 Differential expression for gene, transcript, and exon models of RNA-Seq data. The mean log CPM is plotted against the log fold change in gene (left), alternatively spliced transcript (middle), and exon (right) expression for alcoholics versus controls in the prefrontal cortex. Horizontal blue (light gray in the print version) lines depict a twofold change in expression (increased or decreased) between disease groups and red (dark gray in the print version) dots indicate features with a p-value ≤ 0.05.

transcript or gene product. Altered expression of individual exons may be of substantial interest, especially if these changes coincide with the site of activation, a site of intramolecule docking, or an alcohol-binding site for the fully formed protein substrate.

6. LONG NONCODING RNA

In the absence of forming a functional protein, intracellular molecules can still function as ncRNAs (Mattick & Makunin, 2006). ncRNAs make up a sizeable share of the transcriptional landscape (Carninci et al., 2005; ENCODE Project Consortium et al., 2012, 2007), but the precise function of many noncoding elements remains largely unknown. Defining the diverse biological roles carried out by multiple classes of ncRNAs is a burgeoning aspect of transcriptomics that will likely match or rival the large number and diversity represented by the proteome. Long noncoding RNAs (lncRNAs) represent one of the most abundant classes of nonprotein-coding RNAs in the brain (Jia et al., 2010; Ravasi et al., 2006). Similar to protein-coding transcripts, lncRNAs can be found within specific neuroanatomical regions (Belgard et al., 2011; Mercer, Dinger, Sunkin, Mehler, & Mattick, 2008). A study of human alcoholic brain tissue showed an increase in the expression of the lncRNA *MALAT1* within multiple brain regions (Kryger, Fan, Wilce, & Jaquet, 2012). Overall expression of lncRNAs may be lower than protein-coding transcripts, but can be dynamically regulated in alcoholic brain tissue (Fig. 11.9). Although the role of lncRNAs on alcohol dependence and drug addiction is still unclear, lncRNAs are known to (1) mediate control of epigenetic factors for regulating gene expression (Khalil et al., 2009; Lee, 2012; Wang et al., 2011), (2) act as endogenous competitors (Cesana et al., 2011), (3) regulate alternative splicing events (Barry et al., 2013; Massone et al., 2011; Tripathi et al., 2010), (4) control neuronal development (Pollard et al., 2006), and (5) guide synaptic plasticity (Bond et al., 2009). These diverse roles make it likely that even low-to-moderate changes in lncRNA expression could significantly impact alcohol use disorders and other psychiatric diseases.

7. NOVEL THREE PRIME UNTRANSLATED REGIONS

RNA-Seq can generate rich expression maps for annotated and unannotated regions of the transcriptional landscape (Nagalakshmi et al., 2008). The expression of transcribed RNA features can be extensively

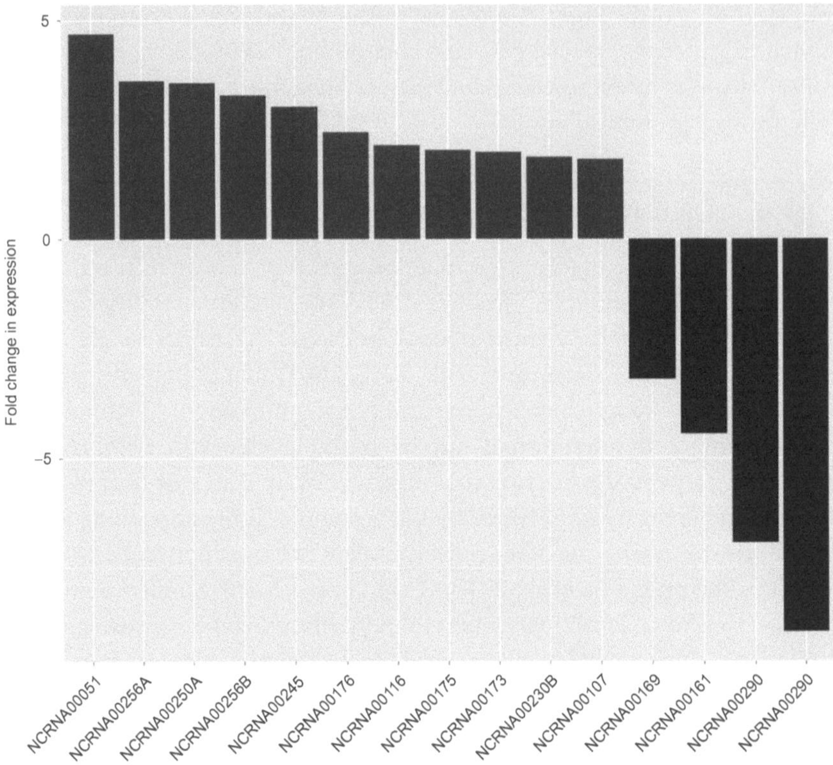

Figure 11.9 Fold change of long noncoding RNAs within the prefrontal cortex. Bar plot of fold change in expression of the top 15 long noncoding RNAs (lncRNAs) between alcoholics and controls within prefrontal cortex. Eleven lncRNAs show increased expression in the prefrontal cortex of alcoholics compared to controls, while four lncRNAs show decreased expression.

regulated across human tissues and cell types (Djebali et al., 2012; Wang et al., 2008), which may involve alternative splicing of exons or pervasive variation within the 3′-UTR of transcripts. Differences in the 3′-UTR of transcripts are known to contribute to expression instability, translation, and act as sites of posttranscriptional regulation (Jackson, 1993). Unbiased transcriptome sequencing of the human brain can identify novel 3′-UTRs for candidate genes (Fig. 11.10). Further characterization of the transcriptome across assorted brain structures, experimental circumstances, and individuals may reveal unique 3′-UTRs or other features for transcribed elements. Probing the neurobiology of novel 3′-UTRs, though time consuming, could eventually expose distinct mechanisms of neuronal function. For example, a short 3′-UTR of *Bdnf* RNA is restricted to neuronal somata,

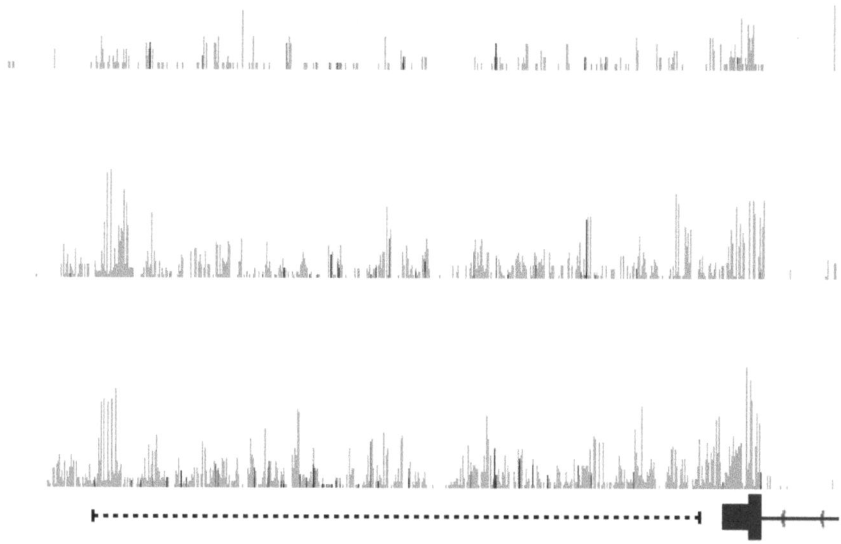

Figure 11.10 Example of a novel three prime untranslated region (3′-UTR) from RNA-Seq data. Snapshot of RNA-Seq counts for three representative samples of a novel long form of a 3′-UTR. Blue (black in the print version) boxes indicate the presence of the last exon and currently annotated 3′-UTR; however, RNA-Seq may detect previously unknown details regarding transcript expression that may impact function.

while a long form of the 3′-UTR is trafficked to the dendrites where it can act upon spine morphology and synaptic transmission (An et al., 2008). Local translation of BDNF, delivered via a long form of the 3′-UTR, can modulate GABAergic transmission (Waterhouse et al., 2012), a well-characterized neurotransmitter system targeted by alcohol and other drugs of abuse (Davies et al., 2003; Harris, Trudell, & Mihic, 2008; Kauer & Malenka, 2007).

8. GENETIC VARIATION AND ALCOHOL DEPENDENCE

Single nucleotide polymorphisms (SNPs) within GABA receptors, and several other candidate genes, are likely contributors in susceptibility to the development of alcohol dependence (Dick & Foroud, 2003). Similar to other psychiatric diseases, alcoholism is influenced by multiple genes with low-to-moderate effect (Sullivan, Daly, & O'Donovan, 2012). Polygenic factors can account for 40–60% of the risk for developing alcohol dependence (Schuckit, 2009); however, SNPs associated with disease usually reside within noncoding regions. Surveying the alcoholic transcriptome for genetic variants further corroborates this assertion (Fig. 11.11). The largest percentage of detected variants is located with sequencing reads mapped to intronic regions, followed by areas located up- or downstream of coding elements and intergenic regions. Introns are typically removed through RNA splicing events, but may be retained within individual isoforms harboring *cis*-acting SNPs controlling their expression (Dieter & Estus, 2010). High-throughput sequencing of human populations, as well as other model systems, is beginning to pinpoint numerous sites of nucleotide variation, that regardless of genomic loci, are capable of gene, alternative splicing, and downstream expression (Gerstein et al., 2010; Graveley et al., 2011; Lappalainen et al., 2013). However, linking any causal points of genetic inference with disease remains a significant challenge in the modern era for quantitative biology. Isolated studies often lack statistical power to definitively link any single SNP, let alone the interaction among multiple SNPs, with disease progression. Strategies are emerging to overcome these types of hurdles and identify unanticipated points of genetic interaction (Pan, 2008; Pandey et al., 2010). Similar to other complex traits, alcoholism is driven by the interaction of countless SNPs and competing environmental influences, which shape the transcriptome and regulate neurobiological functions. Although a number of differences may exist between human DNA and

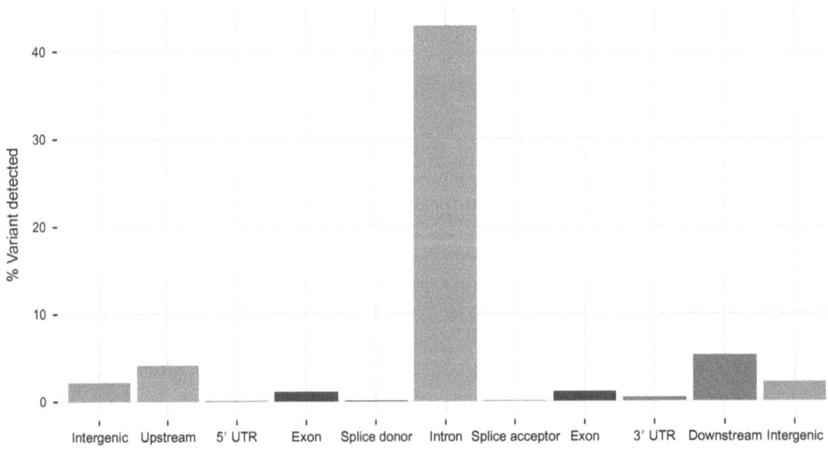

Figure 11.11 Classifying genetic variants within alcoholic prefrontal cortex. Genetic variants detected within alcoholic prefrontal cortex classified by currently annotated gene regions: intergenic, upstream, 5′-UTR, exon, splice site donor, intron, splice site acceptor, 3′-UTR, and downstream. Genetic variants are primarily located in unannotated genomic areas or noncoding/intronic regions.

RNA sequences from the same individual (Li et al., 2011), sequencing the transcriptome in alcoholic brain tissue will continue to provide a valuable resource that represents the multidimensional factors operating in alcohol use disorders.

9. BIOLOGICAL COEXPRESSION NETWORKS

The number of genes implicated in alcohol dependence and other psychiatric illnesses continues to grow, with no single factor being uniquely responsible for the genotype–phenotype relationship. This is not a surprising notion, given that genes and their ensuing proteins do not exist in isolation, but work through coordinated pathways to govern cellular actions. Current canonical pathways, although useful to some extent, are becoming increasingly inadequate to account for the multitude of factors driving cellular behavior and manifesting phenotypes (Califano, Butte, Friend, Ideker, & Schadt, 2012). Using a variety of high-throughput approaches, both expected and unexpected connections can be simultaneously established among multiple cellular substrates to define biological networks for nearly any condition (Barabási & Oltvai, 2004; Vidal, Cusick, & Barabási, 2011). Not all genetic perturbations may be of equal value but may spread their effects across a web of neighboring genes to propagate disease symptomology. Disease-associated genes form an extended network that surrounds highly connected hub genes, which are essential to influence multiple biochemical pathways for development and survival (Goh et al., 2007). Understanding phenotypes across a spectrum of human disorders will require understanding the corresponding network architecture of related diseases.

The human brain transcriptome represents highly organized gene coexpression networks that are consistent across individuals (Hawrylycz et al., 2012; Oldham et al., 2008). Defining gene coexpression patterns for human diseases has revealed convergent molecular profiles (Voineagu et al., 2011), predicted causal systems in neuropathology (Zhang et al., 2013), and unveiled distinct network structures for similar phenotypes (Parikshak et al., 2013). Gene coexpression networks of alcoholic brain tissue, determined with microarray profiling, generated a systemic view of gene expression alterations spanning multiple cell types and brain regions (Ponomarev, Wang, Zhang, Harris, & Mayfield, 2012). A significant portion of transcripts coregulated by chronic alcohol abuse may be conserved within animal models of alcohol consumption (Nunez et al., 2013), permitting an experimentally tractable mode of elucidating gene networks in complex behaviors. Mouse models of alcohol-responsive gene networks can genetically dissect the interrelationship among endophenotypes (Wolen et al., 2012) and clarify networks of candidate genes in alcohol-responsive behaviors (Farris & Miles, 2013). Most network models of acute and chronic alcohol exposure have relied primarily upon microarrays, excluding many available ncRNA substrates. RNA-Seq-derived coexpression networks for alcohol-related phenotypes are still emerging, but should offer greater insight into the whole transcriptome (Giorgi, Del Fabbro, & Licausi, 2013; Iancu et al., 2012). Leveraging the network structure, in addition to differential expression, of the complete transcriptome using RNA-Seq will facilitate a more comprehensive assessment of transcribed features involved in alcoholism and drug conditions.

10. FUTURE DIRECTIONS

Realizing the full potential of RNA-Seq will eventually involve incorporating multiple levels of discrete data types and model systems (Fig. 11.12). The entire complement of RNA molecules, including miRNAs and lncRNAs, exists as a highly orchestrated network, regulated in part by genetic variants or epigenetic phenomena spread throughout the genome. Alternative splicing of mature RNA enables considerable biodiversity of protein products and protein–protein interaction networks. The human proteome (Rual et al., 2005) is far from complete and will likely evolve in parallel with information gleaned from the transcriptome. In the long term, such information will further inform the interpretation of neurophysiological and neuroanatomical studies, including large-scale initiatives like the Human Connectome Project (Van Essen et al., 2013, 2012), in human health. A major challenge will be distilling the vast amount of biological data that bridge multiple scales and also are linked to discrete phenotypes. Focusing on intermediate phenotypes of complex traits may be useful for discovering large, consistent effects exerted by gene networks.

Although it should be emphasized that many of the biological effects seen in neuropsychiatric diseases may be specific to humans, model systems will continue to serve a fundamental role in the post transcriptomic era of modern biology. For example, systematically combining multiple biological networks within a yeast reference population clearly demonstrated that integrating several datasets can improve prediction of causal regulators of complex system behavior (Zhu et al., 2008). Combining information from human and animal models can ascertain core networks affecting disease

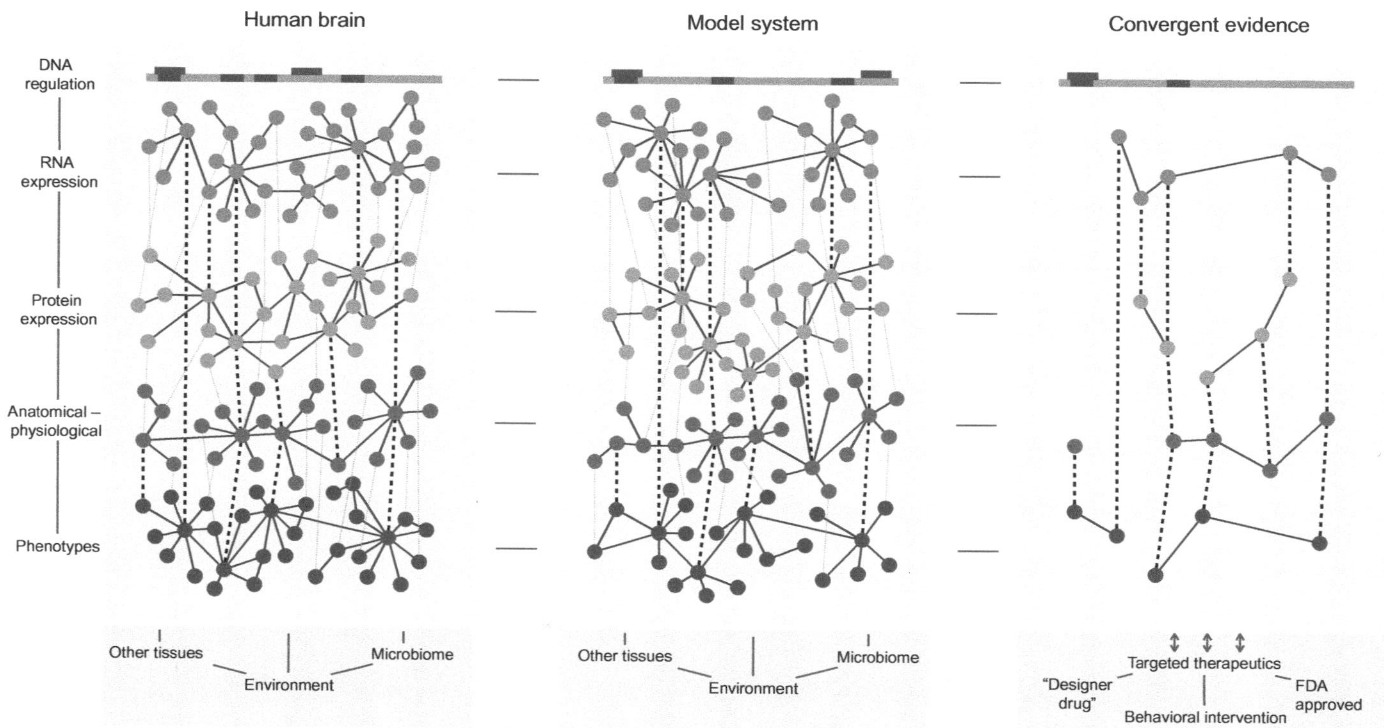

Figure 11.12 Utilizing RNA-Seq in the context of a multiscale systems approach for understanding the neurobiology of alcohol dependence. Differing brain regions of alcoholics and controls can be evaluated using high-throughput sequencing for regulation of DNA and RNA expression. Information from sequencing data can then be layered with current protein data, brain imaging, physiological function, a variety of phenotypic traits (i.e., drinking behavior, withdrawal, craving), and additional influences (non-CNS tissues, human microbiota, and environmental pressures). Pooling resources from both clinical and preclinical sources can clarify points of convergent validity to determine individualized treatment plans incorporating behavioral therapy, current FDA approved compounds, or designer compounds that best target the underlying structure of an individual's disease.

(Emilsson et al., 2008). Alternatively, animal models explicitly created for a desired attribute may be sequenced to find novel causal contributors. Studying alcohol preferring and nonpreferring rats identified a stop codon within the metabotropic glutamate receptor 2 (*Grm2*) that controls protein expression and alcohol-drinking behavior (Zhou et al., 2013). This is just one example of the presumably large collection of variants that will be identified in alcohol consumption, which may eventually intersect with those recognized in human populations. Identifying networks with convergent validity across model organisms and humans (Fig. 11.12) has the potential to isolate systems for therapeutic intervention tailored to the specific needs of the individual.

Whole-transcriptome sequencing has far reaching effects in both clinical and preclinical applications. As a foundation for basic sciences, RNA-Seq continues to impart a deeper appreciation for the vast transcriptional structure of genes and quantification of transcript expression throughout differing cells, tissues, and species. Although still in the early stages of use, RNA-Seq can monitor spatial organization of the transcriptome (Lee et al., 2014) and extract some of the subtle expression differences induced by individual neurons and their microenvironments (Lovatt et al., 2014). With psychiatric illnesses representing one of the most challenging areas of medicine, sophisticated tools such as those furnished by deep sequencing technologies are essential for deciphering all of the converging elements that orchestrate these diseases.

REFERENCES

An, J. J., Gharami, K., Liao, G.-Y., Woo, N. H., Lau, A. G., Vanevski, F., et al. (2008). Distinct role of long 3' UTR BDNF mRNA in spine morphology and synaptic plasticity in hippocampal neurons. *Cell*, *134*(1), 175–187. http://dx.doi.org/10.1016/j.cell.2008.05.045.

Anders, S., McCarthy, D. J., Chen, Y., Okoniewski, M., Smyth, G. K., Huber, W., et al. (2013). Count-based differential expression analysis of RNA sequencing data using R and Bioconductor. *Nature Protocols*, *8*(9), 1765–1786. http://dx.doi.org/10.1038/nprot.2013.099.

Barabási, A.-L., & Oltvai, Z. N. (2004). Network biology: Understanding the cell's functional organization. *Nature Reviews. Genetics*, *5*(2), 101–113. http://dx.doi.org/10.1038/nrg1272.

Barbosa-Morais, N. L., Irimia, M., Pan, Q., Xiong, H. Y., Gueroussov, S., Lee, L. J., et al. (2012). The evolutionary landscape of alternative splicing in vertebrate species. *Science (New York, N.Y.)*, *338*(6114), 1587–1593. http://dx.doi.org/10.1126/science.1230612.

Barry, G., Briggs, J. A., Vanichkina, D. P., Poth, E. M., Beveridge, N. J., Ratnu, V. S., et al. (2013). The long non-coding RNA Gomafu is acutely regulated in response to neuronal activation and involved in schizophrenia-associated alternative splicing. *Molecular Psychiatry*, *19*, 486–494. http://dx.doi.org/10.1038/mp.2013.45.

Belgard, T. G., Marques, A. C., Oliver, P. L., Abaan, H. O., Sirey, T. M., Hoerder-Suabedissen, A., et al. (2011). A transcriptomic atlas of mouse neocortical layers. *Neuron*, *71*(4), 605–616. http://dx.doi.org/10.1016/j.neuron.2011.06.039.

Bond, A. M., Vangompel, M. J. W., Sametsky, E. A., Clark, M. F., Savage, J. C., Disterhoft, J. F., et al. (2009). Balanced gene regulation by an embryonic brain ncRNA is critical for adult hippocampal GABA circuitry. *Nature Neuroscience*, *12*(8), 1020–1027. http://dx.doi.org/10.1038/nn.2371.

Bottomly, D., Walter, N. A. R., Hunter, J. E., Darakjian, P., Kawane, S., Buck, K. J., et al. (2011). Evaluating gene expression in C57BL/6J and DBA/2J mouse striatum using RNA-Seq and microarrays. *PLoS One*, *6*(3), e17820. http://dx.doi.org/10.1371/journal.pone.0017820.

Bullard, J. H., Purdom, E., Hansen, K. D., & Dudoit, S. (2010). Evaluation of statistical methods for normalization and differential expression in mRNA-Seq experiments. *BMC Bioinformatics*, *11*, 94. http://dx.doi.org/10.1186/1471-2105-11-94.

Califano, A., Butte, A. J., Friend, S., Ideker, T., & Schadt, E. (2012). Leveraging models of cell regulation and GWAS data in integrative network-based association studies. *Nature Genetics*, *44*(8), 841–847. http://dx.doi.org/10.1038/ng.2355.

Cao, X., Yeo, G., Muotri, A. R., Kuwabara, T., & Gage, F. H. (2006). Noncoding RNAs in the mammalian central nervous system. *Annual Review of Neuroscience*, *29*, 77–103. http://dx.doi.org/10.1146/annurev.neuro.29.051605.112839.

Carninci, P., Kasukawa, T., Katayama, S., Gough, J., Frith, M. C., Maeda, N., et al. (2005). The transcriptional landscape of the mammalian genome. *Science (New York, N.Y.)*, *309*(5740), 1559–1563. http://dx.doi.org/10.1126/science.1112014.

Cesana, M., Cacchiarelli, D., Legnini, I., Santini, T., Sthandier, O., Chinappi, M., et al. (2011). A long noncoding RNA controls muscle differentiation by functioning as a competing endogenous RNA. *Cell*, *147*(2), 358–369. http://dx.doi.org/10.1016/j.cell.2011.09.028.

Davies, A. G., Pierce-Shimomura, J. T., Kim, H., VanHoven, M. K., Thiele, T. R., Bonci, A., et al. (2003). A central role of the BK potassium channel in behavioral responses to ethanol in C. elegans. *Cell*, *115*(6), 655–666.

Dick, D. M., & Foroud, T. (2003). Candidate genes for alcohol dependence: A review of genetic evidence from human studies. *Alcoholism, Clinical and Experimental Research*, *27*(5), 868–879. http://dx.doi.org/10.1097/01.ALC.0000065436.24221.63.

Dieter, L. S., & Estus, S. (2010). Isoform of APOE with retained intron 3; quantitation and identification of an associated single nucleotide polymorphism. *Molecular Neurodegeneration*, *5*, 34. http://dx.doi.org/10.1186/1750-1326-5-34.

Dillies, M.-A., Rau, A., Aubert, J., Hennequet-Antier, C., Jeanmougin, M., Servant, N., et al. (2013). A comprehensive evaluation of normalization methods for Illumina high-throughput RNA sequencing data analysis. *Briefings in Bioinformatics*, *14*(6), 671–683. http://dx.doi.org/10.1093/bib/bbs046.

Djebali, S., Davis, C. A., Merkel, A., Dobin, A., Lassmann, T., Mortazavi, A., et al. (2012). Landscape of transcription in human cells. *Nature*, *489*(7414), 101–108. http://dx.doi.org/10.1038/nature11233.

Eddy, S. R. (2001). Non-coding RNA genes and the modern RNA world. *Nature Reviews. Genetics*, *2*(12), 919–929. http://dx.doi.org/10.1038/35103511.

Emilsson, V., Thorleifsson, G., Zhang, B., Leonardson, A. S., Zink, F., Zhu, J., et al. (2008). Genetics of gene expression and its effect on disease. *Nature*, *452*(7186), 423–428. http://dx.doi.org/10.1038/nature06758.

ENCODE Project Consortium, Bernstein, B. E., Birney, E., Dunham, I., Green, E. D., Gunter, C., et al. (2012). An integrated encyclopedia of DNA elements in the human genome. *Nature*, *489*(7414), 57–74. http://dx.doi.org/10.1038/nature11247.

ENCODE Project Consortium, Birney, E., Stamatoyannopoulos, J. A., Dutta, A., Guigo, R., Gingeras, T. R., et al. (2007). Identification and analysis of functional elements in 1% of the human genome by the ENCODE pilot project. *Nature*, *447*(7146), 799–816. http://dx.doi.org/10.1038/nature05874.

Farris, S. P., & Miles, M. F. (2013). Fyn-dependent gene networks in acute ethanol sensitivity. *PLoS One*, *8*(11), e82435. http://dx.doi.org/10.1371/journal.pone.0082435.

Garber, M., Grabherr, M. G., Guttman, M., & Trapnell, C. (2011). Computational methods for transcriptome annotation and quantification using RNA-seq. *Nature Methods*, *8*(6), 469–477. http://dx.doi.org/10.1038/nmeth.1613.

Gerstein, M. B., Lu, Z. J., Van Nostrand, E. L., Cheng, C., Arshinoff, B. I., Liu, T., et al. (2010). Integrative analysis of the Caenorhabditis elegans genome by the modENCODE project. *Science (New York, N.Y.)*, *330*(6012), 1775–1787. http://dx.doi.org/10.1126/science.1196914.

Giorgi, F. M., Del Fabbro, C., & Licausi, F. (2013). Comparative study of RNA-seq- and microarray-derived coexpression networks in Arabidopsis thaliana. *Bioinformatics (Oxford, England)*, *29*(6), 717–724. http://dx.doi.org/10.1093/bioinformatics/btt053.

Goh, K.-I., Cusick, M. E., Valle, D., Childs, B., Vidal, M., & Barabási, A.-L. (2007). The human disease network. *Proceedings of the National Academy of Sciences of the United States of America*, *104*(21), 8685–8690. http://dx.doi.org/10.1073/pnas.0701361104.

Graveley, B. R., Brooks, A. N., Carlson, J. W., Duff, M. O., Landolin, J. M., Yang, L., et al. (2011). The developmental transcriptome of Drosophila melanogaster. *Nature*, *471*(7339), 473–479. http://dx.doi.org/10.1038/nature09715.

Hansen, K. D., Irizarry, R. A., & Wu, Z. (2012). Removing technical variability in RNA-seq data using conditional quantile normalization. *Biostatistics (Oxford, England)*, *13*(2), 204–216. http://dx.doi.org/10.1093/biostatistics/kxr054.

Harris, R. A., Trudell, J. R., & Mihic, S. J. (2008). Ethanol's molecular targets. *Science Signaling*, *1*(28), re7. http://dx.doi.org/10.1126/scisignal.128re7.

Hartmann, B., & Valcárcel, J. (2009). Decrypting the genome's alternative messages. *Current Opinion in Cell Biology*, *21*(3), 377–386. http://dx.doi.org/10.1016/j.ceb.2009.02.006.

Hawrylycz, M. J., Lein, E. S., Guillozet-Bongaarts, A. L., Shen, E. H., Ng, L., Miller, J. A., et al. (2012). An anatomically comprehensive atlas of the adult human brain transcriptome. *Nature*, *489*(7416), 391–399. http://dx.doi.org/10.1038/nature11405.

Iancu, O. D., Kawane, S., Bottomly, D., Searles, R., Hitzemann, R., & McWeeney, S. (2012). Utilizing RNA-Seq data for de novo coexpression network inference. *Bioinformatics (Oxford, England)*, *28*(12), 1592–1597. http://dx.doi.org/10.1093/bioinformatics/bts245.

Jackson, R. J. (1993). Cytoplasmic regulation of mRNA function: The importance of the 3' untranslated region. *Cell*, *74*(1), 9–14.

Jia, H., Osak, M., Bogu, G. K., Stanton, L. W., Johnson, R., & Lipovich, L. (2010). Genome-wide computational identification and manual annotation of human long noncoding RNA genes. *RNA (New York, N.Y.)*, *16*(8), 1478–1487. http://dx.doi.org/10.1261/rna.1951310.

Kauer, J. A., & Malenka, R. C. (2007). Synaptic plasticity and addiction. *Nature Reviews. Neuroscience*, *8*(11), 844–858. http://dx.doi.org/10.1038/nrn2234.

Khalil, A. M., Guttman, M., Huarte, M., Garber, M., Raj, A., Rivea Morales, D., et al. (2009). Many human large intergenic noncoding RNAs associate with chromatin-modifying complexes and affect gene expression. *Proceedings of the National Academy of Sciences of the United States of America*, *106*(28), 11667–11672. http://dx.doi.org/10.1073/pnas.0904715106.

Kiss, T. (2002). Small nucleolar RNAs: An abundant group of noncoding RNAs with diverse cellular functions. *Cell*, *109*(2), 145–148.

Kryger, R., Fan, L., Wilce, P. A., & Jaquet, V. (2012). MALAT-1, a non protein-coding RNA is upregulated in the cerebellum, hippocampus and brain stem of human alcoholics. *Alcohol*, *46*(7), 629–634. http://dx.doi.org/10.1016/j.alcohol.2012.04.002.

Lappalainen, T., Sammeth, M., Friedländer, M. R., 't Hoen, P. A. C., Monlong, J., Rivas, M. A., et al. (2013). Transcriptome and genome sequencing uncovers functional variation in humans. *Nature*, *501*(7468), 506–511. http://dx.doi.org/10.1038/nature12531.

Lee, J. T. (2012). Epigenetic regulation by long noncoding RNAs. *Science (New York, N.Y.)*, *338*(6113), 1435–1439. http://dx.doi.org/10.1126/science.1231776.

Lee, J. H., Daugharthy, E. R., Scheiman, J., Kalhor, R., Yang, J. L., Ferrante, T. C., et al. (2014). Highly multiplexed subcellular RNA sequencing in situ. *Science (New York, N.Y.)*, *343*, 1360–1363. http://dx.doi.org/10.1126/science.1250212.

Li, M., Wang, I. X., Li, Y., Bruzel, A., Richards, A. L., Toung, J. M., et al. (2011). Widespread RNA and DNA sequence differences in the human transcriptome. *Science (New York, N.Y.)*, *333*(6038), 53–58. http://dx.doi.org/10.1126/science.1207018.

Lovatt, D., Ruble, B. K., Lee, J., Dueck, H., Kim, T. K., Fisher, S., et al. (2014). Transcriptome in vivo analysis (TIVA) of spatially defined single cells in live tissue. *Nature Methods*, *11*(2), 190–196. http://dx.doi.org/10.1038/nmeth.2804.

Mane, S. P., Evans, C., Cooper, K. L., Crasta, O. R., Folkerts, O., Hutchison, S. K., et al. (2009). Transcriptome sequencing of the microarray quality control (MAQC) RNA reference samples using next generation sequencing. *BMC Genomics*, *10*(1), 264. http://dx.doi.org/10.1186/1471-2164-10-264.

Marioni, J. C., Mason, C. E., Mane, S. M., Stephens, M., & Gilad, Y. (2008). RNA-seq: An assessment of technical reproducibility and comparison with gene expression arrays. *Genome Research*, *18*(9), 1509–1517. http://dx.doi.org/10.1101/gr.079558.108.

Massone, S., Vassallo, I., Fiorino, G., Castelnuovo, M., Barbieri, F., Borghi, R., et al. (2011). 17A, a novel non-coding RNA, regulates GABA B alternative splicing and signaling in response to inflammatory stimuli and in Alzheimer disease. *Neurobiology of Disease*, *41*(2), 308–317. http://dx.doi.org/10.1016/j.nbd.2010.09.019.

Mattick, J. S., & Makunin, I. V. (2006). Non-coding RNA. *Human Molecular Genetics*, *15*(Spec. No. 1), R17–R29. http://dx.doi.org/10.1093/hmg/ddl046.

Mercer, T. R., Dinger, M. E., Sunkin, S. M., Mehler, M. F., & Mattick, J. S. (2008). Specific expression of long noncoding RNAs in the mouse brain. *Proceedings of the National Academy of Sciences of the United States of America*, *105*(2), 716–721. http://dx.doi.org/10.1073/pnas.0706729105.

Morozova, O., & Marra, M. A. (2008). Applications of next-generation sequencing technologies in functional genomics. *Genomics*, *92*(5), 255–264. http://dx.doi.org/10.1016/j.ygeno.2008.07.001.

Mortazavi, A., Williams, B. A., McCue, K., Schaeffer, L., & Wold, B. (2008). Mapping and quantifying mammalian transcriptomes by RNA-Seq. *Nature Methods*, *5*(7), 621–628. http://dx.doi.org/10.1038/nmeth.1226.

Nagalakshmi, U., Wang, Z., Waern, K., Shou, C., Raha, D., Gerstein, M., et al. (2008). he transcriptional landscape of the yeast genome defined by RNA sequencing. *Science (New York, N.Y.)*, *320*(5881), 1344–1349. http://dx.doi.org/10.1126/science.1158441.

Nunez, Y. O., Truitt, J. M., Gorini, G., Ponomareva, O. N., Blednov, Y. A., Harris, R. A., et al. (2013). Positively correlated miRNA-mRNA regulatory networks in mouse frontal cortex during early stages of alcohol dependence. *BMC Genomics*, *14*, 725. http://dx.doi.org/10.1186/1471-2164-14-725.

Oldham, M. C., Konopka, G., Iwamoto, K., Langfelder, P., Kato, T., Horvath, S., et al. (2008). Functional organization of the transcriptome in human brain. *Nature Neuroscience*, *11*(11), 1271–1282. http://dx.doi.org/10.1038/nn.2207.

Oshlack, A., & Wakefield, M. J. (2009). Transcript length bias in RNA-seq data confounds systems biology. *Biology Direct*, *4*, 14. http://dx.doi.org/10.1186/1745-6150-4-14.

Ozsolak, F., & Milos, P. M. (2011). RNA sequencing: Advances, challenges and opportunities. *Nature Reviews. Genetics*, *12*(2), 87–98. http://dx.doi.org/10.1038/nrg2934.

Pan, W. (2008). Network-based model weighting to detect multiple loci influencing complex diseases. *Human Genetics*, *124*(3), 225–234. http://dx.doi.org/10.1007/s00439-008-0545-1.

Pan, Y., Kaiguo, M., Razak, Z., Westwood, J. T., & Gerlai, R. (2011). Chronic alcohol exposure induced gene expression changes in the zebrafish brain. *Behavioural Brain Research*, *216*(1), 66–76. http://dx.doi.org/10.1016/j.bbr.2010.07.017.

Pandey, G., Zhang, B., Chang, A. N., Myers, C. L., Zhu, J., Kumar, V., et al. (2010). An integrative multi-network and multi-classifier approach to predict genetic interactions. *PLoS Computational Biology*, *6*(9). http://dx.doi.org/10.1371/journal.pcbi.1000928.

Parikshak, N. N., Luo, R., Zhang, A., Won, H., Lowe, J. K., Chandran, V., et al. (2013). Integrative functional genomic analyses implicate specific molecular pathways and circuits in autism. *Cell*, *155*(5), 1008–1021. http://dx.doi.org/10.1016/j.cell.2013.10.031.

Pickrell, J. K., Marioni, J. C., Pai, A. A., Degner, J. F., Engelhardt, B. E., Nkadori, E., et al. (2010). Understanding mechanisms underlying human gene expression variation with RNA sequencing. *Nature*, *464*(7289), 768–772. http://dx.doi.org/10.1038/nature08872.

Pietrzykowski, A. Z., Friesen, R. M., Martin, G. E., Puig, S. I., Nowak, C. L., Wynne, P. M., et al. (2008). Posttranscriptional regulation of BK channel splice variant stability by miR-9 underlies neuroadaptation to alcohol. *Neuron*, *59*(2), 274–287. http://dx.doi.org/10.1016/j.neuron.2008.05.032.

Pollard, K. S., Salama, S. R., Lambert, N., Lambot, M.-A., Coppens, S., Pedersen, J. S., et al. (2006). An RNA gene expressed during cortical development evolved rapidly in humans. *Nature*, *443*(7108), 167–172. http://dx.doi.org/10.1038/nature05113.

Ponomarev, I., Wang, S., Zhang, L., Harris, R. A., & Mayfield, R. D. (2012). Gene coexpression networks in human brain identify epigenetic modifications in alcohol dependence. *The Journal of Neuroscience: The Official Journal of the Society for Neuroscience*, *32*(5), 1884–1897. http://dx.doi.org/10.1523/JNEUROSCI.3136-11.2012.

Rapaport, F., Khanin, R., Liang, Y., Pirun, M., Krek, A., Zumbo, P., et al. (2013). Comprehensive evaluation of differential gene expression analysis methods for RNA-seq data. *Genome Biology*, *14*(9), R95. http://dx.doi.org/10.1186/gb-2013-14-9-r95.

Ravasi, T., Suzuki, H., Pang, K. C., Katayama, S., Furuno, M., Okunishi, R., et al. (2006). Experimental validation of the regulated expression of large numbers of non-coding RNAs from the mouse genome. *Genome Research*, *16*(1), 11–19. http://dx.doi.org/10.1101/gr.4200206.

Risso, D., Schwartz, K., Sherlock, G., & Dudoit, S. (2011). GC-content normalization for RNA-Seq data. *BMC Bioinformatics*, *12*, 480. http://dx.doi.org/10.1186/1471-2105-12-480.

Robinson, M. D., & Oshlack, A. (2010). A scaling normalization method for differential expression analysis of RNA-seq data. *Genome Biology*, *11*(3), R25. http://dx.doi.org/10.1186/gb-2010-11-3-r25.

Rogelj, B., Hartmann, C. E. A., Yeo, C. H., Hunt, S. P., & Giese, K. P. (2003). Contextual fear conditioning regulates the expression of brain-specific small nucleolar RNAs in hippocampus. *The European Journal of Neuroscience*, *18*(11), 3089–3096.

Rual, J.-F., Venkatesan, K., Hao, T., Hirozane-Kishikawa, T., Dricot, A., Li, N., et al. (2005). Towards a proteome-scale map of the human protein-protein interaction network. *Nature*, *437*(7062), 1173–1178. http://dx.doi.org/10.1038/nature04209.

Schuckit, M. A. (2009). Alcohol-use disorders. *Lancet*, *373*(9662), 492–501. http://dx.doi.org/10.1016/S0140-6736(09)60009-X.

Soneson, C., & Delorenzi, M. (2013). A comparison of methods for differential expression analysis of RNA-seq data. *BMC Bioinformatics*, *14*, 91. http://dx.doi.org/10.1186/1471-2105-14-91.

Sullivan, P. F., Daly, M. J., & O'Donovan, M. (2012). Genetic architectures of psychiatric disorders: The emerging picture and its implications. *Nature Reviews. Genetics*, *13*(8), 537–551. http://dx.doi.org/10.1038/nrg3240.

Tang, F., Barbacioru, C., Wang, Y., Nordman, E., Lee, C., Xu, N., et al. (2009). mRNA-Seq whole-transcriptome analysis of a single cell. *Nature Methods*, *6*(5), 377–382. http://dx.doi.org/10.1038/nmeth.1315.

Tarazona, S., García-Alcalde, F., Dopazo, J., Ferrer, A., & Conesa, A. (2011). Differential expression in RNA-seq: A matter of depth. *Genome Research*, *21*(12), 2213–2223. http://dx.doi.org/10.1101/gr.124321.111.

Tazi, J., Tazi, J., Bakkour, N., Bakkour, N., Stamm, S., & Stamm, S. (2009). Alternative splicing and disease. *Biochimica et Biophysica Acta*, *1792*(1), 14–26. http://dx.doi.org/10.1016/j.bbadis.2008.09.017.

Trapnell, C., Roberts, A., Goff, L., Pertea, G., Kim, D., Kelley, D. R., et al. (2012). Differential gene and transcript expression analysis of RNA-seq experiments with TopHat and Cufflinks. *Nature Protocols*, 7(3), 562–578. http://dx.doi.org/10.1038/nprot.2012.016.

Tripathi, V., Ellis, J. D., Shen, Z., Song, D. Y., Pan, Q., Watt, A. T., et al. (2010). The nuclear-retained noncoding RNA MALAT1 regulates alternative splicing by modulating SR splicing factor phosphorylation. *Molecular Cell*, *39*(6), 925–938. http://dx.doi.org/10.1016/j.molcel.2010.08.011.

Van Essen, D. C., Smith, S. M., Barch, D. M., Behrens, T. E. J., Yacoub, E., Ugurbil, K., et al. (2013). The WU-Minn Human Connectome project: An overview. *NeuroImage*, *80*, 62–79. http://dx.doi.org/10.1016/j.neuroimage.2013.05.041.

Van Essen, D. C., Ugurbil, K., Auerbach, E., Barch, D., Behrens, T. E. J., Bucholz, R., et al. (2012). The Human Connectome Project: A data acquisition perspective. *NeuroImage*, *62*(4), 2222–2231. http://dx.doi.org/10.1016/j.neuroimage.2012.02.018.

Vidal, M., Cusick, M. E., & Barabási, A.-L. (2011). Interactome networks and human disease. *Cell*, *144*(6), 986–998. http://dx.doi.org/10.1016/j.cell.2011.02.016.

Voineagu, I., Wang, X., Johnston, P., Lowe, J. K., Tian, Y., Horvath, S., et al. (2011). Transcriptomic analysis of autistic brain reveals convergent molecular pathology. *Nature*, *474*(7351), 380–384. http://dx.doi.org/10.1038/nature10110.

Walter, N. A. R., Bottomly, D., Laderas, T., Mooney, M. A., Darakjian, P., Searles, R. P., et al. (2009). High throughput sequencing in mice: A platform comparison identifies a preponderance of cryptic SNPs. *BMC Genomics*, *10*, 379. http://dx.doi.org/10.1186/1471-2164-10-379.

Wang, Z., Gerstein, M., & Snyder, M. (2009). RNA-Seq: A revolutionary tool for transcriptomics. *Nature Reviews. Genetics*, *10*(1), 57–63. http://dx.doi.org/10.1038/nrg2484.

Wang, E. T., Sandberg, R., Luo, S., Khrebtukova, I., Zhang, L., Mayr, C., et al. (2008). Alternative isoform regulation in human tissue transcriptomes. *Nature*, *456*(7221), 470–476. http://dx.doi.org/10.1038/nature07509.

Wang, K. C., Yang, Y. W., Liu, B., Sanyal, A., Corces-Zimmerman, R., Chen, Y., et al. (2011). A long noncoding RNA maintains active chromatin to coordinate homeotic gene expression. *Nature*, *472*(7341), 120–124. http://dx.doi.org/10.1038/nature09819.

Waterhouse, E. G., An, J. J., Orefice, L. L., Baydyuk, M., Liao, G.-Y., Zheng, K., et al. (2012). BDNF promotes differentiation and maturation of adult-born neurons through GABAergic transmission. *The Journal of Neuroscience: the Official Journal of the Society for Neuroscience*, *32*(41), 14318–14330. http://dx.doi.org/10.1523/JNEUROSCI.0709-12.2012.

Wolen, A. R., Phillips, C. A., Langston, M. A., Putman, A. H., Vorster, P. J., Bruce, N. A., et al. (2012). Genetic dissection of acute ethanol responsive gene networks in prefrontal cortex: Functional and mechanistic implications. *PLoS One*, 7(4), e33575. http://dx.doi.org/10.1371/journal.pone.0033575.

Xing, Y., Yu, T., Wu, Y. N., Roy, M., Kim, J., & Lee, C. (2006). An expectation-maximization algorithm for probabilistic reconstructions of full-length isoforms from splice graphs. *Nucleic Acids Research*, *34*(10), 3150–3160. http://dx.doi.org/10.1093/nar/gkl396.

Zhang, B., Gaiteri, C., Bodea, L.-G., Wang, Z., McElwee, J., Podtelezhnikov, A. A., et al. (2013). Integrated systems approach identifies genetic nodes and networks in late-onset Alzheimer's disease. *Cell*, *153*(3), 707–720. http://dx.doi.org/10.1016/j.cell.2013.03.030.

Zhou, Z., Karlsson, C., Liang, T., Xiong, W., Kimura, M., Tapocik, J. D., et al. (2013). Loss of metabotropic glutamate receptor 2 escalates alcohol consumption. *Proceedings of the National Academy of Sciences of the United States of America*, *110*(42), 16963–16968.

Zhu, J., Zhang, B., Smith, E. N., Drees, B., Brem, R. B., Kruglyak, L., et al. (2008). Integrating large-scale functional genomic data to dissect the complexity of yeast regulatory networks. *Nature Genetics*, *40*(7), 854–861. http://dx.doi.org/10.1038/ng.167.

INDEX

Note: Page numbers followed by "*f*" indicate figures and "*t*" indicate tables.

B

C

CONTENTS OF RECENT VOLUMES

Volume 39

Volume 40

Volume 41

Volume 42

Volume 43

Volume 44

Volume 45

Volume 46

Volume 47

Volume 48

Volume 49

Volume 50

Volume 51

Volume 52

Volume 53

Volume 54

Volume 55

Volume 56

Volume 57

Volume 58

Volume 59

Volume 60

Volume 61

Volume 62

Volume 63

Volume 64

Volume 65

Volume 66

Volume 67

Volume 68

Volume 69

Volume 70

Volume 71

Volume 72

Volume 73

Volume 74

Volume 75

Volume 76

Volume 77

Volume 78

Volume 79

Volume 80

Volume 81

Volume 82

Volume 83

Volume 84

Volume 85

Volume 86

Volume 87

Volume 88

Volume 89

Volume 90

Volume 91

Volume 92

Volume 93

Volume 94

Volume 95

Volume 96

Volume 97

Volume 98

Volume 99

Volume 100

Volume 101

Volume 102

Volume 103

Volume 104

Volume 105

Volume 106

Volume 107

Volume 108

Volume 109

Volume 110

Volume 111

Volume 112

Volume 113

Volume 114

Volume 115

FSC
www.fsc.org
MIX
Papier | Fördert gute Waldnutzung
FSC® C083411

Kolibri 360 GmbH
Ferdinand-Jühlke-Straße 7
99095 Erfurt, Deutschland
produktsicherheit@kolibri360.de